# Electronics
# Sourcebook
# for Engineers

George Loveday

CEng MIERE

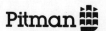
Pitman

PITMAN PUBLISHING
128 Long Acre, London WC2E 9AN

A Division of Longman Group UK Limited

First published in Great Britain under the title *Essential Electronics* 1982
Reprinted 1984, 1985
Second edition published as *Electronics Sourcebook for Engineers* 1986
Reprinted 1987, 1988, 1989, 1991

Printed and bound in Singapore

ISBN 0 273 02667 4

# Preface

*"Any change, even a change for the better, is always accompanied by drawbacks and discomforts"*. So wrote Arnold Bennett, the novelist.

No one would question that Electronics is changing rapidly and the drawback to this is that even the experienced engineer is hard pressed to keep pace with developments. For new students and enthusiasts the problems associated with a rapidly changing subject are even more acute. Electronics must sometimes seem a confusing jungle of loosely related topics—a mixture of devices with strange names (backward diode, Hexfet, etc.), theorems, parameters, circuits, and concepts. What I have attempted in this book is to gather together most of the various interrelated topics into one handy reference book. The intention is to provide a concise explanatory guide to all the important areas, giving only information that is necessary for a reasonable grasp of the subject area.

Naturally it has not been easy deciding what should and should not be included. In a sense, therefore, this is my personal list, based on much experience and involvement with the subject as a lecturer and enthusiast.

My thanks to John Cushion, the editor at Pitman Books, for the layout, helpful criticism, and support in this project.

G. L.

# Contents

# Abbreviations

In electronics the use of abbreviations and acronyms is understandably widespread, mainly for the reason that their use saves space in technical publications and data sheets. For example, it is simpler to write PROM instead of programmable read only memory, or d.v.m. in place of digital voltmeter. PROM is an example of an acronym — a word formed from the initial letters of the other words.

The following is a list, in alphabetical order, of abbreviations used in modern electronics. Note that either capitals or lower case letters are frequently used for the same abbreviation. Integrated circuit, for example, is abbreviated to either IC or i.c. In each case the most common form is shown.

| | |
|---|---|
| Acceptable quality level | AQL |
| Alternating current | a.c. |
| Amplitude modulation | a.m. |
| Analogue-to-digital convertor | ADC |
| Arithmetic logic unit | ALU |
| American standard code for information interchange | ASCII |
| Audio frequency | a.f. |
| Automatic frequency control | AFC |
| Automatic gain control | AGC |
| Automatic test equipment | ATE |
| Beat frequency oscillator | b.f.o. |
| Binary coded decimal | BCD |
| British Standard | BS |
| Cathode-ray oscilloscope | c.r.o. |
| Cathode-ray tube | c.r.t. |
| Central processor unit | CPU |
| Charge coupled device | c.c.d. |
| Collector diffusion isolation | CDI |
| Complementary MOS logic | CMOS |
| Current differencing amplifier | c.d.a. |
| Current mode logic (same as ECL) | CML |
| Decibel | dB |
| Digital panel meter | d.p.m. |
| Digital-to-analogue convertor | DAC |
| Diode transistor logic | DTL |
| Direct current | d.c. |
| Erasable programmable read only memory | EPROM |
| Electromotive force | e.m.f. |
| Emitter coupled logic | ECL |
| Field effect transistor | FET |
|   Junction field-effect transistor | JFET |
|   Insulated-gate field-effect transistor | IGFET |
|   Metal-oxide-silicon field-effect transistor | MOSFET |
| Frequency modulation | f.m. |
| Frequency shift keying | FSK |
| Full-scale deflection | f.s.d. |
| Ground connection | GND |
| Hertz | Hz |
| High frequency | h.f. |
| High threshold logic | HTL |
| High voltage | h.v. |
| Integrated circuit | IC or i.c. |
| Integrated injection logic | $I^2L$ |
| Intermediate frequency | i.f. |

| | |
|---|---|
| Large-scale integration | LSI |
| Light-activated silicon controlled rectifier | l.a.s.c.r. |
| Light-dependent resistor | l.d.r. |
| Light-emitting diode | l.e.d. |
| Medium-scale integration | MSI |
| Metal oxide silicon (device) | MOS |
| Metal-oxide-silicon field-effect transistor | MOSFET MOST |
| Microprocessor | $\mu P$ |
| n-channel MOS | NMOS |
| Negative feedback | n.f.b. |
| Operational amplifier | op-amp or o.p.a. |
| Oscillator | osc. |
| p-channel MOS | PMOS |
| Phased locked loop | PLL |
| Phase modulation | p.m. |
| Potential difference | p.d. |
| Printed circuit board | p.c.b. |
| Programmable logic array | PLA |
| Programmable read-only memory | PROM |
| Pulse amplitude modulation | p.a.m. |
| Pulse code modulation | p.c.m. |
| Pulse repetition frequency | p.r.f. |
| Quality assurance | QA |
| Quality control | QC |
| Random-access memory | RAM |
| Read-only memory | ROM |
| Reference | ref. |
| Resistor transistor logic | RTL |
| Silicon | Si |
| Silicon controlled rectifier (thyristor) | SCR |
| Small-scale integration | SSI |
| Switched mode power unit | SMPU |
| Transistor transistor logic | TTL |
| Transmission gate logic | TGL |
| Unijunction transistor | UJT |
| Universal asynchronous receiver transmitter | UART |
| Very high frequency | v.h.f. |
| Very low frequency | v.l.f. |
| Vertical metal oxide (field effect transistor) | VMOS |
| Visual display unit | VDU |
| Voltage controlled oscillator | v.c.o. |

## Acceptor Impurity

A trivalent material (one that has 3 valence electrons) which is used to dope intrinsic (pure) semiconductor to create p-type semiconductor. Each atom of the acceptor impurity can only form covalent bonds with three electrons from adjacent silicon atoms. Thus a vacancy or "hole" is set up in the crystal lattice wherever an acceptor atom exists. Holes are positive charge carriers which travel through the semiconductor by accepting electrons.

Typical doping levels in practical semiconductor devices are as low as 1 part in $10^8$, i.e. one atom of the impurity to 100 million atoms of silicon. Common acceptor impurity materials are: Boron, Gallium, Indium.

▶ *Donor impurity* (this creates n-type)
▶ *Semiconductor theory*

## Access Time

This refers to the speed with which the content of any location within a memory can be made available. It is the time interval between the instant that an address is sent to the memory and the instant that the data stored at that address location is presented at the output.

Random access, where any location in the store can be reached in the same time as any other, is the fastest method. This is why RAMs are used as the central working stores of computers. Backing stores, such as magnetic disc, drum and tape, use accessing methods that are cyclic or serial. The access time then varies for different store locations, and an average time for access is then quoted, i.e. the time for one half revolution in a cyclic system. Typical figures are given in Table A1.

**Table A1**  *Stores and access times*

| Type of store | Accessing method | Access time |
|---|---|---|
| Core | random | 1 $\mu$s |
| Semiconductor | | |
|   bipolar | random | 50 ns |
|   MOS | random | 200 ns |
| Drum | cyclic | 10 ms |
| Disc | cyclic | 30 ms |
| Tape | serial | several minutes |
| Magnetic bubbles | random | 10 to 100 $\mu$s |

▶ *Address*   ▶ *Memory*

**Table A2**  *Typical accuracies of common measuring instruments*

| Instrument | Accuracy |
|---|---|
| Moving coil multimeter | $\pm1\%$ f.s.d. d.c. voltage and current<br>$\pm2\%$ f.s.d. a.c. voltage and current<br>$\pm3\%$ zero to midscale (ohms)<br>$\pm5\%$ midscale to $\frac{2}{3}$ f.s.d. (ohms)<br>$\pm10\%$ $\frac{2}{3}$ f.s.d. to f.s.d. (ohms) |
| Digital multimeter: general purpose $3\frac{1}{2}$ digits | $\pm0.3\%$ of reading $\pm1$ digit for a.c., d.c. and ohms |
| Cathode ray oscilloscope | $\pm3\%$ amplitude and time |

## Accuracy

The degree to which a measurement, an indication, or a conversion is in agreement with the true value. In other words, the "error" or "range of error" between the indicated and the correct value. For example, if a voltmeter indicates 9.9 V when the true value is 10 V, then the accuracy of the meter can be said to be $-1\%$ at 10 V. In practice the accuracy of an analogue multimeter is quoted as a percentage of full-scale deflection.

The accuracy of any instrument depends to a large extent upon its regular calibration, when its indication will be compared against a known standard. Many measuring situations do not require a high degree of accuracy, and performance parameters such as resolution, sensitivity, input impedance (loading effect) and repeatability may be more important.

▶ *Calibration*   ▶ *Measurement*

## Active Device

Those components within a circuit that have gain or which direct the flow of current, i.e. transistors, thyristors, diodes and valves. Thus an ACTIVE PULLUP is the transistor in TTL logic gates which replaces the load resistor.

▶ *Transistor transistor logic*

## Active Filter

A term used to describe a filter circuit that is constructed of both active and passive elements (fig. A1). The active element is usually an i.c.

**a) Sallen and key low pass**

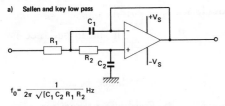

$$f_0 = \frac{1}{2\pi \sqrt{[C_1 C_2 R_1 R_2]}} \text{ Hz}$$

*Example:* Variable low pass (2.5 kHz to 25 kHz)

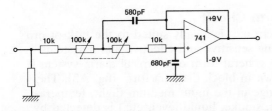

**b) Sallen and key high pass**

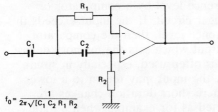

$$f_0 = \frac{1}{2\pi \sqrt{[C_1 C_2 R_1 R_2]}}$$

*Example:* Variable high pass (250 Hz to 3 kHz)

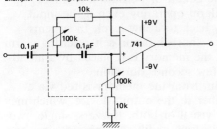

**c) Circuits based on the twin-T filter**

1 kHz band pass

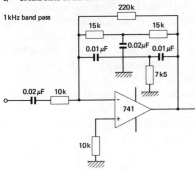

*Figure A1* **Active filters: basic circuits with examples**

---

1 kHz band stop (notch)

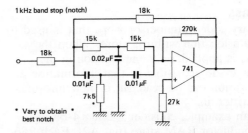

\* Vary to obtain \* best notch

**d) Notch filter with variable Q-factor**

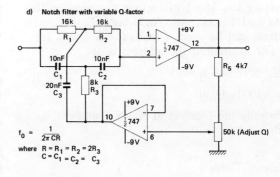

$$f_0 = \frac{1}{2\pi CR}$$

where $R = R_1 = R_2 = 2R_3$
$C = C_1 = C_2 = C_3$

**e) Band pass filter using multi-feedback**

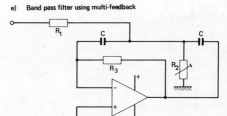

Simpler circuit than twin-T type. $R_2$ adjusts centre frequency.

$$f_0 = \frac{1}{2\pi C} \sqrt{\left[\frac{R_1 + R_2}{R_1 R_2 R_3}\right]} \qquad Q = \frac{1}{2} \sqrt{\left[\frac{R_3 (R_1 + R_2)}{R_1 R_3}\right]}$$

op-amp and the passive components are $R$ and $C$. The latter are connected to the inputs of the op-amp and in the feedback loop to produce the required frequency-dependent transfer function.

All the common types of filter — low pass, high pass, band pass, and band stop (notch) — can be constructed [▶ *Filter*]. Active filters are usually of the second-order type, which means that there is an $\omega^2$ term in the equation describing the transfer function. This gives sharp cut-off characteristics of 12 dB per octave compared with 6 dB per octave for simple *RC* filters. Therefore filters can be made with excellent performance from relatively low-cost components. Inductors, which would be expensive at audio frequencies, are not required because the active element, the op-amp, can be used to simulate inductive reactance.

## Address

A binary coded number or word that is used to specify a location within a store. When the address is applied, only one location in the store is made accessible so that information at that location can be read out or new information written in.

As an example, consider a $256 \times 4$-bit semiconductor RAM store (fig. A2). Each 256-bit store has $16 \times 16$ locations arranged in 16 rows and 16 columns. An 8-bit binary coded address can be used, 4 leads for X and 4 for Y. For example if the address is

X   0011   ($= 3$)

Y   1001   ($= 9$)

then only the location at $X_2$, $Y_8$ will be addressed.

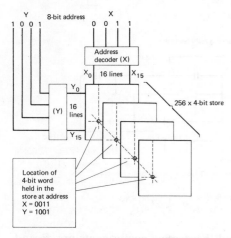

*Figure A2*   **8-bit binary coded address**

## Admittance

Admittance is the ratio of r.m.s. current to r.m.s. voltage for sinusoidal signals. The symbol for admittance is $Y$ and the units are Siemens.

Thus in an a.c. circuit   $Y = \dfrac{I}{V}$ Siemens

Note that admittance is the reciprocal of impedance, i.e.

$$Y = \frac{1}{Z} = \frac{1}{R + jX}$$

$$Y = \frac{R}{R^2 + X^2} - j\frac{X}{R^2 + X^2}$$

$$= G + jB$$

where conductance $G = R/|Z|^2$

and susceptance $B = -j\dfrac{X}{|Z|^2}$ if the impedance is inductive

and $B = +j\dfrac{X}{|Z|^2}$ if the impedance is capacitive.

▶ *Parameter*   ▶ *Field effect transistor*

## Alarm Circuit

Electronics is ideally suited to the task of providing sensitive and effective alarm systems. The general form of a typical alarm system is shown in block diagram form (fig. A3). The change in the input medium (light, temperature, smoke, liquid level, etc.) is detected by a suitable transducer which converts it into an electrical signal. This signal is then compared with a reference level by a comparator or Schmitt trigger circuit. If the input exceeds this preset reference level, then the comparator gives an output which sets the memory device. A memory is often used, especially in situations where the input may return to a lower value or where short duration changes are being detected as for example when a light beam is interrupted. The memory, which may be a bistable circuit or perhaps a thyristor, keeps the alarm operating even if the input falls rapidly back to a low level. In systems where only a short-duration self-resetting alarm is required, the memory is replaced by a monostable of a few seconds duration.

The output from the memory is fed via a drive amplifier to the output device, which may be audible, visual or both. Buzzers, single, two-tone or multi-tone audible warning devices and sirens are all commercially available, the frequency to gain maximum attention being between 1 kHz and 3 kHz. Flashing lights at about 2 Hz are also used and, as shown later in a circuit example, the low-frequency oscillator driving the lamp can be used to gate an audio oscillator driving a loudspeaker. This gives a "bleeping" alarm.

The general requirements for alarm circuits can be listed as

1) Good long-term stability.

2) High reliability, for the circuit may remain unused for long periods before input conditions demand an alarm output.

3) Attention-gaining alarm output.

4) Low power consumption in the quiescent no-alarm state.

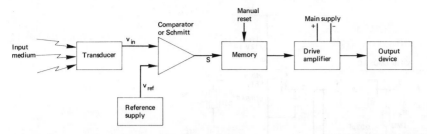

*Figure A3*  Block diagram of typical alarm system

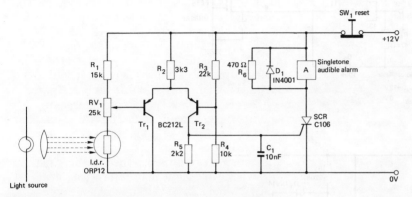

*Figure A4*  Alarm based on the interruption of a light beam

5) Provision of manual test and alarm-reset facilities.

Some of the most effective designs are those which are relatively simple, for in general the lower the component count, the better the reliability. Long-term stability depends to a great extent upon the type of input transducer and its siting. If the transducer is remote from the system, care must be taken with the connecting leads to avoid the problem of electrical noise affecting the system operation. Twisted pairs of wires feeding a differential amplifier and/or input filters must be used. The range of transducers that can be used is very wide (Table A3, p. 16).

**1** *Audible alarm operated by the interruption of a light beam* (fig. A4).

Transistors $Tr_1$ and $Tr_2$ form a differential switch circuit with the reference voltage at $Tr_2$ base set to approximately +4 V by the potential divider $R_3$ and $R_4$. While the light beam falls on the l.d.r. the voltage on $Tr_1$ base (set by $RV_1$) will be less than 4 V ensuring that $Tr_1$ conducts and that $Tr_2$ is off. If the light beam is obstructed, the resistance of the l.d.r. rises,

causing the base voltage of $Tr_1$ to increase above 4 V. $Tr_2$ turns on and the voltage across $R_5$ goes sufficiently positive to trigger the thyristor. In this circuit the thyristor is the memory element because, once it is triggered, it will remain conducting. In this way, the alarm, a single tone at 2.6 kHz, will sound continuously after the light beam is blocked momentarily. The alarm can be reset by operating $SW_1$.

The unit could be modified to be used for a liquid level alarm. While a prism is in air, the light beam is passed to the l.d.r. When the prism is immersed in liquid, of similar refractive index to the glass prism, the light beam no longer reaches the l.d.r. and the alarm will sound.

**2** *Liquid-level alarm* (fig. A5)

This simple example shows how one CMOS i.c. and two VMOS power FETs can be used to give a relatively high-power flashing light and a "bleeping" audible output. The i.c., a quad two input NAND type 4011B, is connected to form two oscillator circuits. Normally both oscillators are held off since test points (a) and (b) will

5

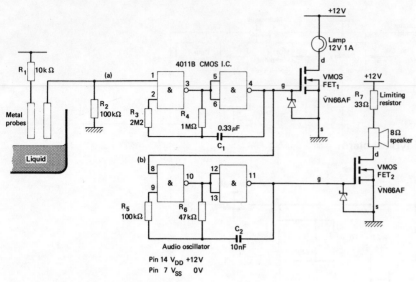

*Figure A5* Liquid-level alarm

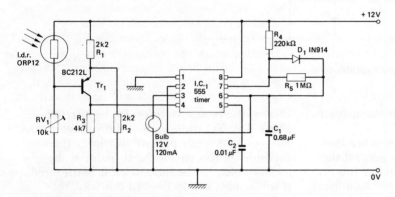

*Figure A6* Dark-activated hazard alarm using a 555 timer

both be at logic 0, i.e. very near zero volts. If the level of the liquid, which must be a conducting type, rises to reach the metal probes, then point (a) goes positive to logic 1, allowing the low-frequency oscillator to run. This puts a low-frequency square wave drive to the gate of $VMOS_1$ and at the same time enables the audio oscillator to give bursts of 2 kHz square wave drive to $VMOS_2$. A big advantage of VMOS devices is that their very high input impedance allows them to be directly connected to the outputs of CMOS gates.

Thus when the liquid reaches the probes, the lamp flashes on and off at a frequency determined by $R_4$ and $C_1$, and at the same time a bleeping audio output at a frequency deter-

mined by $R_6$ and $C_2$ is given from the loudspeaker. Since CMOS and VMOS devices are used the current taken from the +12 V supply during the alarm's off-state is negligible.
▶ *VMOS power FET*   ▶ *CMOS*
▶ *Oscillator (square wave)*

**3** *Dark-activated hazard alarm* (fig. A6)
This uses a 555 timer i.c. wired as a low-frequency astable. While light falls on the l.d.r. its resistance remains relatively low and $Tr_1$ is nonconducting. The voltage on pin 4 of the 555 will be at zero volts which prevents the circuit from oscillating. At a low light level, pre-selected by $RV_1$, $Tr_1$ will conduct and the voltage across $R_3$ will rise positive. The

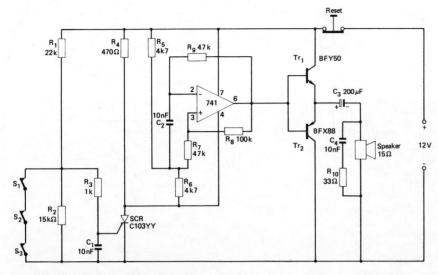

*Figure A7* Security alarm system

capacitor $C_1$ is now free to charge and the circuit will oscillate causing the bulb to flash on and off. The on-time is determined as $C_1$ charges via $R_4$ and $D_1$ and the off time by $C_1$ and $R_5$. With the values shown, the frequency is approximately 1 Hz and the duty cycle 0.1 to 1. The advantage of using a low duty cycle is that the mean current taken from the supply is kept relatively low.

Note that there is no memory section and that the bulb will cease flashing and be off when the ambient light level is sufficient to again cause $Tr_1$ to switch off. Simple modifications can be made to

*a)* Provide a higher output by using the output signal from pin 3 via resistor to switch a power transistor or relay.

*b)* Change the function to a frost or low-temperature alarm by replacing the l.d.r. with a thermistor.

*c)* Obtain an audio output by changing the values of $R_4$, $R_5$ and $C_1$ and connecting a loudspeaker via series resistor and electrolytic capacitor from pin 3 to ground.

▶ *Timer*

**4** *Security alarm* (fig. A7)

This is an alarm that provides an effective audio output if any one of the input switches is operated. In alarm systems such as this the switches should be normally closed types so that any failure of a switch will be detected. The switches could be pressure-sensitive reed types operated by small permanent magnets, or microswitches. The operation is quite straightforward. While the switches $S_1$, $S_2$ and $S_3$ remain closed, the SCR is off and the voltage supplied to the op-amp oscillator circuit is zero. If any one of the switches is operated, even momentarily, the SCR will be switched into a forward conducting state. Nearly the full supply voltage will then be applied to the oscillator which will then generate square waves at about 1.5 kHz.

$Tr_1$ and $Tr_2$ wired as complementary emitter followers provide the necessary drive to the loudspeaker. Note that, once triggered, the SCR remains in the forward conducting state and the alarm can only be silenced by operating the Reset.

**5** *Gas alarm* (fig. A8)

A gas sensor, such as the RS307-733, is a device used to detect the presence of natural gas, propane, iso-butane, or methane, and its output can be used to switch an alarm signal before a dangerous concentration is reached. It consists of two platinum wire elements, one of which is coated with special materials for sensing gas and the other compensates for variations in temperature and humidity.

To operate the device, a supply voltage of 3 V at a current of 140 mA is used. This heats both wires. When a small concentration of gas occurs, the resistance of the sensing platinum wire element will fall. This change of resistance

7

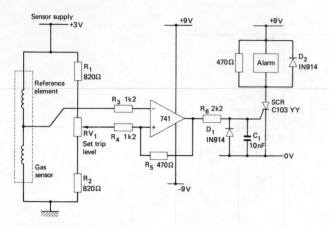

Figure A8 Gas alarm

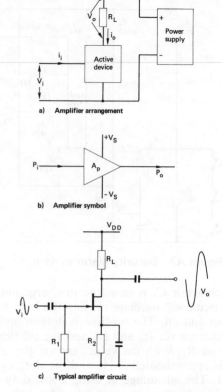

a) Amplifier arrangement

b) Amplifier symbol

c) Typical amplifier circuit

Figure A10 Amplifier arrangement, with amplifier symbol and typical circuit (using a FET)

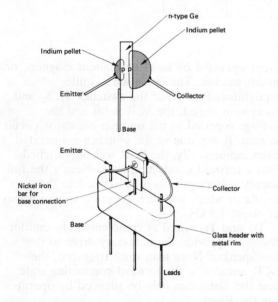

Figure A9 Alloy-junction transistor, with mounting method

can be detected by a bridge and comparator circuit as shown and then used to trigger a thyristor to sound the alarm.

## Alloy-junction Transistor

The germanium alloy-junction process was the first commercially successful method for making transistors (fig. A9). Although a few a.f. power transistors are still being manufactured in this way, the method is mostly of historical interest since it has been superseded by the planar process.

In the alloy-junction method, transistors are made individually. Two pellets of indium, which is an acceptor impurity, are fixed on opposite sides of a thin wafer of n-type germanium (about $50\ \mu m$ thick). The small assembly is held in a jig and heated to about 500°C (which is above the melting point of indium but below that of germanium) to create two p-regions as shown. These will be the emitter and collector with the thin n-space between them (typically $10\ \mu m$) forming the base region. Leads are then added by soldering or welding, and the transistor is sealed into a can.

The process is relatively costly with low yield and can only be used for manufacturing low-frequency transistors. This is because the cut-off frequency $f_T$ of a transistor depends to a great extent on the width of the base region. Typically, $f_T \simeq 1$ MHz for alloyed transistors.

n-p-n transistors were manufactured using p-type germanium and lead-antimony pellets.

▶ Transistor    ▶ Planar process

8

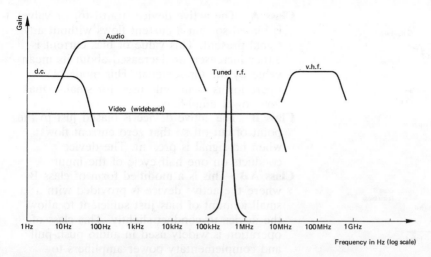

*Figure A11* Classification of amplifiers by frequency response

## Alphanumerics

A set of symbols consisting of the alphabet characters, the digits 0 to 9, and the other special characters on the typewriter keyboard.

## Ampere

The unit of electric current. Electric current is measured in terms of the total electric charge passing a given point in a conductor in unit time.

$$1 \text{ AMPERE} = 1 \text{ COULOMB PER SECOND}$$

This is equivalent to the flow of about $6 \times 10^{18}$ electrons per second.

*Definition* (BS 3763): The unit of electric current called the ampere is that constant current which, if maintained in two parallel rectilinear conductors of infinite length, of negligible circular cross-section, and placed at a distance of one metre apart in a vacuum, would produce between these conductors a force equal to $2 \times 10^{-7}$ newton per metre length.

## Amplifier

Amplifiers are one of the most common of all electronic building blocks. By definition an amplifier is any circuit that provides gain. It receives a low-power input which controls, via an external supply, a larger amount of power at the output.

An amplifier arrangement consists of some active device (transistor, FET, or valve) with biasing components, a source of power, and a load (fig. A10). The input signal is used to control the current flowing through the active device. For example, with a FET in common source mode, the input voltage between gate and source ($V_{GS}$) will control the current flowing from drain to source ($i_{DS}$). Since the output current flows in the load it will develop a voltage across the load so that

$$P_o = V_o i_o \text{ watts}$$
while $\quad P_i = V_i i_i \text{ watts}$

Therefore

$$\text{POWER GAIN } A_p = P_o/P_i$$

In many cases an amplifier may be designed primarily for voltage or current gain:

$$\text{VOLTAGE GAIN } A_v = V_o/V_i$$
$$\text{CURRENT GAIN } A_i = i_o/i_i$$

These are all expressions of gains as ratios; it is usually more convenient to express gain in logarithmic units [▶ *Decibel*]:

$$A_p = 10 \log (P_o/P_i) \text{ dB}$$
$$A_v = 20 \log (V_o/V_i) \text{ dB}$$
$$A_i = 20 \log (i_o/i_i) \text{ dB}$$

provided that the input and output impedances are identical

**9**

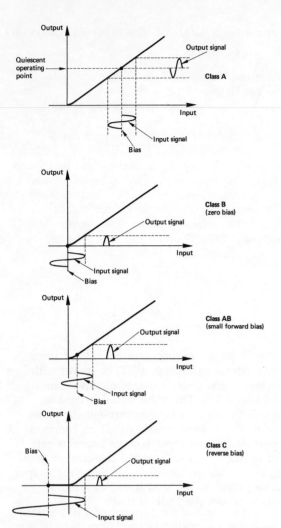

*Figure A12* Biasing arrangements of amplifiers

## Classification of amplifiers

A wide variety of types exists. They are usually described under one or a combination of the following headings.

**1** *Intended use*: power, voltage or current gain.

**2** *Frequency response* (fig. A11)
  d.c. (from zero frequency).
  Audio (15 Hz to 20 kHz).
  Tuned r.f. (narrow band with centre frequency from tens of kHz to hundreds of megahertz).
  Video or pulse (wideband d.c. to 10 MHz).
  v.h.f. (up to thousands of megahertz).

**3** *Method of operation*: which means the biasing arrangement that determines the position of the quiescent operating point (fig. A12).

**Class A**  The active device (transistor or valve) is biased so that a current flows without any signal present. This value of bias current is either increased or decreased about its mean value by the input signal. This mode of operation is commonly used for small-signal low-power amplifiers.

**Class B**  The active device is biased just to the point of cut-off so that zero current flows when no signal is present. The device conducts on one half cycle of the input.

**Class AB**  This is a modified form of class B where the active device is provided with a small amount of bias just sufficient to allow the device to conduct slightly. This class of operation is widely used in audio push-pull and complementary power amplifiers to avoid non-linearity at the cross-over point.

**Class C**  The active device is reverse-biased beyond the point of cut-off so that it only conducts when the amplitude of one half cycle of the input exceeds a relatively large value. This method is used in pulsed and r.f. power amplifiers.

## Single-stage amplifiers

All multi-stage amplifiers are built up by connecting single stages to achieve the desired gain and impedance match. Most active devices used in amplifiers can be connected in three configurations as follows.

**1** *Bipolar transistor*
Common Emitter (CE)  Most commonly used since it provides the highest gain.
Common Base (CB)  Used mainly as v.h.f. and u.h.f. amplifiers. More stable than the CE because only a very small capacitance links input and output. The base tends to act as a screen between the emitter (input) and the collector (output).
Common Collector (CC)  Usually called "emitter follower". Voltage gain just less than unity, high input impedance and low output impedance. Therefore widely used for impedance matching.

**2** *Unipolar transistor* (FET)
Common Source (CS)  Medium voltage gain. Low noise. Very high input impedance.
Common Gate (CG)  Used in some cascode circuits and u.h.f. amplifiers.
Common Drain (CD)  More usually called "source follower". Voltage gain just less than unity, very high imput impedance and low

output impedance. Used for impedance matching.

The table beginning on page 12 shows circuits and appropriate formulae. Typical values of small-signal parameters at 1 kHz are for example:

| BC107<br>Si n-p-n<br>transistor | $h_{ie} = 3k6\,\Omega$<br>$h_{re} = 1.8 \times 10^{-4}$<br>$h_{fe} = 280$<br>$h_{oe} = 24\,\mu S$ | $I_C = 2\,mA$<br>$V_{CE} = 5\,V$<br>$f = 1\,kHz$ |
|---|---|---|
| 2N3819<br>n-channel<br>JFET | $y_{fs} = 2\,mS$<br>$y_{os} = 50\,\mu S$ | $V_{DS} = 15\,V$<br>$V_{GS} = 0\,V$<br>$f = 1\,kHz$ |

▶ *Transistor* (*parameters*)  ▶ *Equivalent circuit*

In the table, $R_L$ is assumed to be 3 kΩ and $R_S$, the generator impedance, 50 Ω.

### Multi-stage amplifiers

The majority of amplifiers employ negative feedback of some kind to stabilise the gain against variations in device parameters, power supply, and temperature, and to modify input and output impedance. In addition n.f.b. reduces internal generated noise and gives wider bandwidth [▶ *Negative feedback*].

A few examples of amplifiers are as follows.

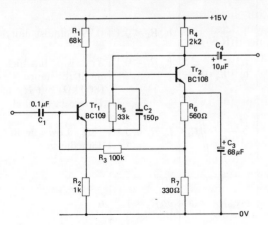

**Figure A13**  Two-stage voltage amplifier

**1**  *Two-stage voltage amplifier* (fig. A13)
A pair of transistors is used with negative feedback via $R_5$ from $Tr_2$ collector to $Tr_1$ emitter. A portion $\beta$ of the output is fed back to oppose the input signal. This feedback fraction is

$$\beta = R_2/(R_2 + R_5)$$

Since the loop gain $A_o\beta \gg 1$, the a.c. gain of the circuit is

$$A_c \simeq 1/\beta = (R_2 + R_5)/R_2 = 34$$

An additional d.c. feedback loop via $R_3$ stabilises the quiescent operating point and, since $R_3$ is in parallel with the input impedance of $Tr_1$, the input impedance at signal frequencies is approximately 100 kΩ. The frequency response is from about 15 Hz to 35 kHz. $C_1$ and $R_{in}$ limit the low-frequency response, and $C_2$ in parallel with $R_5$ limit the high frequencies.

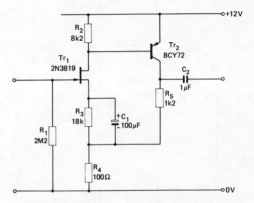

**Figure A14**  FET input pre-amplifier

**2**  *FET input pre-amplifier* (fig. A14)
One of the important features of FETs is that under normal operating conditions they have very high values of input impedance and also a lower noise figure than bipolar transistors.

An n-channel JFET (2N3819) is used as a common source amplifier in the input stage and this is directly coupled to $Tr_2$ a p-n-p transistor in common emitter. Negative feedback is applied via $R_5$ giving a voltage gain of approximately 12. Since the input impedance over the useful frequency range is equal to $R_1$, any high impedance source can be directly coupled to $Tr_1$ gate, i.e. transducers such as crystal microphones and pick-ups. The frequency response is from a few hertz up to about 400 kHz.

Even higher input impedance can be obtained by a technique called bootstrapping. [▶ *Bootstrapping*].
**3**  *Pulse amplifier* (fig. A15)
This arrangement, with voltage gain set by $RV_1$, uses a differential pair $Tr_1$, $Tr_2$. The

| | | Reasonably exact | Approximate | Assuming | Typical value |
|---|---|---|---|---|---|
| **CE** | Current gain | $A_i = \dfrac{+h_{fe}}{1 + h_{oe}R_L}$ | $+h_{fe}$ | $h_{oe}R_L \leqslant 0.1$ | 280 |
| | Voltage gain | $A_v = \dfrac{-h_{fe}R_L}{(h_{ie} + R_S)(1 + h_{oe}R_L)}$ | $\dfrac{-h_{fe}R_L}{h_{ie}}$ | $h_{oe}R_L \leqslant 0.1$ $h_{ie} \gg R_S$ | −233 (minus sign indicates inversion) |
| | Input $Z$ | $Z_i = h_{ie} + h_{re}A_iR_L$ | $h_{ie}$ | $h_{re} \rightarrow 0$ | 3 k6 This is in parallel with $R_1$ and $R_2$ |
| | Output $Z$ | $Z_o = R_L \,\bigg/\!\!\bigg/\, \dfrac{1}{h_{oe} + \dfrac{h_{fe}h_{re}}{h_{ie} + R_S}}$ | $R_L$ | $h_{oe}R_L \leqslant 0.1$ | 3 kΩ |
| **CB** | Current gain | $A_i = \dfrac{h_{fb}}{1 + h_{ob}R_L}$ | $h_{fb}$ | $h_{ob}R_L \leqslant 0.1$ | 0.99 $h_{fb} = \dfrac{h_{fe}}{1 + h_{fe}}$ |
| | Voltage gain | $A_v = \dfrac{h_{fb}R_L}{(R_S + h_{ib})(1 + h_{ob}R_L)}$ | $\dfrac{R_L}{R_S + h_{ib}}$ | $h_{ob}R_L \leqslant 0.1$ | 48 $h_{ib} = \dfrac{h_{ie}}{1 + h_{fe}}$ |
| | Input $Z$ | $Z_i = h_{ib} + h_{rb}A_iR_L$ | $h_{ib}$ | $h_{rb} \rightarrow 0$ | 12.8 Ω |
| | Output $Z$ | $Z_o = R_L \,\bigg/\!\!\bigg/\, \dfrac{1}{h_{of} - \dfrac{h_{fb}h_{rb}}{h_{ib} + R_S}}$ | $R_L$ | $h_{ob}R_L \leqslant 0.1$ | 3 kΩ |
| **CC (emitter follower)** | Current gain | $A_1 = \dfrac{h_{fc}}{1 + h_{oc}R_L}$ | $1 + h_{fe}$ | $h_{oc}R_L \leqslant 0.1$ | 281 $h_{oc} \simeq h_{oe}$ |
| | Voltage gain | $A_v = 1 - \dfrac{h_{ie}}{h_{ie} + A_iR_L}$ | $1 - \dfrac{h_{ie}}{R_{in}}$ | $h_{oc}R_L \leqslant 0.1$ | 0.99 almost unity |
| | Input $Z$ | $Z_i = h_{ie} + A_iR_L$ | $h_{ie} + (1 + h_{fe})R_L$ | $h_{oc}R_L \leqslant 0.1$ | Transistor has high input resistance (800 kΩ) but $R_1$ and $R_2$ are in parallel. |
| | Output $Z$ | $Z_o = R_L \,\bigg/\!\!\bigg/\, \dfrac{1}{h_{oc} + \dfrac{1 + h_{fe}}{h_{ie} + R_S}}$ | $\dfrac{R_S + h_{ie}}{1 + h_{fe}} \,\bigg/\!\!\bigg/\, R_L$ | $h_{oc}R_L \leqslant 0.1$ | 12Ω Low output impedance. |
| **CS** | Voltage gain | $A_v = \dfrac{-y_{fs}R_L}{1 + y_{os}R_L}$ | $-y_{fS}R_L$ | $y_{oS}R_L \ll 1$ | 6 |
| | Input $Z$ | — | $R_1 /\!/ R_2$ | — | Depends on values typically in MΩ |
| | Output $Z$ | $Z_0 = \dfrac{R_L}{1 + y_{oS}R_L}$ | $R_L$ | $y_{oS}R_L \ll 1$ | 3 kΩ |

Note // means "in parallel with"

| | | | | | |
|---|---|---|---|---|---|
| <br>CD<br>(source follower) | **Voltage gain** | $Av = \dfrac{y_{fs}R_L}{1+R_L(y_{fs}+y_{oS})}$ | $\dfrac{R_L}{R_L+(1/y_{fs})}$ | $R_L(y_{fs}+y_{oS}) \ll 1$ | 0.99 almost unity |
| | **Input $Z$** | — | $R_1//R_2$ | — | Very high, depends on values |
| | **Output $Z$** | $Z_0 = \dfrac{R_L+y_{oS}R_L^2}{1+R_L(y_{oS}+y_{fs})}$ | $R_L \left\Vert \dfrac{1}{y_{fs}}\right.$ | $y_{oS} \to 0$ | 1.2 kΩ |
| <br>CG | **Voltage gain** | $A_v = \dfrac{y_{fs}R_L}{1+y_{oS}(R_L+R_S)+y_{fs}R_S}$ | $\dfrac{R_L}{(1/y_{fs})+R_S}$ | $y_{oS}(R_L+R_S)\ll 1$ | 5.5 |
| | **Input $Z$** | $Z_{in} = \dfrac{1}{y_{oS}+y_{fs}}+(R_S+R_L)$ $+\dfrac{y_{oS}}{y_{oS}+y_{fs}}+\dfrac{R_Sy_{fs}}{y_{oS}+y_{fs}}$ | $R_S+\dfrac{1}{y_{fs}}$ | $y_{oS}\ll y_{fs}$ <br> $\dfrac{y_{oS}(R_L+R_S)}{\ll 1}$ | 550 Ω |
| | **Output $Z$** | $Z_0 = \dfrac{R_L}{1+\dfrac{y_{oS}(R_L+R_S)}{1+y_{fs}R_S}}$ | $R_L$ | $\dfrac{y_{oS}(R_L+R_S)}{\ll 1}$ | 3 kΩ |

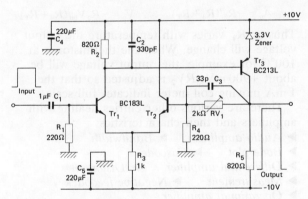

*Figure A15*  Pulse amplifier

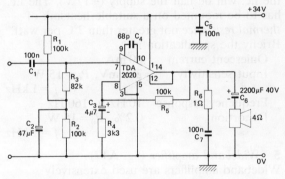

*Figure A16*  Audio power amplifier

differential amplifier is useful since it possesses the capability of rejecting common mode signals and it forms the basis of most op-amp designs. In this circuit a portion of the output is fed back to $Tr_2$ base which is the inverting input of the differential pair. Thus this feedback opposes the input signal on $Tr_1$ base. The feedback fraction is

$$\beta = R_4/(R_4+RV_1)$$

and therefore the pulse gain is

$$A_v \simeq (R_4+RV_1)/R_4 \quad \text{a maximum of 10}$$

**4**  *Audio power amplifier* (fig. A16)
There are many varieties of audio amplifiers and especially i.c. types [▶ *Audio amplifier*].

Here a monolithic audio i.c. type TDA 2020 is wired with a single supply and is capable of 15 W output with a distortion of 0.2%. The i.c. can be used with a symmetrical power supply ($\pm V_S$) in which case the loudspeaker can be directly coupled to pin 14. However in the event of an amplifier fault or one side of the power unit failing, the loudspeaker would be damaged. By using a single supply and a coupling capacitor $C_6$, the speaker is provided with

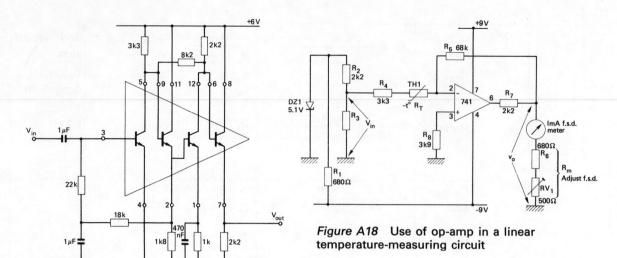

**Figure A17** Wideband amplifier using RCA CA3018 i.c.

**Figure A18** Use of op-amp in a linear temperature-measuring circuit

protection. The non-inverting input (pin 7) is biased to half the supply voltage by $R_1$ and $R_2$ and 100% d.c. feedback is provided by $R_5$. Therefore the quiescent output voltage from the i.c. will be half the supply (+17 V). The i.c. has to be mounted on a suitable heatsink of *thermal resistance* not greater than 2°C per watt. Briefly the specification is

| | |
|---|---|
| Quiescent current | 60 mA |
| Input sensitivity | 260 mV, $P_0 = 15$ W, $f = 1$ kHz |
| Frequency response | 10 Hz to 160 Hz |
| Distortion | 0.2%, $P_0 = 15$ W, $f = 1$ kHz |

**5** *Wideband amplifier* (fig. A17)
Wideband amplifiers are used extensively in t.v., radar and measuring instruments such as c.r.o.s where the requirement is to produce an amplified version of input pulses which may have fast rise and fall times. If $t$ = pulse rise time (10% to 90%), then the maximum frequency (−3 dB) required from the amplifier is

$$f_{max} = 1/2t$$

Thus if the pulse rise time is 50 ns, then $f_{max}$ is 10 MHz. This example based on an i.c. has a bandwidth of 800 Hz to 30 MHz and a gain of 49 dB ± 1 dB.

**6** *Op-amp used in linear temperature measuring circuit* (fig. A18)
A GM472 bead-type thermistor is used in this circuit as the temperature-sensing element. A voltage of about −50 mV is developed across

$R_3$ by the potential divider $R_2$ and $R_3$. The 741 op-amp is connected as an inverting amplifier with negative feedback via $R_5$. The voltage gain of the amplifier is

$$A_v = -R_5/(R_4 + R_T) \qquad \therefore V_o = -R_5 V_{in}/(R_4 + R_T)$$

Thus as $R_T$ varies with temperature the output voltage will change. When the thermistor is at 100°C for example, the output voltage will be about +900 mV. $RV_2$ is adjusted so that the 1 mA moving coil meter indicates full scale.

There are various other entries dealing with amplifiers and their characteristics:
▶ *Audio amplifier*    ▶ *Bandwidth*
▶ *Bode plot*    ▶ *Cascode*
▶ *Differential amplifier*    ▶ *Distortion*
▶ *Measurement*    ▶ *Negative feedback*
▶ *Operational amplifier*

## Amplitude Distortion

Distortion which may occur in an amplifier as a result of non-linearity (fig. A19). For example, if the output of an amplifier is 10 V for an input of 0.1 V but only 18 V (instead of 20 V) for an input of 0.2 V, then amplitude distortion exists. The amount of distortion can usually be minimised by underdriving an amplifier system, for it is near maximum output that the greatest non-linearity generally occurs.

Amplitude distortion is also referred to as "non-linear distortion" and will cause harmonics of the input frequencies to be present at the output. In practice the total harmonic

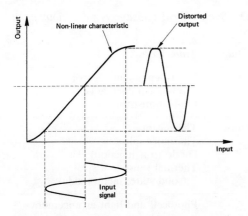

**Figure A19** Amplitude distortion

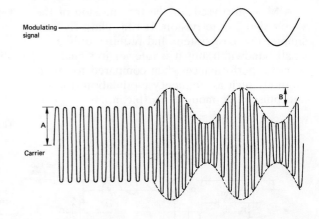

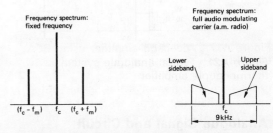

**Figure A20** Carrier amplitude-modulated by fixed-frequency audio sine wave

distortion (THD) is quoted for audio systems at stated power outputs.

▶ *Distortion*

## Amplitude Modulation

The process by which a low-frequency signal is impressed onto a higher-frequency wave by varying the amplitude of the higher-frequency wave while leaving its frequency unchanged. The lower frequency is called the modulating signal and the higher frequency the carrier. This method is used in a.m. radio transmissions to enable the audio frequencies to be transmitted long distances, by superimposing them onto a radio frequency carrier which can then be radiated from an aerial. Radio frequencies are defined as those at which coherent electromagnetic radiation of energy is useful for communication purposes. The lowest radio frequency is about 10 kHz and the range extends up to hundreds of GHz.

In fig. A20 an r.f. carrier is shown amplitude-modulated by a fixed-frequency audio sine wave. The percentage modulation is

$$\frac{B}{A} \times 100\%$$

and the modulated carrier consists of three frequencies, all r.f., as follows:

$f_c$    the carrier

$f_c + f_m$    the upper side frequency

$f_c - f_m$    the lower side frequency

If the carrier is represented by

$$e_c = A \sin \omega_c t \quad \text{where} \quad \omega_c = 2\pi f_c$$

and the modulating signal by

$$e_m = B \sin \omega_m t \quad \text{where} \quad \omega_m = 2\pi f_m$$

then the amplitude of the modulated carrier can be expressed as

$$(A + B \sin \omega_m t) \sin \omega_c t$$

Expanding this gives

$$A \sin \omega_c t + B \sin \omega_m t \sin \omega_c t$$
$$= A \sin \omega_c t + \tfrac{1}{2}B \cos(\omega_c - \omega_m)t - \tfrac{1}{2}B \cos(\omega_c + \omega_m)t$$
$$= A \sin \omega_c t \quad \text{carrier}$$
$$\quad + \tfrac{1}{2}B \sin[(\omega_c - \omega_m)t + \tfrac{1}{2}\pi] \quad \text{lower s.f.}$$
$$\quad + \tfrac{1}{2}B \sin[(\omega_c + \omega_m)t - \tfrac{1}{2}\pi] \quad \text{upper s.f.}$$

If the carrier is modulated with a complex waveform, or a range of audio signals, then a range of frequencies will be present to form upper and lower sidebands. In a.m. radio, since by international agreement stations are spaced 9 kHz apart, the maximum audio frequency that can be used to modulate the carrier is 4.5 kHz. Note that both sidebands are transmitted although only one is demodulated in the receiver.

**15**

A.M. is also used for the transmission of the vision signal in television. It is a relatively simple, low cost system and requires only a small bandwidth, but it is inferior in signal to noise performance when compared to other methods such as frequency modulation (f.m.) and pulse code modulation (p.c.m.).

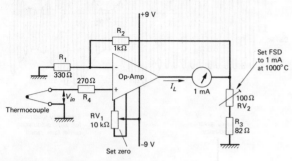

*Figure A21* Typical analogue system: thermocouple amplifier

## Analogue Signal and Circuit

An analogue signal is one that is variable and can, therefore, take any value between some defined limits. A good example of an analogue system would be the amplification of the small e.m.f. from a thermocouple to a level sufficient to give an indication of temperature on a 1 mA movement meter. A linear or analogue i.c., in this case an op-amp, is used as shown in fig. A21 to increase the thermocouple output voltage. As the temperature being measured by the thermocouple varies, so a small change in thermocouple e.m.f. takes place — this is an analogue signal. If this signal were to be coded into a set of pulses for discrete levels, then a digital signal would be generated.

## Analogue-to-Digital Convertor

An ADC is the essential link between the analogue and digital portions of a system. Many inputs, especially those from sensors or transducers, will appear first as analogue signals and these must be coded into digital information before they can be processed, analysed or stored within the digital circuits. The ADC takes the input signal, samples it, and then produces a coded digital word that corresponds to the level of that portion of the analogue signal being examined. The digital output can be either serial (one bit at a time) or parallel with all coded bits presented simultaneously.

**Table A3**  *Some typical transducers for various input mediums*

| Input medium | Transducer |
| --- | --- |
| Light level | Light-dependent resistor (photoconductive cell) Photo-transistor |
| Temperature | Thermistor Thermocouple Diode (p-n junction) Thermal sensor (solid state, metal oxide) |
| Smoke | Photocell and light arrangement |
| Gas | Gas sensor (specially coated platinum wire) |
| Position | Limit switch Magnet and reed relay Potentiometer L.v.d.t. (linear variable differential transformer) Light-activated switch Reflected light-operated opto-coupler |
| Liquid level | Metal probes Thermistor (cools when immersed) Float and potentiometer |
| Intruder | Pressure-sensitive switches Magnet and reed relay Ultrasonic beam Light beam and photocell |

There are many ways of performing analogue-to-digital conversion depending upon the required conversion speed and accuracy. Methods range from the slow and inexpensive to the ultra fast and relatively costly. The most commonly used types of ADC are

*Open loop systems*
Voltage to frequency
Voltage to pulse width
Simultaneous conversion

*Closed loop systems*
Single ramp and counter
Dual ramp
Successive approximation

Note that in many cases the analogue input must be held steady during the conversion process.
▶ *Sample and hold*

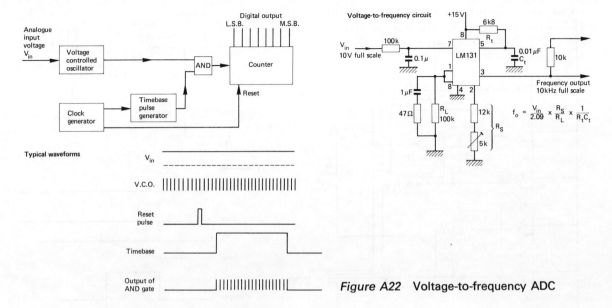

Figure A22 Voltage-to-frequency ADC

## 1 Voltage-to-frequency ADC (fig. A22)

A simple system that can be used where high accuracy is not required. The analogue input is applied to a voltage-controlled oscillator (v.c.o.). This produces an output frequency which is a precise linear function of the analogue input. The counter is first reset and then the time base produces a fixed-time duration pulse to open the AND gate. The v.c.o. output is then applied to the counter for this time period. At the end of the time pulse, the AND gate is closed and the contents of the counter, presented as binary digits on parallel lines, will be proportional to the analogue input signal.

I.C.s are available for the v.c.o. that give high dynamic frequency range and good accuracy. In the example an LM 131 i.c. is connected to give an output frequency from 10 Hz to 10 kHz for an analogue input voltage from zero to 10 V full scale. The typical linearity is ±0.03%.

## 2 Voltage-to-pulse-width ADC (fig. A23)

Here the analogue input voltage is used to control the width of the output pulse from a monostable. A trigger pulse from the control circuit starts the monostable and the output pulse width is then proportional to the applied input voltage. The monostable pulse opens the gate to allow the stable fixed-frequency clock pulses to be counted. At the end of the monostable pulse, the content of the counter will be proportional to the input.

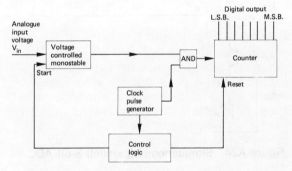

Figure A23 Voltage-to-pulse-width ADC

## 3 Simultaneous conversion (fig. A24)

Sometimes referred to as the parallel method because all the bits for the digital representation are determined simultaneously. This method is therefore the fastest available. The analogue input is applied to a parallel bank of voltage comparators, each of which responds to a discrete level of input voltage.

A circuit example to convert analogue input into only 3-bit digital information gives a good idea of the complexity involved. A constant current source $Tr_1$ supplies 1 mA to a chain of resistors $R_1$ to $R_7$. These set up the levels at which the seven comparators switch: 0.5 V, 1 V, 1.5 V, 2 V up to 3.5 V. If the analogue input just exceeds 1.5 V, then the outputs of comparators A, B and C will be logic 0, while outputs of comparators D, E, F and G will remain high at logic 1. The outputs of the

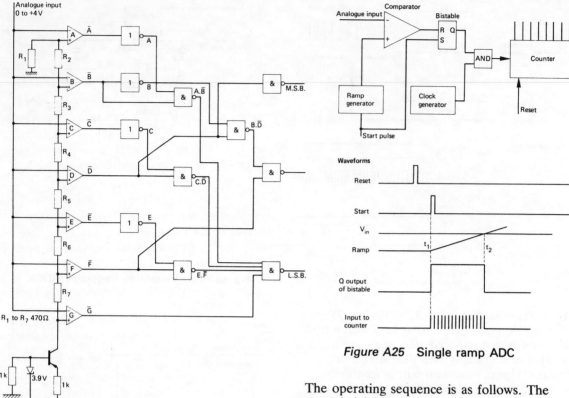

Figure A24  Simultaneous (parallel) 3-bit ADC

Figure A25  Single ramp ADC

comparators are then converted into usable digital form by logic gates.

For $n$ bits of binary information the method requires $(2^n - 1)$ comparators. For example, to give a 6-bit digital word at the output the comparators alone total 63. Naturally the logic to perform the conversion then becomes very complex. This is the main disadvantage of the method. However now that LSI circuits are available, for the ultra fast convertors, this method is being increasingly used.

6 to 12 bits in 50 to 250 ns.

**4  Single ramp and counter** (fig. A25)
The analogue input voltage is compared with a linear ramp. While the input is greater than the ramp from time $t_1$, the counter accumulates clock pulses. As soon as the ramp just exceeds the input at time $t_2$, the counter is stopped. The number of clock pulses counted is then proportional to the analogue input.

The operating sequence is as follows. The counter is initially reset to zero. Then at time $t_1$ the start pulse triggers a linear ramp generator and at the same time sets the bistable. The $Q$ output of the bistable rises high and allows pulses from the stable clock generator to be fed to the counter. While the analogue input is greater than the ramp, pulses from the clock generator will continue to be accumulated by the counter. When the ramp voltage just exceeds the value of the analogue input, the comparator output switches high (time $t_2$) and resets the bistable. The AND gate closes and the digital output from the counter is then proportional to the analogue input.

The method is very useful for medium accuracy and slow speed. Limiting factors are linearity of the ramp, input offset in the comparator and frequency accuracy of the clock pulse generator.

12 to 16 bits in 20 to 200 ms.

**5  Dual ramp and counter** (fig. A26)
This is very similar to the single ramp technique but gives far superior performance. The non-linearities of the integrator are self-correcting and there is no need for a highly stable clock generator. The system is used

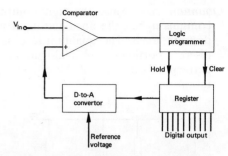

**Figure A27** Successive approximation ADC

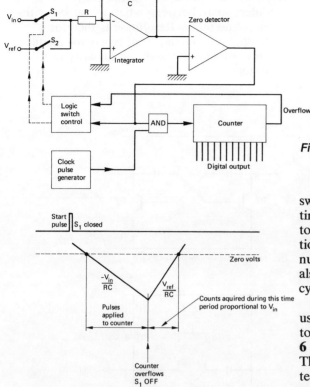

**Figure A26** Dual ramp ADC

extensively in digital voltmeters. The basic operation is as follows.

Initially both switches $S_1$ and $S_2$ are off and the counter is reset. Following this, the control logic operates $S_1$ to apply $V_{in}$, the analogue input, to the integrator. The output of the integrator will be a negative-going ramp with a slope of $-V_{in}/RC$. At the instant this ramp goes through zero, going negative, the output of the zero detector switches high to allow clock pulses to be fed to the counter. The counter then acquires clock pulses until it reaches maximum count $2^n$, which will be 4096 for a 12 bit unit. As the counter overflows and returns to zero, the control logic switches $S_1$ off and $S_2$ on. This applies the reference voltage to the input of the integrator. The output of the integrator now ramps, from the previous negative value proportional to $V_{in}$, in the opposite direction towards zero with a slope of $+V_{ref}/RC$. The counter continues to count until the integrator output again crosses zero. At this point the output of the zero detector

switches low and the AND gate closes. The time taken for the reference ramp $(+V_{ref}/RC)$ to reach zero from the negative value is proportional to the input voltage. Therefore the number of counts acquired during this time is also proportional to the analogue input. The cycle then repeats.

Since the components of the integrators are used for both ramps, any non-linearities tend to be self-correcting.

**6** *Successive approximation* (fig. A27)
The system requires a logic programmer, register, and a digital-to-analogue convertor. I.C.s are available for this and the method gives a reasonable compromise between circuit complexity and speed. A typical conversion time for 12 bits may be as fast as 25 $\mu$s. At the start of the conversion the most significant bit is applied to the D-to-A convertor. The output from the D-to-A is compared with the analogue input by the comparator. If the D-to-A output is larger than $V_{in}$, the 1 is removed from that bit and placed in the next most significant bit for comparison. This process continues until a point of balance is reached, that is when $V_{in}$ is just greater than the output from the D-to-A convertor. When this is achieved the digital output from the register is equivalent to the analogue input.

▶ *Digital-to-analogue convertor*
Explanation of some of the terms used in A-to-D convertors.

*Conversion time* The time interval between the command being given to perform the conversion and the appearance at the output of the complete digital equivalent of the analogue input.

*Conversion rate* The frequency at which conversions can be made. This has to take into account both conversion and recovery times.

**Quantum level** Since the digital output contains $n$ bits, the analogue input has $2^n$ discrete levels. The quantum level is the discrete portion of the analogue input that corresponds to 1 bit of the digital representation.

**Resolution** The ability of the convertor to differentiate between adjacent values of the analogue input. Resolution is therefore limited by the number of bits in the digital output and noise and non-linearity in the system.

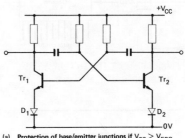

(a)  Protection of base/emitter junctions if $V_{CC} > V_{EBO}$

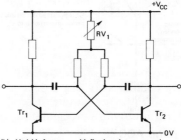

(b)  Variable frequency with fixed mark-to-space ratio

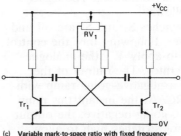

(c)  Variable mark-to-space ratio with fixed frequency

**Figure A28a  Basic astable multivibrator**
Neglecting the effect of $V_{BE}$ and $V_{CE(SAT)}$
$f \approx 1/0.7(C_1R_1 + C_2R_2)$
If $C_1R_1 = C_2R_2 = CR$, then $f = 1/1.4CR$.
Max. frequency typically 1 MHz.
Min. frequency typically 1 Hz.
Timing resistor should be $R_1 = 10R_{C1}$,
$R_2 \approx 10R_{C2}$

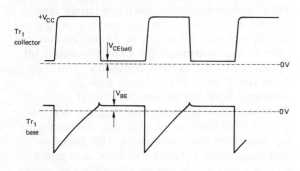

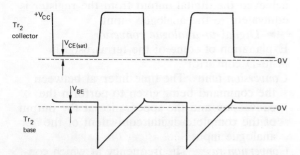

**Figure A28b  Waveforms in basic astable multivibrator**

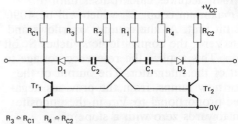

$R_3 \triangleq R_{C1}$    $R_4 \triangleq R_{C2}$

d)  Improvement to leading edge rise time

**Figure A29  Modifications to astable**

*Accuracy* The degree to which the digital output is a true representation of the input. The sources of error are quantization, non-linearity, noise, and short-term drift. Typical accuracy is $\pm 0.05\% \pm \frac{1}{2}$ LSB.

*Precision* The repeatability of successive measurements. This is mainly limited by the small quantization error that must be present. If an analogue input falls in the mid-point of a quantum level then successive measurements can be $\pm 1$ LSB.

## Angstrom Unit

Used as a unit of wavelength particularly for visible and ultra violet waves.

$$1 \text{ Angstrom unit } (\text{Å}) = 10^{-10} \text{ m}$$
$$= 10^{-8} \text{ cm}$$
$$= 10^{-4} \text{ } \mu\text{m}$$

For example the wavelength of the visible light given off from an LED could be quoted as 655 nm or 6550 Å.

## Astable Multivibrator

A circuit that has two partially-stable states and which oscillates continuously to produce a square or pulse waveform at its output (figs. A28, A29). The term multivibrator, applied also to the monostable and bistable circuits, is used because the output signal is rich in harmonics.

▶ *Clock pulse generator*

## Asymmetrical

Usually used to describe a waveform that has a mark which is unequal to the space, i.e. a waveform that has a mark-to-space ratio that does not equal unity.

## Asynchronous Operation

A digital system or part of a system that is not controlled by an external clock. In an asynchronous system the signal to start the next operation is the previous operation. A good example of this is a ripple-through (asynchronous) counter (fig. A30). The output of each flip-flop, on its negative edge, triggers the next. Such counters are basically simple to construct but, as with other non-synchronised systems, a delay builds up as shown and the changes of state at the three outputs on count 8 are not at the same time. This can cause false outputs or glitches when the counter is decoded.

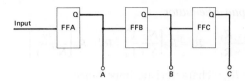

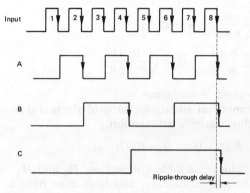

*Figure A30* Ripple-through asynchronous counter

## Attenuator

An attenuator is a network of resistors which when introduced into a circuit will reduce the voltage, current, or power, by a fixed ratio. They are used extensively in measuring instruments, signal generators and communication systems.

A potential divider (fig. A31) is an example of a simple attenuator.

$$v_o = v_i\left(\frac{R_2}{R_1 + R_2}\right)$$

and attenuation $\quad N = \dfrac{v_i}{v_o} = \dfrac{R_1 + R_2}{R_2}$

*Figure A31* Potential divider as attenuator

Such a circuit can be used for matching a signal from a high impedance source to one of lower impedance. However it is usually required that the introduction of an attenuator into a circuit does not affect the existing impedance relationships, i.e. the attenuator has the same characteristic input and output impedance.

## 1  T-pad attenuator

$$N = \frac{R_0 + R_1}{R_0 - R_1} \quad R_1 = R_0\left[\frac{N-1}{N+1}\right] \quad R_2 = R_0\left[\frac{2N}{N^2-1}\right]$$

where $R_0$ = characteristic impedance
$= \sqrt{[R_1^2 + 2R_1 R_2]}$.

## 2  $\pi$-attenuator

$$N = \frac{R_2 - R_0}{R_2 + R_0} \quad R_1 = R_0\frac{N^2-1}{2N}$$

$$R_2 = R_0\frac{N+1}{N-1} \quad R_0 = \sqrt{\frac{R_1 R_2^2}{R_1 + 2R_2}}$$

## 3  Bridged-T attenuator

A symmetrical attenuator particularly useful in providing variable attenuation.

$$R_1 = R_0(N-1) \quad R_2 = R_0/(N-1)$$

To achieve variable attenuation, $R_1$ and $R_2$ must be ganged together and both must have a logarithmic law.

For the T-attenuator, with $R_0 = 50\,\Omega$, the following table gives values for $R_1$ and $R_2$ for attenuation ($N$) in dB.

| Attenuation in dB | $R_1\,\Omega$ | $R_2\,\Omega$ |
|---|---|---|
| 2 | 5.73 | 215 |
| 3 | 8.55 | 142 |
| 4 | 11.3 | 105 |
| 6 | 16.6 | 66.9 |
| 8 | 21.5 | 47.3 |
| 10 | 26 | 35.1 |
| 12 | 29.9 | 26.8 |
| 16 | 36.3 | 16.3 |
| 20 | 40.9 | 10.1 |
| 30 | 46.9 | 3.17 |
| 40 | 49 | 1 |

If for example a T-attenuator of 50 $\Omega$ impedance and 60 dB attenuation is required, then 3 T-sections of 20 dB each can be cascaded (fig. A32).

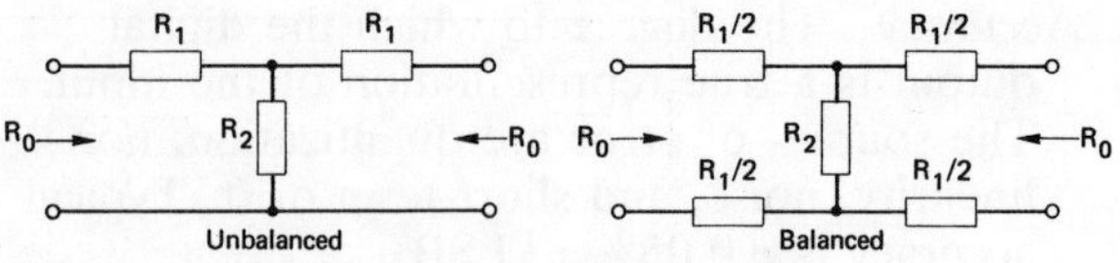

## Audio Amplifier

These are systems designed to amplify signals in the range of 15 Hz up to 20 kHz. A typical system would consist of a pre-amplifier, a tone control circuit and a power amplifier. Fig. A33 shows one channel of a stereo system capable of 10 watts per channel with a total harmonic distortion of less than 0.1%. The pre-amplifier accepts signals from a magnetic cartridge and

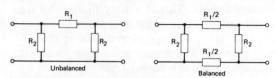

**Figure A31a  T-pad attenuator**

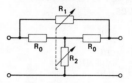

**Figure A31b  $\pi$-attenuator**

**Figure A31c  Bridged-T attenuator**

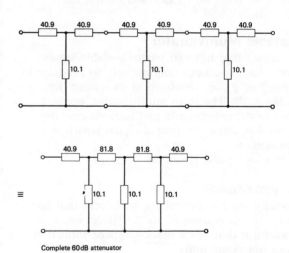

**Figure A32**

the amplified signals are passed to a standard tone control network which provides a bass control of about ±15 dB at 100 Hz and a treble control of ±15 dB at 10 kHz. The power output stage uses a TDA 2030 i.c. which has built-in short circuit protection. Several i.c.s are available for audio applications, Pin-out diagrams are shown in fig. A34 for the i.c.s used in the circuit.

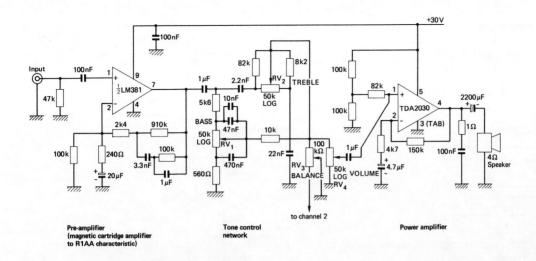

*Figure A33* Audio amplifier system

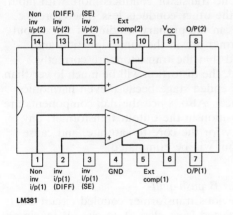

LM381

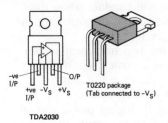

TDA2030

*Figure A34* I.C.s for audio application

The important requirements of audio power amplifiers are

*a*) High power output with good efficiency. Maximum power must be transferred to the load and heat loss from the amplifier minimised.

*b*) Low values of harmonic and intermodulation distortion. This implies that the amplifier has good linearity and negligible cross-over distortion.

*c*) Flat response over the audio range.

*d*) High sensitivity. The r.m.s. voltage required at the input to give maximum power output. This should usually be less than 1 V.

*e*) Protection circuits: to protect the power amplifier in case of an output short circuit and to protect the speaker in the event of an amplifier fault.

The pre-amplifier must be capable of amplifying low-level signals with low noise. The LM 381 in the example has an internal power supply decoupling circuit giving 120 dB supply rejection so that 100 Hz "ripple" or noise from the supply should not affect the system.

Configurations for power amplifiers using discrete components are shown in fig. A35.

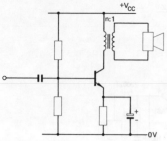

1　Class A single ended

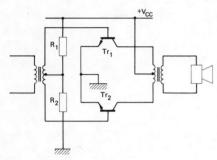

2　Class A push-pull

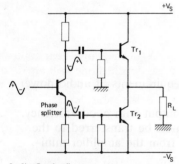

3　Class B push-pull

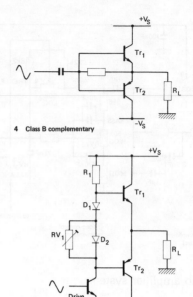

4　Class B complementary

5　Class AB complementary

Figure A35　Power amplifier configurations

**1　Class A single ended**

Transformer coupling is essential to match the speaker to the relatively high output resistance of the transistor.

$$n = \sqrt{\frac{R'_L}{R_L}}$$

where $n$ = turns ratio of the transformer
$R_L$ = resistance of speaker
$R'_L$ = optimum load.

Conversion efficiency is

$$\eta = \frac{\text{Signal power to load}}{\text{d.c. power supplied}} \times 100\%$$

Theoretical maximum 50%.

Disadvantages: poor efficiency, high levels of distortion.

**2　Class A push-pull**

While one transistor conducts more with input signal, the other conducts less. Therefore a.c. currents in the primary winding of the output transformer are in the same direction and, provided that the transistors are correctly matched, the distortion will be much lower than a single ended stage because even harmonics will cancel. Also, since the d.c. components are in opposition in the output transformer, any tendency for the core to saturate, and cause distortion, will be eliminated.

**3　Class B push-pull**

The previous transformer coupled circuit can be converted from class A to class B by simply removing the bias resistor $R_1$. The transistors are then cut off. One will conduct when the input signal goes positive and the other on the input's negative half cycle.

Fig. 3 shows that the two transformers can be eliminated by using a power supply with a centre tap to ground (giving $\pm V_S$) and by providing drive from a phase splitter.

Advantages of class B: higher power output, high efficiency (theoretical maximum of 78.5%), negligible power loss at no signal conditions.

**Table A4** *Audio i.c. amplifier data*

*Dual pre-amplifiers*

| Type | Voltage gain | Supply voltage | Channel separation | Input resistance |
|---|---|---|---|---|
| LM 381 | 110 dB | 40 V max | 60 dB | 100 kΩ |
| PA 239 | 68 dB | 16 V max | 90 dB | 100 kΩ |

*Output amplifiers*

| Type | Power output | Supply voltage | Distortion |
|---|---|---|---|
| LM 380 | 3 W/8Ω | 8 to 22 V | — |
| TAA 611 | 3.3 W/8Ω | 6 to 16 V | — |
| TAA 621 | 4 W/16Ω | 6 to 24 V | — |
| TAA 641A | 2.2 W/4Ω | 6 to 12 V | — |
| TBA 800 | 2.5 W/16Ω | 30 V max | — |
| TBA 810 | 7 W/4Ω | 120 V max | — |
| TBA 820 | 2 W/8Ω | 3 to 16 V | — |
| SN 76008 | 10 W/4Ω | 26 V max | 1% at 6 W |
| TCA 830 | 4.2 W/4Ω | 4 to 20 V | |
| TCA 940 | 10 W/4Ω | 24 V max | |
| TDA 2010 | 10 W/4Ω | ±14 V | 1% at 10 W |
| TDA 2020 | 15 W/4Ω | ±14 V | 1% at 15 W |
| TDA 2030 | 21 W/8Ω | ±6 to ±18 V | 0.1% at 13 W |

The main disadvantage of class B is that cross-over distortion will occur unless large amounts of negative feedback are used and/or the transistor is supplied with a small amount of forward bias. This bias overcomes the base/emitter threshold voltage of the transistors and puts the amplifier into Class AB.

### 4/5 Complementary output stages

These use a matched pair of n-p-n and p-n-p power transistors in either the CE or as shown in the CC configuration. When the input goes positive, $Tr_1$ conducts to supply current to the load, and on the input's negative half cycle $Tr_2$ conducts. Cross-over distortion will exist unless some forward bias is supplied.

Fig. 5 shows a typical class AB complementary circuit where forward bias is provided to both transistors by diodes $D_1$ and $D_2$. The amount of bias has to be carefully adjusted to avoid the possibility of both transistors conducting heavily. The volt drop across the two diodes is set to about 1 V which is sufficient to allow only a few milliamps to flow through the output transistors. In addition the diodes provide some degree of thermal stability. As the output transistors warm up, the required $V_{BE}$ of each will fall by approximately 2 mV/°C. If

the diodes are mounted on the same heat sink, then the volt drop across each diode will also fall by 2 mV/°C thus ensuring that the amount of bias provided stays at the same level.

### Auto-transformer

This is a transformer that consists of one electrically continuous winding with either fixed taps or a moveable slide (fig. A36). It is used where part of the winding is common to both secondary and primary and has the advantage of small physical size and lower losses than the double winding arrangement. The disadvantage is that there is no isolation between the secondary and primary windings.

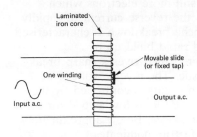

*Figure A36* Auto-transformer

## Availability

The purpose of maintenance is to provide the user with maximum use, or maximum up-time. An instrument may have excellent reliability, i.e. have a low chance of failure during operation, but if and when a failure does occur the repair time (or down-time) must be short to keep availability high.

$$\text{Availability} = \frac{\text{MTBF}}{\text{MTBF} + \text{MTTR}}$$

where MTBF is the mean time between failures and MTTR is the mean time to repair any fault. Availability can be expressed as a fraction or as a percentage.

*Example*: An instrument has an MTBF of 1000 hours and an MTTR of 100 hours.

$$\text{Availability} = 1000/(1000 + 100)$$
$$= 0.91 \text{ or } 91\%$$

If the average repair time increased to say 500 hours, then the availability of the instrument would fall.

$$\text{New value of availability} = 1000/(1000 + 500)$$
$$= 0.67 \text{ or } 67\%$$

▶ *Maintenance*

## Avalanche Breakdown

Avalanche breakdown occurs in semiconductor p-n junctions when the voltage across the junction in the reverse direction is large enough to give the few free electrons sufficient energy to dislodge, by impact, valence electrons. These dislodged electrons will themselves be accelerated to remove still more electrons which therefore causes the reverse current to rapidly increase. Avalanche breakdown is characterised by a sudden and rapid build up of hole/electron pairs which gives a sharp breakdown characteristic. The reverse voltage at which breakdown occurs is controlled by the doping level and can be as low as 6 V for some voltage regulator diodes. Below approximately 6 V the zener effect predominates.

▶ *Zener and voltage regulator diode*

## Back EMF

If the current flowing through an inductor is varied it will set up a changing magnetic field which induces a voltage across the inductor. This voltage is called the back e.m.f. and it tends to oppose the change in current. Thus when a current in an inductor is suddenly switched off, the collapsing magnetic field generates a voltage across the inductor which tends to maintain the current. If the change in current is very rapid, the back e.m.f. voltage may be several hundred volts.

$$\text{Back e.m.f. } e = -L\frac{di}{dt}$$

For example suppose a transistor is used to switch a relay coil as shown in fig. B1. When

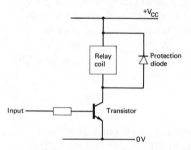

*Figure B1*

the transistor is switched off, the current in the relay coil will fall to zero in a few microseconds and a very large back e.m.f. will be generated, taking the collector of the transistor perhaps several hundred volts positive. To avoid damage to the transistor a suppression circuit must be wired across the relay coil. In most cases a silicon diode is used.

## Backlash

In most mechanical linkages, such as the gears in a control system, there will exist a small amount of free movement. If the gears are rotated clockwise, then stopped, and moved counter-clockwise, there will be a small movement in the gears before they mesh. This is called backlash.

The same term is sometimes applied to electronic trigger circuits, such as the Schmitt, to describe the hysteresis effect. Backlash in an electronic circuit would mean the difference between the upper and lower trip points. For example a trigger circuit that switches on at 5 V (upper trip point) may not switch back to

its previous state until the input is below 3 V. In this case the backlash in the circuit would be 2 V.

▶ *Schmitt trigger circuit*

## Backward Diode

A semiconductor diode that is used in the reverse mode as a detector of microwave signals, typically up to frequencies of 40 GHz. It is basically a very heavily doped p-n junction, giving an ultra thin depletion layer (fig. B2). The diode therefore goes into breakdown in the reverse direction for only small reverse voltages. In the forward direction it behaves as a normal diode, a germanium device requiring 200 mV forward bias for conduction. The doping level, although very high, is not sufficient to give tunnel diode characteristics.

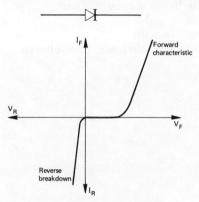

*Figure B2*   Backward diode

## Band Limiting

This refers to the use of filters to limit the frequency content of a signal. In systems where there is a restricted bandwidth it is essential to limit the maximum and the minimum frequencies of the input signal. For example in frequency division multiplexing each message channel is allocated a particular band of frequencies. The input message modulates a fixed carrier and, to prevent overlap, the resulting modulated signal must be band-limited by a band pass filter before transmission.

## Bandwidth

The bandwidth of an electronic system refers to the capability of the system to either amplify, transmit or process signals of different frequency. It is the frequency range of input

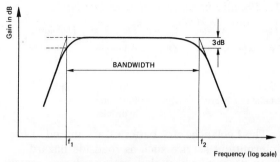

*Figure B3*   Bandwidth

signals to which the system responds. For example an audio power amplifier may be required to amplify signals within the frequency range of 20 Hz up to 20 kHz. Between these two frequencies the gain of the amplifier should be substantially flat. The 20 Hz and 20 kHz frequencies are called the lower and upper cut-off frequencies and represent the frequencies at which the power output of the amplifier has fallen to 50% of its midband value, in other words the output is 3 dB down. The bandwidth is therefore the difference between the upper and lower cut off frequencies (fig. B3).

Typical bandwidth requirements of transmission systems are as follows.

| | |
|---|---|
| Telegraphy (Morse code at 25 words/min) | 100 Hz |
| Telephony (speech) | 300 Hz to 3400 Hz (4 kHz is taken as the bandwidth of one telephone channel) |
| A.M. radio (double side-band) | 9 kHz |
| F.M. radio | 200 kHz |
| T.V. video signals (625 lines) | 5.5 MHz |

## Battery

A battery is a combination of cells connected together to provide a useful source of electrical energy. Apart from their obvious use in portable instruments there are several other important applications of batteries as power sources; these include:

   Roadside hazard lamps
   Location beacons
   Alarm systems
   Standby and emergency power supplies.

## 1  Primary batteries — non-rechargeable

| Type | Anode | Cathode | Electrolyte | Depolariser |
|---|---|---|---|---|
| Leclanché | Carbon | Zinc | Ammonium chloride | Manganese dioxide |
| Menotti (dry) | Carbon | Zinc | Zinc sulphate | Copper sulphate |
| Mallory | Mercury | Zinc | Potassium hydroxide | — |

The Leclanché dry batteries are intended for intermittent service such as roadside hazard lamp flashers, where they recover during rest periods by the action of the depolariser. The mercury cell is more suitable for continuous operation, has excellent regulation and good storage properties.

## 2  Secondary battery systems — rechargeable

| Type | Anode | Cathode | Electrolyte | |
|---|---|---|---|---|
| Lead-acid | Lead peroxide | Lead | Sulphuric acid | Must be charged from a constant-voltage source |
| Nickel-iron | Nickel hydroxide and nickel | Iron and mercury | Potassium hydroxide | |
| Nickel-Cadmium | Nickel hydroxide | Cadmium | | Must be charged from a constant-current source |

Applications of secondary battery systems:
Lead-acid: standby emergency installations.
Ni-Cad (sintered cells): extremely rugged and give long service life; relatively high self-discharge current and should therefore not be left for long period without being recharged.
Ni-Cad (mass plate cells): physically small and rugged; they have low self-discharge currents and are therefore ideal for portable equipment.
Suitable charging circuits for batteries are shown in fig. B4.
The choice of a battery system depends upon many factors.

1) Dimensions and weight
2) Required capacity in ampere hours
3) Voltage and regulation during discharge
4) End point voltage
5) Type of discharge, i.e. continuous, intermittent or pulsed
6) Storage and shelf life
7) Performance with environmental effects.

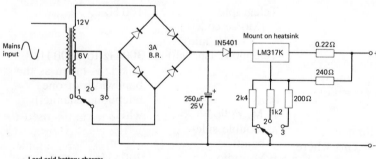

**Lead-acid battery charger**

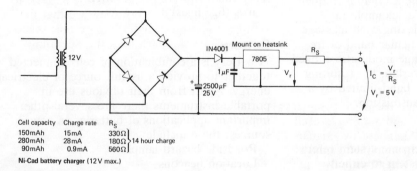

| Cell capacity | Charge rate | $R_S$ | |
|---|---|---|---|
| 150mAh | 15mA | 330Ω | |
| 280mAh | 28mA | 180Ω | 14 hour charge |
| 90mAh | 0.9mA | 560Ω | |

**Ni-Cad battery charger (12 V max.)**

$$I_C = \frac{V_r}{R_S}$$

$$V_r = 5V$$

*Figure B4*  **Battery-charging circuits**

**Table B1** *Representative ratings of batteries*

| Battery type | Voltage charged | Voltage discharged | Capacity | Shelf life at 20°C | Storage time | Discharge | Charging |
|---|---|---|---|---|---|---|---|
| Ni-Cad (sintered cells) | 1.27 V 8.90 V | 1 V 7 V | 180 mAh 1.2 Ah } | 40 days | 5 years | 700 cycles* | Constant current source. Cells are charged in series. |
| Ni-Cad (mass plates) | 1.27 V | 1 V | 150 mAh 150 mAh } | 10 months | 5 years | 300 cycles | Same as for sintered cells but charging rate must be lower. |
| Lead acid | 12.9 V | 10.2 V | 5 Ah | 18 months | 5 years | 200 cycles | Constant voltage parallel or series when batteries have same Ah capacity and state of charge. |

\* The number of energy cycles is constant, i.e. for a battery quoted at 700 cycles and which is only discharged by say 10%, then the total cycle life is $\frac{100}{10} \times 700 = 7000$ cycles.

## Baud

The unit of signalling speed used in telegraphy.

1 Baud = 1 pulse per second

Named after the French telegraph engineer J. M. Baudot (1845–1903) who constructed the first successful teleprinter.

## Beat Frequency Oscillator

When two sine waves of different frequency are mixed, the resulting output will contain components of the sum and difference frequencies as well as the original frequencies, i.e.

$f_1$ and $f_2$
$f_1 + f_2$ sum frequency
$f_1 - f_2$ difference frequency

The difference and the sum can be said to be produced by the two frequencies $f_1$ and $f_2$ "beating" together. This principle can be used in signal generators to give a wide range of output frequencies. For example by using the difference between two high frequency signals the output frequency can be made variable from practically zero to a high value.

A stable crystal-controlled oscillator produces the reference frequency $f_1$ and a variable oscillator is used for $f_2$ (fig. B5). These two signals are mixed and the resultant amplified signal is passed through a low pass filter that only allows the difference $(f_1 - f_2)$ to appear at the output.

## Bel

A logarithmic unit used to express the ratio of two values of power.

Power gain $= \log_{10} (P_o/P_i)$ bels

The bel is rather a large unit and in practice tenths of a bel, called decibels, are more commonly used. Then

Power gain $= 10 \log_{10} (P_o/P_i)$ dB

▶ *Decibel*

## Bias

A bias voltage or current is one that is used to fix the operating point on the characteristic of an active electronic device. Take, for example, a class A small-signal transistor amplifier. Some d.c. value of forward bias must be applied to the base/emitter junction to ensure that the transistor is conducting. The base bias sets up a collector current which fixes the operating

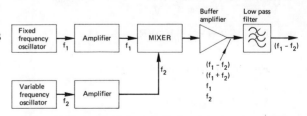

*Figure B5* Principle of beat frequency oscillator

**29**

point on the linear portion of the transistor characteristic. This allows the output signal to swing equally about this quiescent point with minimum distortion. Amplifiers are classified by the type of bias used [▶ *Amplifiers*]. Some class A bias circuits are as follows.

## Bipolar transistors

**1** *Simple fixed bias* (fig. B6a)
$V_{CE}$ should be $\frac{1}{2}V_{CC}$ approximately.

$$V_{CE} = V_{CC} - I_C R_C$$

$$I_B = I_C/h_{FE}$$

$$R_B = (V_{CC} - V_{BE})/I_B$$

This circuit provides no stabilisation of the operating point.

**2** *Collector feedback bias* (fig. B6b)
An improvement over **1** since some degree of stabilisation is provided. If the collector voltage falls for some reason, the base current also falls which reduces the collector current and therefore tends to increase the collector voltages.

If $I_C \gg I_B$

$$V_{CE} \simeq V_{CC} - I_C R_C$$

$$I_B = I_C/h_{FE}$$

$$R_B \simeq [V_{CC} - (I_C R_C + V_{BE})]/I_B$$

**3** *Potential divider bias* (fig. B6c)
Both circuits **1** and **2** can be improved by the addition of an emitter resistor. However for best stabilisation the potential divider bias is recommended.

Make $I_1 \geqslant 10 I_B$ and the ratio $R_1$ to $R_2$ typically 6:1.

$$I_B = I_C/h_{FE(min)}$$

$$I_C \simeq I_E = V_E/R_E$$

where $V_E = V_B - V_{BE}$.

$V_E$ should be small compared to $V_{CE}$.

$$V_{CE} = V_{CC} - (I_C R_C + V_E)$$

*Example*   $V_{CC} = 10$ V, $R_C = 3$ k9, $h_{FE(min)} = 100$
$V_E = 1$ V, $V_{CE} = 4.5$ V

$$I_C = [V_{CC} - (V_{CE} + V_E)]/R_C = 4.5/3 \text{ k9}$$
$$= 1.15 \text{ mA} \quad (I_C)$$

$$R_E = V_E/I_E = 1/1.15 \text{ mA} = 869.6 \ \Omega$$
$$\equiv 820 \ \Omega \text{ n.p.v.} \quad (R_E)$$

$$V_B = V_E + V_{BE} = 1.6 \text{ V}$$

$$I_B = I_C/h_{FE(min)} = 1.15 \text{ mA}/100 = 11.5 \ \mu A$$

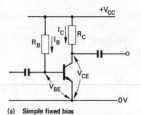
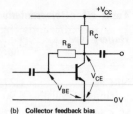

(a)   Simple fixed bias      (b)   Collector feedback bias

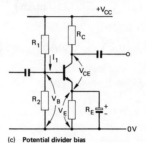

(c)   Potential divider bias

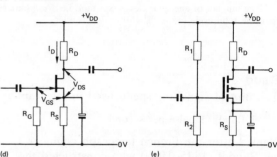

(d)                                    (e)

**Figure B6**  Bias circuits for bipolar transistors and field effect transistors

Therefore, make the current through the potential divider $I_1 = 120 \ \mu A$. Then

$$R_2 = V_B/I_1 = 1.6/120 \ \mu A = 13.33 \text{ k}\Omega$$
$$\equiv 12 \text{ k}\Omega \text{ n.p.v.} \quad (R_2)$$

$$R_1 = (V_{CC} - V_B)/I_1 = 8.4/120 \ \mu A = 70 \text{ k}\Omega$$
$$\equiv 68 \text{ k}\Omega \text{ n.p.v.} \quad (R_1)$$

The potential divider bias is an example of negative feedback. The bias voltage is fixed and held relatively constant by $R_1$ and $R_2$ and the required collector current is set by $R_E$. Suppose the collector current increases (transistor replaced by one with higher gain), then $V_E$ will rise slightly but, since $V_B$ is fixed, the resulting base/emitter voltage $V_{BE}$ will fall. Thus the forward bias to the transistor decreases, which tends to keep the collector current constant and to hold the operating point fixed.

An electrolytic capacitor must be fitted as shown to ensure that a.c. signals do not appear on the emitter. The capacitor is referred to as the bypass or decoupling capacitor.

## Field effect transistors

**1** *Automatic bias* (fig. B6d)
JFETs or depletion mode MOSFETS.

$$-V_{GS} = I_D R_S \qquad V_{DS} = V_{DD} - I_D R_D$$

As the drain current flows, a voltage is set up across $R_S$ that causes the source to rise positive. Since with a FET virtually no gate current flows, the voltage at the gate connected by $R_G$ to ground is nearly zero. In this way an automatic reverse bias voltage is set up between gate and source which sets the drain current. In practice $R_S$ may have to be a potentiometer (1 kΩ) to allow for device parameter spread.

**2** *Improved-bias JFETs or MOSFETs* (fig. B6e)  Uses potential divider bias to swamp variations in $V_{GS}$ between devices.

Nearly all devices have to have some bias, and this naturally represents a shunting effect on the circuits input impedance. Special techniques are employed to reduce the effects of bias component on input impedance.

▶ *Bootstrapping*

## Binary Code

A binary code is a sequence of binary digits that bears a unique relationship to the decimal integers. For example, using the simplest binary code, the 8421 code, the decimal number 14 is encoded as follows:

$$14_{10} = 1110 = 1(8) + 1(4) + 1(2) + 0(1)$$

$$7_{10} = 0111 = 0(8) + 1(4) + 1(2) + 1(1)$$

There are basically two types of binary code — weighted and non-weighted.

In a weighted code only the decimal numbers from 0 to 9 are encoded. When a larger decimal number has to be encoded in a weighted code, each digit is encoded separately to give what is termed a Binary Coded Decimal word (BCD).

Thus, $14_{10}$ is represented in 8421 BCD as

    0001   0100
     (1)    (4)

and $239_{10}$ in 8421 BCD is

    0010   0011   1001
     (2)    (3)    (9)

The 8421 BCD code is commonly used in digital instruments such as counters and d.v.m.s but there are several other popular codes in use as shown in Table B2.

**Table B2** *Binary codes*

| Decimal no. | 8421 | 4221 | Excess 3 | Gray | 2 out of 5 |
|---|---|---|---|---|---|
| 0 | 0000 | 0000 | 0011 | 0000 | 00011 |
| 1 | 0001 | 0001 | 0100 | 0001 | 00101 |
| 2 | 0010 | 0010 | 0101 | 0011 | 00110 |
| 3 | 0011 | 0011 | 0110 | 0010 | 01001 |
| 4 | 0100 | 1000 | 0111 | 0110 | 01010 |
| 5 | 0101 | 0111 | 1000 | 1110 | 01100 |
| 6 | 0110 | 1100 | 1001 | 1010 | 10001 |
| 7 | 0111 | 1101 | 1010 | 1011 | 10010 |
| 8 | 1000 | 1110 | 1011 | 1001 | 10100 |
| 9 | 1001 | 1111 | 1100 | 1000 | 11000 |

The 4221 and excess 3 codes are used in computation. This is because the complement for 4221 and excess 3 can be easily obtained by inverting each bit. In computers the process of subtraction is usually performed by adding the complement of the subtrahend, i.e.

    $0010 = 2_{10}$ in 4221

and its complement $= 1101 = 7_{10}$ in 4221

(The number plus its nines-complement always equals 9.)

The Gray code, which is non-weighted, has the useful characteristic that only 1 bit changes at a time. This is particularly useful in position encoders because it reduces errors and malfunctions.

The 2 out of 5 is an example of an error-detecting code. It has an additional bit that allows the number to be checked to see if it is correct after transmission or processing. The bit that is added makes an even number of 1s in the coded word. If an error is made in 1 bit, a test for an even number of 1s will fail. This is called PARITY CHECKING.

## Binary Number System and Binary Arithmetic

Binary numbers are those with a base of 2. (Decimals have a base of 10 and octals a base of 8.) Binary numbers contain only two digits 0 and 1 [▶ *Bit*]. For example

| Decimal | | Binary |
|---|---|---|
| 1 | = | 1 |
| 2 | = | 10 |
| 3 | = | 11 |
| 4 | = | 100 |
| | etc. | |
| 9 | = | 1001 |
| 10 | = | 1010 |

The binary number system is used extensively in digital circuits because the bits 0 and 1 can be reliably represented by switches which are either on (conducting) or off (non-conducting).

## Conversion from binary to decimal

Write down the binary number and then, starting from the right, multiply each bit by its corresponding weight. The result is added to give the decimal equivalent.

*Example 1*

$$2^4\ 2^3\ 2^2\ 2^1\ 2^0 \quad \text{weight}$$

$$1\ 1\ 0\ 1\ 0$$

$$(1 \times 2^4) + (1 \times 2^3) + (0 \times 2^2) + (1 \times 2^1) + (0 \times 2^0)$$

$$16\ +\ 8\ +\ 0\ +\ 2\ +\ 0\ = 26$$

*Example 2*

$$2^5\ 2^4\ 2^3\ 2^2\ 2^1\ 2^0 .\ 2^{-1}\ 2^{-2}$$

$$1\ 0\ 1\ 1\ 0\ 1\ .\ 1\quad 1$$

$$32\ 0\ 8\ 4\ 0\ 1\ ..5\quad .25\ = 45.75$$

## Conversion from decimal to binary

Divide the decimal number successively by 2 until only 1 remains. At each division the remainder is noted. The final remaining 1 followed by the previous remainders in reverse order is the equivalent binary number.

*Example* 45 converted to binary.

2/45
2/22 remainder 1
2/11 remainder 0
2/ 5 remainder 1
2/ 2 remainder 1
/ 1 remainder 0
→ 1

Binary number is 101101

## Rules of binary arithmetic

| Addition | Subtraction | Multiplication |
|---|---|---|
| $0+0=0$ | $0-0=0$ | $1 \times 1=1$ |
| $0+1=1$ | $1-0=1$ | $1 \times 0=0$ |
| $1+0=1$ | $1-1=0$ | $0 \times 0=0$ |
| $1+1=10$ | $0-1=1$ | $0 \times 1=0$ |
| | with a borrow | |
| | of 1 | |

*Examples*

**1** 101 add to 10011 (i.e. $5+19$)

$$
\begin{array}{r}
10011\ + \\
101 \\
\hline
11000 = 24
\end{array}
$$

**2** 110 subtracted from 10100 (i.e. $20-6$)

$$
\begin{array}{r}
10100\ - \\
110 \\
\hline
1110 = 14
\end{array}
$$

**3** 1001 multiplied by 101 (i.e. $9 \times 5$)

$$
\begin{array}{r}
1001\ \times \\
101 \\
\hline
1001 \\
0000 \\
1001 \\
\hline
101101 = 45
\end{array}
$$

*Division* Using long division, the divisor is subtracted from the dividend and a 1 is placed in the quotient. If the subtraction cannot be performed, then 0 is placed in the quotient.

*Example* Divide 110000 by 100 (i.e. $48 \div 4$)

$$
\begin{array}{r}
1100 \\
100\overline{)110000} \quad \text{answer 1100} \\
100 \\
\hline
100 \\
100 \\
\hline
00
\end{array}
$$

## Bipolar Device

Any active component in a system that depends for its operation on both types of charge carrier is termed bipolar. In other words, a bipolar transistor or i.c. is one that utilises both holes (+ve charge carriers) and electrons (−ve charge carriers) for current flow.

Mostly the term is used to differentiate between the more common types of transistor and field effect transistors. FETs are called UNIPOLAR devices.

▶ *Transistor*

## Bistable

A bistable is a circuit that has two possible stable states and which can be triggered into one of these states by the application of a suitable short duration pulse (fig. B7). Once triggered, the bistable holds its new state and stores the information after the input command ceases. Bistables, sometimes called flip-flops or

latches, are the building blocks of sequential logic circuits such as counters, dividers, and shift registers, and are also used extensively in computer stores.

A variety of bistable types exist: R–S, clocked R–S, T, D, J–K.

By considering the simplest type, the R–S latch, it is then relatively easy to build up an understanding of the more complex types.

**1**  *R–S bistable*, made from two cross-coupled transistor switches or two cross-coupled gates as shown. In circuit (*a*), because of the values chosen for the resistors, when one transistor is conducting with its collector at $V_{CE(sat)}$ then the other must be held off. Thus if one output is low (0.1 V) the other must be high ($V_{CC}$). The two outputs are called $Q$ and $\bar{Q}$ and the two inputs are known as $S$ for SET and $R$ for RESET.

If a pulse is applied to the Set input taking it momentarily to logic 0, then the $Q$ output will be forced to assume a logic 1 state. This is because $Tr_2$ will be turned off and the feedback within the circuit will be force $Tr_1$ to conduct taking $\bar{Q}$ to logic 0. The state of the bistable can now only be changed by applying a pulse (to logic 0) to the Reset input.

An important point to note is that if both $S$ and $R$ inputs are made 0 simultaneously then the resulting states of $Q$ and $\bar{Q}$ will be indeterminate. In other words the circuit may go into either state. Therefore the simultaneous application of $S$ and $R$ pulses is not allowed.

The action of a bistable can be described by a truth table, but because it has a memory circuit, the state of the circuit before the input pulse is applied must be included in the table in the input side. The previous state is written as $Q_n$ and the final or resulting state as $Q_{n+1}$. The truth table for an R–S bistable is

| R | S | $Q_n$ | $Q_{n+1}$ |
|---|---|---|---|
| 0 | 0 | 0 | * indeterminate |
| 0 | 0 | 1 | * indeterminate |
| 0 | 1 | 0 | 0 ⎫ a 0 on $R$ with $S$ at 1 |
| 0 | 1 | 1 | 0 ⎭ will cause the $Q$ output to go to 0 |
| 1 | 0 | 0 | 1 ⎫ a 0 on the $S$ with $R$ at 1 |
| 1 | 0 | 1 | 1 ⎭ will cause the $Q$ output to go to 1 |
| 1 | 1 | 0 | 0 ⎫ no change |
| 1 | 1 | 1 | 1 ⎭ |

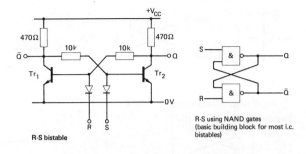

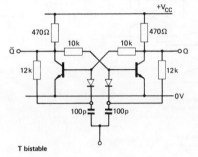

*Figure B7a*  **R–S and T bistables**

The R–S can be converted to a *T type* as shown. Here the $RC$ networks with the diodes form a pulse steering circuit to pass the negative edge of the T-waveform to the transistor that is conducting. Each negative edge therefore causes the circuit to change state. The circuit will divide the input by 2 and it is referred to as a binary divider. The T is not often used since the function can be easily carried out by the D or J–K bistables.

The R–S can be implemented using NOR gates, in which case the bistable is set and reset by logic 1s. The truth table will be the complement of the one given for the R–S implemented with NAND gates.

**2**  *Clocked R–S bistable*  The addition of a clock input to a bistable gives the possibility of synchronous operation. The circuit is set and reset by logic 1s but this can only occur while the clock pulse input is high. This gives what is termed *level clocking*. Level clocking has the

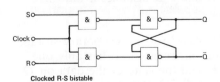

*Figure B7b*  **Clocked R-S bistable**

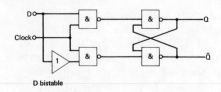

D bistable

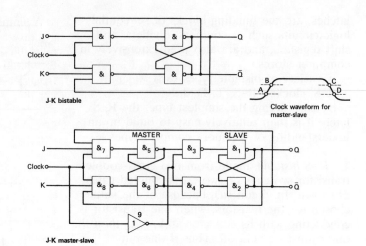

J-K bistable

Clock waveform for master-slave

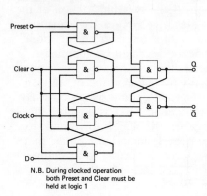

N.B. During clocked operation both Preset and Clear must be held at logic 1

J-K master-slave

**Figure B7d** J-K bistable

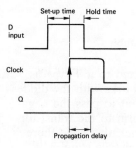

Set-up time    Hold time

D input

Clock

Q

Propagation delay

**Figure B7c** D bistable

| Clock | $D$ | $Q_n$ | $Q_{n+1}$ |
|-------|-----|-------|-----------|
| H | 0 | 0 | 0 |
| H | 0 | 1 | 0 |
| H | 1 | 0 | 1 |
| H | 1 | 1 | 1 |

A typical D bistable in i.c. form is the 7474 which is +ve edged triggered. The input signal to the $D$ must be applied before the clock's leading edge and it must be held therefore for a short time to allow correct operation. For the 7474 the set-up time is typically 20 ns and the hold time 5 ns.

**4  J–K bistable**  One of the most versatile and widely used bistables. It has the advantage in that it cannot be forced into an indeterminate state. The truth table is

| $J$ | $K$ | $Q_n$ | $Q_{n+1}$ | |
|-----|-----|-------|-----------|---|
| 0 | 0 | 0 | 0 | no change |
| 0 | 0 | 1 | 1 | |
| 0 | 1 | 0 | 0 | when $J=0$, $K=1$, |
| 0 | 1 | 1 | 0 | output goes to 0 |
| 1 | 0 | 0 | 1 | when $J=1$, $K=0$, |
| 1 | 0 | 1 | 1 | output goes to 1 |
| 1 | 1 | 0 | 1 | when $J=K=1$, output |
| 1 | 1 | 1 | 0 | always complements |

disadvantage that changes in the inputs while the clock is high will cause changes in the state of the bistable. An improvement is EDGE TRIGGERING where the input information can only be transferred to change the state of the bistable on one edge of the clock waveforms. Most i.c. bistables are edge triggered. With the clocked R–S the output will still be indeterminate if 1s are applied at the same time to $R$ and $S$ inputs while the clock is high. The D and J–K bistables avoid this problem

**3  D bistable**  Used for the temporary storage of data and in shift registers. It has only one input called $D$ and the state of the data on $D$ is transferred to the $Q$ output when the clock goes high (level triggered). The truth table is

The J–K achieves this operation because of the additional feedback from $Q$ to $K$ gate and $\bar{Q}$ to $J$ gate. Only one input gate can be enabled at any one time. To avoid timing problems — referred to as race hazards — the

J–K is usually implemented in MASTER-SLAVE form.

In the master-slave J–K, as the clock goes high at point (A), gates 3 and 4 close isolating the slave from the master. The $Q$ and $\bar{Q}$ outputs cannot change state while the clock is high. At point (B) on the waveform, gates 7 and 8 open allowing the $J$ and $K$ input data to change the state of the master.

On the trailing edge of the clock waveform at point (C), gates 7 and 9 close disconnecting the $J$ and $K$ inputs from the master, and then finally at point (D) gates 3 and 4 open and the slave assumes the same state as the master. Thus the output of a J–K master-slave changes state on the trailing edge of the clock waveform.

▶ *Counters and dividers*    ▶ *Shift register*

## Bit

The binary unit of information used in digital systems and computers. It is taken from the first letter and the last two letters of BINARY DIGIT.

Several bits comprise a binary word [▶ Word].

## Block Diagram

These types of diagram are invaluable for fault location on systems or as an aid in the understanding of how a system or instrument operates. Instead of studying the full circuit, all the unnecessary detail is removed, and whole portions of the system are just considered as functioning blocks. For example, a circuit used to produce a fixed reference frequency can be represented by a block called "crystal oscillator"; all the components for an amplifier can be replaced by a block labelled "wideband amplifier" and so on. The result is a much simplified and clearer view of the system.

*Figure B8* Example of a block diagram. The corresponding full circuit diagram (of a motor speed controller) is shown in Fig. M36b (p. 167)

A simple example is shown in fig. B8. Here a motor speed control circuit is reduced to four main blocks and it is not necessary to know all the detail to appreciate the operation. A fixed-frequency 400 Hz pulse generator supplies trigger pulses to a monostable. The output of the monostable controls the switching action of the power switch via a drive circuit. By adjusting the width of the output pulses from the monostable more or less power can be supplied to the motor to alter its speed.

Block diagrams, by reducing the detail, can be very useful for fault finding on systems. By measurement and deduction it is possible to rapidly narrow down the search for a fault to *one* block within the system. Then either that block can be replaced or further fault finding procedures carried out to locate the failed component.

▶ *Fault diagnosis*

## Blocking Oscillator

A useful and effective circuit for producing short-duration pulses with fast edges at a low duty cycle (fig. B9). One application is as a clock pulse generator in digital systems. Low-power transistors can be used to generate relatively high-power pulses because the duty cycle is low. The mean power dissipated is small because the transistor is only on for a short period of the cycle. The positive feedback of the oscillator is via the small pulse transformer between the collector and base circuits. When power is first applied, $Tr_1$ will be off until the voltage across $C$ rises to just above 0.6 V. $Tr_1$ starts to conduct and collector current flows through the primary winding of $T_1$. Because of the positive feedback, the changing collector current induces an e.m.f. in the secondary that forces the transistor to conduct further. Very rapidly the transistor turns on and saturates, its collector voltage falling to about 0.1 V ($V_{CE(sat)}$). After a short time the collector

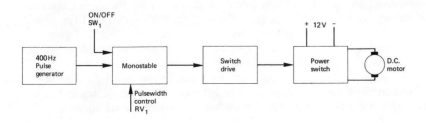

**Table B3**  *Bistables*

| | Type no. | Description | Clock triggering method | Other inputs |
|---|---|---|---|---|
| **TTL logic** | 7470 | Single J-K | Level triggered | Preset and clear |
| | 7471 | Single J-K master-slave | −ve edge | Preset |
| | 7472 | Single J-K master-slave | −ve edge | Preset and clear |
| | 7473 | Dual J-K | Level triggered | Clear |
| | 7474 | Dual D | +ve edge | Preset and clear |
| | 7474LS | Dual (low power schottky) | +ve edge | |
| | 7474S | Dual (schottky) | +ve edge | |
| | 7475 | Quad D latch | No clock provided | — |
| | 7475LS | Dual (low power schottky) | No clock provided | |
| | 7476 | Dual J-K | −ve edge | Preset and clear |
| | 7476LS | Dual (low power schottky) | −ve edge | |
| | 74174 | Hex D | | Clear |
| | 74175 | Quad D | | Clear |
| **ECL** | 10130 | Dual D | +ve edge | Direct set |
| | 10131 | Dual D master-slave | +ve edge | Direct set |
| | 10135 | Dual J-K master-slave | +ve edge | Direct set |
| **CMOS** | 4013 | Dual D | +ve edge | Set direct, clear direct |
| | 4027 | Dual J-K | +ve edge | Set direct, clear direct |
| | 4043 | Quad R-S (tri-state output) | | |
| | 40174 | Hex D | | reset |
| | 40175 | Quad D | | reset |

current stops changing, either because of the limit imposed by the transistor's current gain or because the core of the transformer saturates. When this happens the magnetic field in the transformer collapses and the base e.m.f. falls. The transistor begins to turn off with its collector voltage rising towards $+V_{CC}$. The feedback rapidly drives the transistor off and the base waveform goes negative, the peak reverse voltage across $C$ being approximately $nV_{CC}$ where $n$ is the turns ratio of the transformer. The transistor remains off while $C$ charges exponentially via $R$ towards $+V_{CC}$. As soon as the voltage across $C$ exceeds $+0.6$ V, the transistor again conducts and the cycle is repeated. A diode is wired across the transformer primary to suppress the back e.m.f. caused when the transistor turns off.

Blocking oscillators can also be used to provide sawtooth generators as shown, or monostable and bistable circuits.

## Bode Plot

This plot or diagram is a graphical method for showing how the gain and phase shift of an amplifier vary with frequency. The magnitude of the gain and the amount of phase shift introduced are both plotted against the log of the signal frequency. A typical plot for an a.c. coupled amplifier is shown in the first figure (B10a). At low frequencies, because of the coupling capacitors, the gain is very low and the phase angle of the output relative to the input is leading by 90°. As the signal frequency is increased, the magnitude rises by 20 dB/decade and the phase angle changes by 45°/decade until $f_1$ is reached. At $f_1$, which is referred to as the "corner frequency" or "low-frequency cut-off point", the magnitude of the gain is $1/\sqrt{2}$ or 3 dB less than its mid-band value and the phase lead has fallen to 45°. Over the useful frequency range, or bandwidth, the gain and the phase shift of the amplifier remain fairly constant. At $f_2$, the upper cut-off frequency, the gain is 3 dB down and the phase shift lags by 45°. For this amplifier the gain then falls by 20 dB/decade and the phase angle between output and input increases by −45°/decade to a maximum of −90°. The fall-off in gain and the resulting phase lag are caused by circuit and stray capacitances in

| Clock frequency MHz | |
| --- | --- |
| Min | Typical |
| 15 | 35 |
| 25 | 30 |
| 15 | 20 |
| 15 | 20 |
| 15 | 25 |
| 25 | 33 |
| 75 | 90 |
| Prop. delay D to Q 16 ns | |
| Prop. delay D to Q 16 ns | |
| 15 | 20 |
| 25 | 30 |
| Update frequency 35 MHz max. | |
| Update frequency 35 MHz max. | |

| 3 ns delay | |
| --- | --- |
| 125 | 160 |
| 140 | |
| 5 | 9 at 5 V |
| 4 | 8 at 5 V |
| 5 | 9 at 5 V |
| 5 | 9 at 5 V |

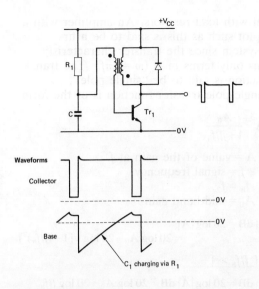

**Waveforms**

Collector ............0V

Base ............0V

$C_1$ charging via $R_1$

**Used as a sawtooth generator**

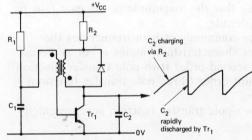

$C_2$ charging via $R_2$

$C_2$ rapidly discharged by $Tr_1$

*Figure B9*  Blocking oscillator

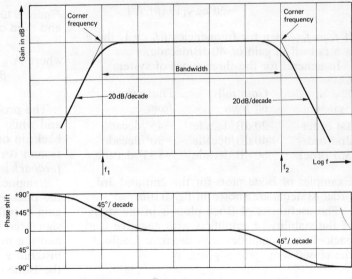

*Figure B10a*  Bode plot for a first-order a.c. coupled amplifier

37

parallel with load resistors. An amplifier with a Bode plot such as this is said to be a first-order system since the transfer characteristic contains only terms in $\omega$ ($\omega = 2\pi f$). The transfer function is said to be "single pole".

A single-pole transfer function is of the form

$$A = \frac{A_0}{1 + jf/f_h}$$

where $A$ = value of the gain at $f$
$\quad\quad f$ = signal frequency
$\quad\quad f_h = f_2$
$\quad\quad A_0$ = mid-band gain

$$|A|\,dB = 20 \log |A|$$
$$= 20 \log A_0 - 20 \log \sqrt{[1 + (f/f_h)^2]}$$

Thus if $f/f_h \gg 1$

$$|A|\,dB = 20 \log |A|\,dB = 20 \log A_0 - 20 \log f/f_h$$

showing that the magnitude of the gain falls by 20 dB/decade.

More commonly at high frequencies the transfer characteristic contains $\omega^2$ or $\omega^3$ terms giving second-order (two-pole transfer function) and third-order (three-pole transfer function) systems.

A two-pole transfer function is represented by

$$A = \frac{A_0}{[1 + j(f/f_{h1})] \cdot [1 + j(f/f_{h2})]}$$
$$|A|\,dB = 20 \log A_0 - 20 \log \sqrt{[1 + (f/f_{h1})^2]}$$
$$- 20 \log \sqrt{[1 + (f/f_{h2})^2]}$$

If $f_{h1} = f_{h2}$, then for frequencies $f/f_{h2} \gg 1$, there is a fall off in gain of 40 dB/decade.

In general for the three types of system:

| System | Gain roll-off from | Phase shift from |
|--------|--------------------|------------------|
| 1st order | 20 dB/decade | $-45°$/decade |
| 2nd order | 40 dB/decade | $-90°$/decade |
| 3rd order | 60 dB/decade | $-135°$/decade |

Examples of Bode plots for the 2nd and 3rd order systems are shown in fig. B10$b$.

The usefulness of Bode plots is in the study of the stability of amplifiers with negative feedback. For an amplifier with negative feedback the closed loop gain is given by

$$A_{CL} = \frac{A_{OL}}{1 + A_{OL}\beta}$$

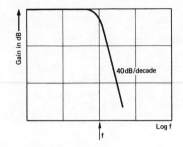

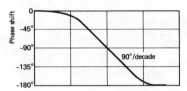

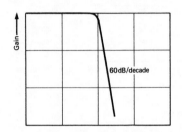

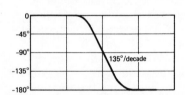

**Figure B10b**  Bode plots for second-order (top) and third-order (bottom) systems

where $A_{OL}$ = open loop gain (assumed +ve)
$\quad\quad \beta$ = fraction gain of the feedback network.

The product $A_{OL}\beta$ is called the loop gain, and while this is positive we get negative feedback; in other words the gain of the amplifier system is reduced and stabilised [▶ *Negative feedback*].

Imagine though, at some high frequency, that the phase shift approaches 180° while the loop gain is still above unity. If this happens the signal fed back will be in phase with the input, and the overall feedback will be positive giving instability and possibly oscillations. By plotting the Bode diagram for the open loop gain of the amplifier, that is the gain before feedback is connected, the amount of negative feedback

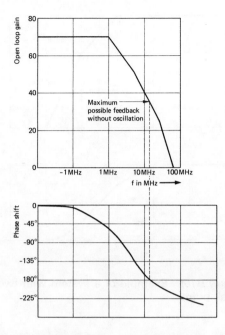

**Figure B10c**

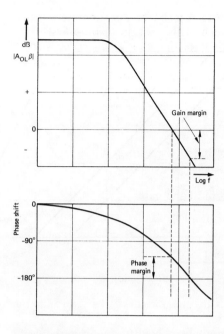

**Figure B10d**

that can be applied and give a stable system can be evaluated. For example, in fig. B10*c* the maximum feedback that can be applied $(1 + A_{OL}\beta)$ is 34 dB. In practice it would not be possible to apply 34 dB since it would be inevitable that the 180° phase shift would be exceeded by minor deviations in the amplifier or external circuit wiring. Thus the limiting phase has to be chosen as less than 180° by some definite margin to give a stable system. A satisfactory circuit should have a minimum phase margin of 45°; this will give less than 3 dB peaking, which means that only 20 dB of feedback can be safely applied to this amplifier.

An alternative approach in stability investigations is to plot the Bode diagram for the loop gain $A_{OL}\beta$ (fig. B10*d*). If it can be shown that the magnitude of $A_{OL}\beta$ is less than unity when the phase shift of $A_{OL}\beta$ is 180°, then the amplifier with the negative feedback loop connected will be stable [▶ *Nyquist diagram*].

The *gain margin* on the Bode plot of $|A_{OL}\beta|$ is the value of the magnitude of $A_{OL}\beta$ in dB at the frequency at which the phase shift is 180°. To give a stable system this must be negative, a typical value being $-10$ dB. The *phase margin* in degrees is the difference between 180° and the phase shift of $A_{OL}\beta$ when the magnitude of $A_{OL}\beta$ is unity. This

should be at least 45° as stated previously. The meaning of these margins is illustrated in fig. B10*d*.

Suppose the Bode plot of an amplifier shows possible instabilities; there are then several methods to improve stability. These are basically shaping the response of the open loop characteristics by inserting a lag, lead, or both to give a break at a much lower frequency (see fig. B10*e*) or to modify the feedback network to include reactive elements.

▶ *Negative feedback* ▶ *Amplifier*

## Boolean Algebra

This type of algebra was invented in the 19th century by George Boole as a system for the mathematical analysis of logic. It is based on logical statements that are either true or false and is therefore a powerful tool in the design and analysis of digital logic circuits.

In Boolean algebra instead of writing "both switches A *and* B must be operated to achieve one output" we can write

$$F = A \cdot B$$

The AND function is represented by the symbol ·

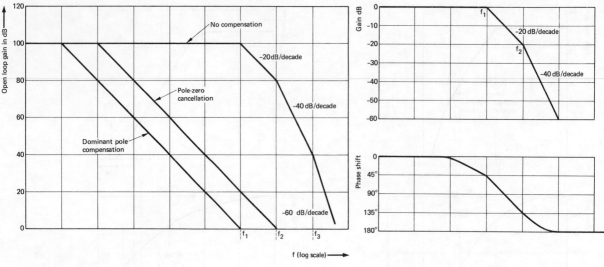

*Figure B10e*

*Figure B10f* **Example of a system with two break points**

Similarly the statement "either switch C *or* D can be operated to achieve an output" can be written

$$F = C + D$$

The OR function is represented by the symbol +

The NOT function is represented by a bar over the switch:

$F = \bar{A}$ means do *not* operate switch $A$ to obtain an output.

From this the NOR function is

$$F = \overline{A + B} \qquad \text{(NOT-OR)}$$

and the NAND function is

$$F = \overline{A \cdot B} \qquad \text{(NOT-AND)}$$

Examples of Boolean logic functions:

$F = A \cdot B + \bar{C}$ means operate $A$ and $B$ together, or not $C$, to get an output at $F$.

$F = A \cdot (B + \bar{C})$ means operate $A$ and, $B$ or not $C$, to get an output at $F$.

Confusion over the last statement can be avoided by removing the brackets:

$$F = A \cdot (B + \bar{C}) = A \cdot B + A \cdot \bar{C}$$

meaning operate $A$ and $B$, or $A$ and not $C$.

The identities and rules of Boolean algebra are shown in fig. B11. The use of Boolean algebra in the simplification of logic can be seen in the following examples

**1** $F = (A + B) \cdot (A + C) \cdot \bar{B}$

Step 1: combine bracket.

$$F = (A \cdot A + A \cdot C + B \cdot A + B \cdot C) \cdot \bar{B}$$

Step 2: use absorption rule.
Since $A \cdot A + A \cdot C + B \cdot A = A$

$$F = (A + B \cdot C) \cdot \bar{B}$$

Step 3: remove bracket.

$$F = A \cdot \bar{B} + \bar{B} \cdot B \cdot C$$

Step 4: from identities
Since $\bar{B} \cdot B = 0$

$$F = A \cdot \bar{B} + 0$$
$$F = A \cdot \bar{B}$$

**2** $F = \overline{A + \overline{B \cdot C} \cdot \overline{(A \cdot D)}}$

This is simplified using De Morgans theorems:

$$F = \bar{A} \cdot \overline{\overline{B \cdot C}} \cdot \overline{\overline{(A \cdot D)}}$$
$$F = \bar{A} \cdot B \cdot C \cdot (\bar{A} + \bar{D})$$

Removing brackets:

$$F = \bar{A} \cdot B \cdot C + \bar{D} \cdot B \cdot C \cdot \bar{A}$$
$$F = B \cdot C \cdot (\bar{A} + \bar{A}D) = B \cdot C \cdot \bar{A}$$

**40**

| Expression | Function | BS Symbol | Alternative symbols |
|---|---|---|---|
| $A + B$ | OR | $F = A + B$ | |
| $A \cdot B$ | AND | $F = A \cdot B$ | |
| $\bar{A}$ | NOT (INVERTOR) | $F = \bar{A}$ | |
| $\overline{A + B}$ | NOR | $F = \overline{A + B}$ | |
| $\overline{A \cdot B}$ | NAND | $F = \overline{A \cdot B}$ | |
| $A \oplus B$ | EXCLUSIVE-OR | $F = A \oplus B$ | |

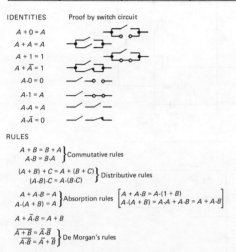

IDENTITIES    Proof by switch circuit

$A + 0 = A$

$A + A = A$

$A + 1 = 1$

$A + \bar{A} = 1$

$A \cdot 0 = 0$

$A \cdot 1 = A$

$A \cdot A = A$

$A \cdot \bar{A} = 0$

RULES

$\left. \begin{array}{l} A + B = B + A \\ A \cdot B = B \cdot A \end{array} \right\}$ Commutative rules

$\left. \begin{array}{l} (A + B) + C = A + (B + C) \\ (A \cdot B) \cdot C = A \cdot (B \cdot C) \end{array} \right\}$ Distributive rules

$\left. \begin{array}{l} A + A \cdot B = A \\ A \cdot (A + B) = A \end{array} \right\}$ Absorption rules $\left[ \begin{array}{l} A + A \cdot B = A \cdot (1 + B) \\ A \cdot (A + B) = A \cdot A + A \cdot B = A + A \cdot B \end{array} \right]$

$A + \bar{A} \cdot B = A + B$

$\left. \begin{array}{l} \overline{A + B} = \bar{A} \cdot \bar{B} \\ \overline{A \cdot B} = \bar{A} + \bar{B} \end{array} \right\}$ De Morgan's rules

**Figure B11** Identities and rules of Boolean algebra

The validity of a Boolean statement can also be checked out by drawing up a truth table. This table is simply a list of all the possible states of the inputs together with the resulting output. For example the truth table for $F = A \cdot \bar{B}$ is

| Input states | | | Output |
|---|---|---|---|
| $A$ | $B$ | $\bar{B}$ | $F = A \cdot \bar{B}$ |
| 0 | 0 | 1 | 0 |
| 0 | 1 | 0 | 0 |
| 1 | 0 | 1 | 1 |
| 1 | 1 | 0 | 0 |

This is just another way of stating that the output from the circuit will only be 1 when $A = 1$ and $B = 0$. Table B4 shows the truth tables for the commonly used combinational logic gates for two inputs. The tables can easily be extended for three or more inputs but tend to become rather large.

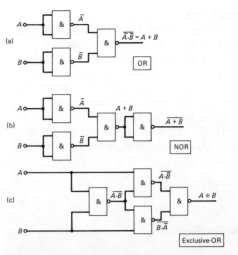

**Figure B12** Use of NAND gates to realise OR, NOR and Exclusive-OR

**Table B4** *Truth tables for various combinational operations*

| Inputs | | Output | | | | |
|---|---|---|---|---|---|---|
| | | OR | NOR | AND | NAND | Exclusive-OR |
| $A$ | $B$ | $A + B$ | $\overline{A + B}$ | $A \cdot B$ | $\overline{A \cdot B}$ | $A \oplus B$ |
| 0 | 0 | 0 | 1 | 0 | 1 | 0 |
| 0 | 1 | 1 | 0 | 0 | 1 | 1 |
| 1 | 0 | 1 | 0 | 0 | 1 | 1 |
| 1 | 1 | 1 | 0 | 1 | 0 | 0 |

The exclusive-OR circuit has the function

$$F = A \cdot \bar{B} + \bar{A} \cdot B$$

It is a circuit that is often used and the ways in which it is achieved with the various logic types is shown in a separate section.

Quite often the designer has to use mostly NAND-type or NOR-type gates, because these are more available or cheaper in the logic family chosen. The De Morgan rules can be used to show how the other logic operations can be realised in practice when only NAND or NOR gates are available.

De Morgan's rules, easily proved by truth table, are

$$\overline{A + B} = \bar{A} \cdot \bar{B} \quad \text{and} \quad \overline{A \cdot B} = \bar{A} + \bar{B}$$

It follows that

$$\overline{\bar{A} \cdot \bar{B}} = \bar{\bar{A}} + \bar{\bar{B}} = A + B \quad \text{the OR function}$$

and

$$\overline{\bar{A} + \bar{B}} = \bar{\bar{A}} \cdot \bar{\bar{B}} = A \cdot B \quad \text{the AND function}$$

## Bootstrapping

A rather quaint but apt term used to describe how a resistor in a circuit can be made to appear as a much higher value to a.c. signals. Suppose an emitter follower circuit is required with high a.c. input impedance. An unmodified circuit has a rather low input impedance because of the shunting effect of the bias resistors. To overcome this, bootstrapping is used (fig. B13). A capacitor, with a value large

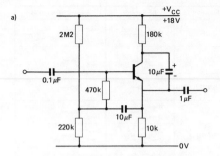

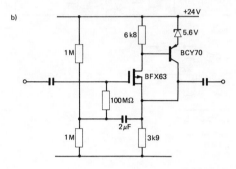

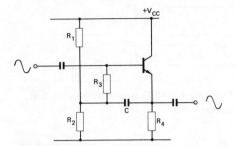

**Figure B13**  Bootstrapping an emitter follower

enough to act as almost a short circuit at the lowest frequency required, is connected from the output to the bottom end of the bias resistor $R_3$. This gives positive feedback but, because the voltage gain of the emitter follower is less than unity, the circuit will not oscillate. However $R_3$ will act as a much higher value of resistor to a.c. signals. As one end of $R_3$ changes in voltage, the other end moves by almost the same amount in the same direction. The effective a.c. signal across $R_3$ is thus very small and it appears that $R_3$ is pulling itself up by its own bootstraps.

The effective a.c. value of $R_3$ is given by

$$R_{3(B)} = R_3/(1 - A_v)$$

For example if $R_3 = 100$ k$\Omega$ and $A_v = 0.99$, then $R_{3(B)} = 10$ M$\Omega$.

In this way the shunting effect of the bias resistor is reduced.

Note that the value of $C$ should be greater than $10/f_1R_3$ where $f_1$ is the lowest signal frequency required.

Other examples of bootstrapping are shown in fig. B.14. In circuit ($a$) the shunting effect of the transistor collector resistance $r_c$ is also reduced by feeding back a signal from the emitter to the collector. Thus circuit has an input impedance of greater than 500 k$\Omega$ at frequencies down to 50 Hz.

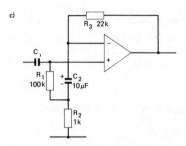

**Figure B14**  Examples of bootstrapping

The MOSFET circuit shown in ($b$) has an input impedance greater than $5 \times 10^9$ and a voltage gain of 0.98.

Bootstrapping can also be used effectively with a.c. coupled op-amp circuits as shown in ($c$). $R_1$ is bootstrapped by the feedback signal to give a high value of input impedance over the useful frequency range. With the values shown the input impedance will be in excess of 10 M$\Omega$.

## Breakdown Testing

Any device will go into breakdown and pass a large reverse current if the voltage across it exceeds some specified value. If the value of breakdown voltage has to be measured then the test current must be limited to a safe value. In semiconductors, breakdown is characterised

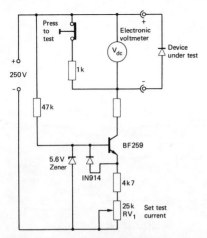

Figure B15 Breakdown testing of a circuit using a high-voltage transistor

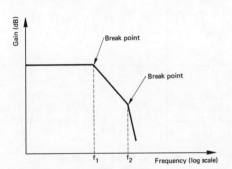

Figure B16 Break points

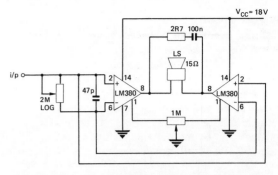

Figure B17 An audio bridge amplifier (basic circuit)

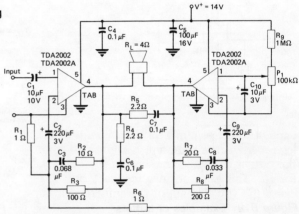

Figure B18 Full circuit of audio bridge amplifier: 15 W circuit

by a sudden and rapid increase in current [▶ *Avalanche breakdown*]. Thus to prevent damage a constant current source must be used. A simple circuit using a high-voltage transistor is shown in fig. B15. It can be used to non-destructively test $V_{(BR)R}$ in diodes, $V_Z$ in zeners, and breakdown voltages such as $V_{CEO}$ in transistors. $Tr_1$ is a constant-current generator with its base voltage held constant by the 5.6 V zener. The transistor current can be adjusted by $RV_1$ from 200 $\mu$A up to a maximum of 1 mA which should be low enough not to cause failure with most devices.

When the test switch is pushed, the voltage across the device under test will rise to its breakdown value where the current will then be limited. The voltage across the device can be read using a high impedance meter such as a digital multimeter.

## Break Point

Suppose the gain of an amplifier has a roll-off at high frequencies of 20 dB/decade (6 dB/octave); then the break point is the point at which the 20 dB/decade slope intersects the value of the mid-band gain (fig. B16). At this point, also referred to as the "corner frequency" or "upper cut-off frequency" the gain will be 3 dB down.

Note than an amplifier may have more than one break point as the roll off in gain changes from 20 dB/decade to 40 dB/decade and then to possibly 60 dB/decade.

▶ *Bode plot*     ▶ *Frequency response*

## Bridge Amplifier

A connection method used in audio and servo power amplifiers to achieve an increase in output power (fig. B17). Two amplifiers are used

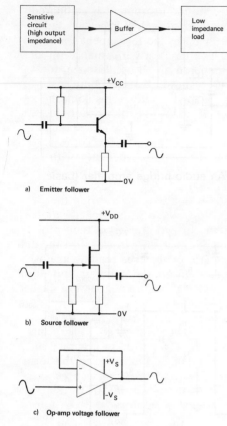

a) Emitter follower

b) Source follower

c) Op-amp voltage follower

*Figure B19* Examples of buffers

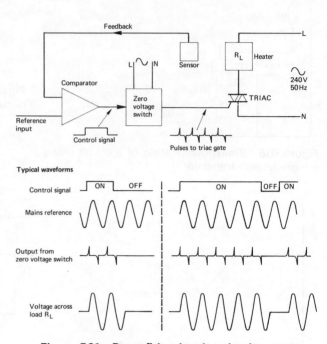

Typical waveforms

*Figure B20* Burst firing in triac circuit

and the load is connected between the outputs. As the output of one amplifier goes positive the other goes equally negative giving twice the voltage swing across the load. The overall output power can be increased by a factor of two or more over a single amplifier.

▶ *Audio amplifier*

## Buffer

An electronic buffer is a circuit that takes a signal from a sensitive high impedance output device and delivers it to a relatively heavy load (fig. B19). The buffer has near unity gain and very high input impedance so that it presents only a light load to the sensitive circuit. Its output impedance must be very low so that it can drive the low impedance load. Buffers are sometimes referred to as matching circuits. The emitter follower, source follower, and the unity gain follower using an op-amp are all good examples of buffer circuits.

Buffers are also used in logic, for example the CD 4050 is a hex non-inverting CMOS buffer that has a higher drive capability than other CMOS gates.

▶ *Bootstrapping*    ▶ *Darlington*

## Burn-in

If a large number of electronic components, straight from production, is put on test at maximum rating, then some will fail early on. The failure rate will fall until a constant failure rate figure is reached where failures are due to chance alone [▶ *Maintenance*].

The devices that fail early are mostly those that have some built-in defect or are inherently weak. By putting all the devices from production on a short duration test at full rating and high temperature these "weak" devices will be sorted out and not sold. The test time is referred to as burn-in and may range in time from a few hours up to say 100 hours depending on the component type.

## Burst Firing

A technique used in triac and thyristor a.c. power control circuits as an alternative to phase control (fig. B20). Instead of the power device being triggered on to conduct for a portion of each mains half cycle, it is switched on for multiples of whole cycles. For example, 1 cycle in every 10 to give low power in the load or 9 cycles in every 10 for high power. The triac or thyristor is fired in bursts and it is triggered on at the point where the a.c. mains supply is passing through zero. The advantage of this is that there will be practically no interference signals generated when the power device switches. In contrast the phase control method can generate large amounts of radio frequency interference as the device may be required to switch on just when the mains is at its peak value.

Power controllers using burst firing can only be used for loads such as heaters which have a relatively slow response time. The bursts of mains cycles applied are smoothed out by the load. The technique cannot be used for lamp dimmers or motor speed control.

▶ *Phase control*  ▶ *Thyristor*
▶ *Zero crossing detector*

## Byte

A digital word with a length of 8 bits, this is the most commonly used data unit in computing.

A NIBBLE is generally considered as $\frac{1}{2}$ byte and is a word of only 4 bits.

KBYTE is a term used to indicate that the number of bytes being used is 1024.

▶ *Bit*

## Calibration

This means checking the accuracy of an instrument's indication or output against a known standard. The characteristics of any instrument will gradually drift with time and, to get the best results from test equipment, regular and careful calibration is essential. Most electronic manufacturers carry out calibration checks on their test equipment at 90 day intervals, or even more frequently on critical pieces of test gear.

The purpose of calibration is to ensure that the instrument, whether it is used for measuring or generating test signals, is performing within its specified accuracy [▶ *Accuracy*]. The value indicated by the instrument is compared with a known standard and, if necessary, adjustments are made to the instrument's internal circuits to achieve the best performance on all ranges. The timebase on an oscilloscope, for example, can be checked by injecting a test signal from a stable oscillator. This could be a crystal-controlled oscillator with divider stages to allow all ranges of the timebase to be calibrated. The accuracy of most oscilloscopes is typically ±3% so the crystal oscillator with an accuracy of probably better than ±0.1% is more than adequate.

Sometimes it is not possible to bring the accuracy of an instrument within specification. A calibration or correction chart is then produced that shows either the error against the dial or scale indication, or the true reading against the dial or scale indication, or the amount that must be added or subtracted against each dial or scale indication.

Suppose a signal generator is carefully checked over the range 100 Hz to 1 kHz using a digital frequency meter with an accuracy of ±0.1%. The data can be presented in the form of a table, or the percentage error can be plotted graphically as shown (fig. C1). Calibration of signal generator:

| Dial indication | Measured output | Error | % error | Correction |
|---|---|---|---|---|
| (Hz) | (Hz) | (Hz) | | (Hz) |
| 100 | 101.6 | +1.6 | +1.6% | −1.6 |
| 200 | 202.2 | +2.2 | +1.1% | −2.2 |
| 300 | 298.8 | −1.2 | −0.4% | +1.2 |
| 400 | 392.4 | −7.6 | −1.9% | +7.6 |
| 500 | 487.5 | −12.5 | −2.5% | +12.5 |
| 600 | 579.6 | −20.4 | −3.4% | +20.4 |
| 700 | 678.3 | −21.7 | −3.1% | +21.7 |
| 800 | 783.3 | −16.5 | −2.06% | +16.5 |
| 900 | 895.5 | −4.5 | −0.5% | +4.5 |
| 1000 | 997.4 | −2.6 | −0.25% | +2.6 |

Keeping records of calibration is essential and a copy of any correction chart should also be attached to the instrument to which it refers.

Modern digital instruments such as the digital voltmeter and digital frequency meter, which possibly have built-in automatic calibration cycles, are ideal for calibrating instruments such as oscilloscopes, analogue multimeters, and signal generators. But if these are not available other methods can be used, as indicated in Table C1.

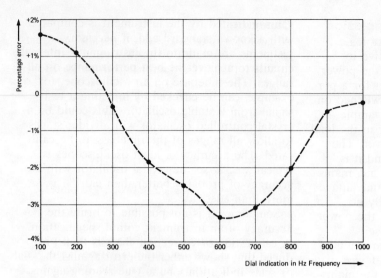

Figure C1   Correction graph

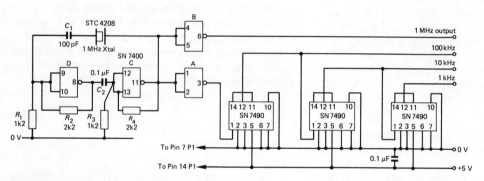

Figure C2   Frequency standard unit

**Table C1**   *Methods of calibrating instruments*

| Instrument | Calibration methods | |
|---|---|---|
| C.R.O. | Timebase: | Crystal-controlled oscillator and divider (fig. C2). |
| | Amplitude: | Switch to d.c. input and use precision voltage source. Alternatively apply known-amplitude square wave. |
| Analogue multimeter | Current ranges: | Place in series with standard meter. |
| | Voltage range: | Use precision voltage source or compare in parallel with standard meter. |
| Signal generator | Amplitude: | Measure with a.c. digital voltmeter at 1 kHz. |
| | Frequency: | Apply to one channel of dual beam c.r.o. while the other is fed from crystal calibrator or use Wien Bridge (audio frequencies only). |

## Canonical Form of a Logic Expression

Suppose the truth table for a 3-input logic circuit is as shown:

|   | Inputs |   |   | Output |
|---|---|---|---|---|
|   | A | B | C | F |
| 0 | 0 | 0 | 0 | 0 |
| 1 | 0 | 0 | 1 | 1 |
| 2 | 0 | 1 | 0 | 0 |
| 3 | 0 | 1 | 1 | 1 |
| 4 | 1 | 0 | 0 | 0 |
| 5 | 1 | 0 | 1 | 1 |
| 6 | 1 | 1 | 0 | 1 |
| 7 | 1 | 1 | 1 | 1 |

Figure C3

From this table, $F = 1$ when the inputs $A \cdot B \cdot C$ are in the following conditions:

001    011    101    110    111

The canonical form for this is written as

$$F = \sum (1, 3, 5, 6, 7)$$
$$= \bar{A} \cdot \bar{B} \cdot C + \bar{A} \cdot B \cdot C + A \cdot \bar{B} \cdot C$$
$$+ A \cdot B \cdot \bar{C} + A \cdot B \cdot C$$

Simplifying

$$F = \bar{A} \cdot C(\bar{B} + B) + A \cdot C(\bar{B} + B) + A \cdot B \cdot \bar{C}$$
$$= \bar{A} \cdot C + A \cdot C + A \cdot B \cdot \bar{C}$$
$$= C + A \cdot B$$

The canonical notation is very useful in the simplification of logic expressions.

*Example* Suppose $F = \sum (12, 13, 14)$.

$$12 = 1100$$
$$13 = 1101$$
$$14 = 1110$$
$$\therefore \quad F = A \cdot B \cdot \bar{C} \cdot \bar{D} + A \cdot B \cdot \bar{C} \cdot D + A \cdot B \cdot C \cdot \bar{D}$$
$$= A \cdot B \cdot (\bar{C} \cdot \bar{D} + \bar{C} \cdot D + C \cdot \bar{D})$$
$$= A \cdot B \cdot (\bar{C} + C \cdot \bar{D})$$
$$= A \cdot B \cdot (\bar{C} + \bar{D})$$
$$F = A \cdot B \cdot (\overline{C \cdot D})$$

## Capacitor

A capacitor is created by two parallel conducting plates separated by an insulating DIELECTRIC. The familiar formula for capacitance $C$ is

$$C = \varepsilon_0 \varepsilon_r \frac{A}{d} \quad \text{(neglecting any edge effects)}$$

where $\varepsilon_0$ is the permittivity of free space
$\varepsilon_r$ is the relative permittivity
$A$ is the area of the plates
$d$ is the distance between the plates, i.e. the thickness of the dielectric.

To achieve any reasonable value of capacitance the area of the plates must be large, the relative permittivity high, and the dielectric thickness small. For the manufacturer this means that the conductors and the dielectric must be thin in order to create a component that has a reasonably small volume. The capacitance-to-volume ratio is important since the space available for a capacitor on a p.c.b. for example is usually limited. With many types (plastic, paper) long strips of thin conducting foils separated by a thin dielectric are rolled together to make the capacitor (fig. C4). Values up to a few microfarads are then possible. Another constraint is that the thin insulating

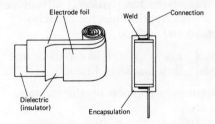

(a) EXTENDED FOIL TYPE

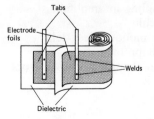

(b) BURIED FOIL TYPE

*Figure C4* **Basic capacitor**

**Table C2** *Capacitors classified by dielectric*

| Dielectric | Permittivity |
|---|---|
| *Ceramic* | |
| a) Low-loss types | 7 |
| b) Temperature compensating | 90 |
| c) High permittivity | |
| (high k) | 1000 → 50 000 |
| *Mica* (silver mica) | 4 to 6 |
| *Paper* (waxed), gradually being | |
| replaced by polypropylene | 4 |
| *Plastic film* | |
| Polycarbonate | 2.8 |
| Polystyrene | 2.4 |
| Polyester | 3.3 |
| Polypropylene | 2.25 |
| *Aluminium Oxide* electrolytics | 7 to 9 |
| *Tantalum Oxide* electrolytics | 27 |

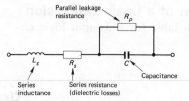

**Figure C5** Equivalent circuit for a capacitor

dielectric must be able to withstand reasonable d.c. voltages without breakdown, so the dielectric strength is also important. Often the *CV* product (capacitance times voltage) is used as a measure of the efficiency of a capacitor type as it gives the total charge *Q* that can be stored. Electrolytic capacitors have the highest *CV* ratio available.

From the preceding paragraph we can see that the characteristics tend to be highly dependent on the type of dielectric material. Modern capacitors can be broadly classified by dielectric.

Each of these types has particular advantages and areas of use. Applications for capacitors are widespread and include

Welding } Energy stored in capacitor and
Photoflash } then discharged rapidly

Spark suppression contacts on thermostats, relays, etc.
Reservoir and smoothing filters in power supplies.
Decoupling and coupling in amplifiers.
Tuning resonant circuits.
Timing elements for monostables, delay circuits, and multivibrators.
Filters and waveform shaping and oscillators.
Power factor correction.
Motor start and run, etc.

The equivalent circuit for a capacitor (fig. C5) shows that it consists of a series inductance, a series resistance (equivalent to losses

in the dielectric), and a capacitance with a parallel leakage resistance. Neglecting the leakage for the moment, the rest form a series resonant circuit where

$$Z = \sqrt{[R_S^2 + (X_L^2 - X_C^2)]}$$

Below RESONANCE the impedance is capacitive, at resonance it is resistive, and above resonance it becomes inductive. It is obviously important to keep the inductance low in order to increase the working frequency range of the capacitor. Low inductance connections are used during manufacture for this purpose. Typical values of resonant frequency are also included in Table C3. Above resonance the impedance rises as the inductive reactance predominates, and the capacitor loses its effectiveness.

Other important characteristics for a capacitor can be derived from the series resistance losses and the leakage resistance. Looking again at the equivalent circuit, for frequencies below resonance we can neglect the effect of the series inductance. A phasor diagram can be drawn showing the effect of the series resistance losses (fig. C6). From this

Loss angle = $\delta$  where $\tan \delta = R_S / X_C$
Phase angle = $\phi$  where $\cos \phi = R_S / Z$
Impedance $Z = \sqrt{[X_C^2 + R_S^2]}$

The loss angle $\delta$ is a measure of the size of the series resistance compared to the capacitive reactance. For a capacitor $\delta$ must obviously be small. The DISSIPATION FACTOR (d.f.) or tan $\delta$, quoted at a particular frequency (50 Hz for electrolytic capacitors, 1 kHz for other types) is a measure of how "lossy" a capacitor is:

Dissipation factor = $R_S / X_C = \tan \delta$

Values range from as low as $2 \times 10^{-4}$ for polystyrene capacitors (at 1 kHz) to above 0.3 for large-value aluminium electrolytics (at 50 Hz).

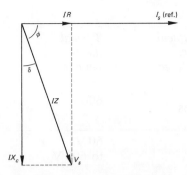

Figure C6 Phasor diagram of a capacitor
$\delta$ = loss angle   $\phi$ = phase angle
$\tan\delta = R_S/X_C$ (dissipation factor)
$\cos\phi = R_S/Z$

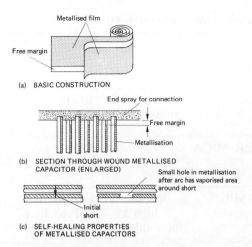

(a) BASIC CONSTRUCTION

(b) SECTION THROUGH WOUND METALLISED
    CAPACITOR (ENLARGED)

(c) SELF-HEALING PROPERTIES
    OF METALLISED CAPACITORS

Figure C7 Metallised capacitors

## Capacitor construction

Basically the constructions of many capacitor types is very similar. The major types are as follows.

**1  Paper capacitors**
Thin sheets of paper, impregnated with oils or waxes to prevent moisture being absorbed and also to increase the dielectric strength, are wound with thin aluminium foils as shown in (figs. C4 and C7). Contact to the metal foils is made either by welding on tabs giving what is termed "buried foil" design, or by extending the foil at either end. With the extended foil type, the end connection is made by soldering or welding an end cap to the exposed foil. Finally the capacitor is sealed inside a metal can, or encapsulated in resin.

**2  Plastic film capacitors**
These are very similar in construction to paper capacitors and both foil and metallised types are produced. In the foil type a number of thin films of plastic material are interleaved with aluminium foils and rolled into a coil by a winding machine. The coil is then fitted with end caps and encapsulated either in resin or in insulating lacquer. Polystyrene film/foil capacitors were the first plastic capacitors to be manufactured and exhibit excellent stability, high insulation resistance, and low temperature coefficient. They are manufactured to fairly close tolerances, but tend to be rather bulky because metallisation of the polystyrene film is not possible since it has a rather low melting point. The plastic dielectric capacitor that is

used extensively in electronic circuits for non-critical applications is the metallised polyethylene, i.e. terphthalate film. They are more commonly referred to as polyester types.

**3  Mica capacitors**
Mica is a naturally occurring substance which, because of its plate-like crystal structure, can be laminated into very thin sheets. It is a very stable material with a high permittivity; thus capacitors made from it give good performance. Silver electrodes are metallised directly onto the sheets of mica and several of these are stacked together to make the complete capacitor. The assembly is then either moulded in plastic or dipped in resin. The latter have higher reliabilities since less stress is applied to the capacitor during the sealing process.

**4  Ceramic capacitors**
These can roughly be divided into two classes: the low-loss low-permittivity types and the high-permittivity types (high k). The low-loss types are usually made from steatite which is a natural mineral. It is finely ground, compressed and then heated to about 900°C to remove impurities. After being reground it is then reformed at about 1300°C. Ceramic capacitors are made in disc, tubular and rectangular plate form. For example, a thin plate is metallised on both sides and connecting leads are soldered to the metallisation. The body is then given several coats of insulating lacquer.

High permittivity ceramics (high k) have the advantage of achieving a relatively large

**Table C3**  *Capacitor characteristics*

| Type | | Range | Tolerance | Typical a.c. voltage | Typical d.c. voltage |
|---|---|---|---|---|---|
| PAPER | foil metallised | 10 nF to 10 μF | ±10% | 250 V/500 V rms | 600 V |
| SILVER MICA | | 5 pF to 10 nF | ±0.5% | — | 60 V to 600 V |
| CERAMIC | low-loss | 5 pF to 10 nF | ±10% | 250 V | |
| | high k | 5 pF to 1 μF | ±20% | | 60 V to 10 kV |
| | monolithic | 1 nF to 47 μF | ±10% | | 60 V to 400 V |
| POLYSTYRENE | | 50 pF to 0.5 μF | ±1% | 150 V | 50 V to 500 V |
| POLYESTER | foil | 100 pF to 10 nF | ±5% | 400 V rms | 400 V |
| | metallised | 1 nF to 2 μF | | | |
| POLYPROPYLENE | | 1 nF to 100 μF | ±5% | 600 V | 1250 V |
| ALUMINIUM | plain foil | 1 μF to 22 000 μF | −20% | Polarised | 6 V |
| ELECTROLYTIC | etched foil | 1 μF to 100 000 μF | +50% | Polarised | to 100 V |
| TANTALUM | foil | 1 μF to 1000 μF | ±10% | Polarised | 1 V |
| ELECTROLYTIC | solid/wet | 1 μF to 2000 μF | ±5% | Polarised | 50 V |

capacitance in a small volume. The common material used is barium titanate. The permittivity can be as high as 10 000. Such capacitors are useful for general purpose coupling and decoupling where fairly wide variations in capacitance value due to temperature, frequency, voltage and time can be tolerated.

A relative newcomer is the so called monolithic ceramic or "block" type. Originally it was intended as a chip component that could be attached direct to film circuits, headers or p.c.b.'s. In this application it had the advantage that the inductance is low because no leads are required. Because of its success this type is now sold as an encapsulated unit with connecting leads attached. The component is made of alternate layers of thin ceramic dielectric (medium to high k) and electrodes. These are compressed and fired to form a monolithic block having the appearance and properties of a single piece of ceramic. A value of say 4.7 μF with a working voltage of 50 V d.c. has a size of only 12 mm × 12 mm × 5 mm. These types are now competing with small electrolytics for general purpose coupling and decoupling applications.

## 5  Electrolytics
These types have one of the highest $CV$ products of all capacitors and are commonly used as reservoir and smoothing elements in power units and for coupling and decoupling in a.f. amplifiers. The large value of capacitance is achieved by the fact that the dielectric formed by the electrolytic action is extremely thin, only a few nanometres.

| Temperature coefficient | $f_R$ | tan δ | Leakage resistance | Stability | Typical application |
|---|---|---|---|---|---|
| 300 p.p.m./°C | 0.1 MHz | 0.005 0.01 | $10^{10}$ Ω $10^9$ Ω | Fair | Motor start and run Mains interference suppression |
| 100 p.p.m./°C | 10 MHz | 0.0005 | $10^{11}$ Ω | Excellent | Tuned circuits Filters |
| ±30 p.p.m./°C Varies Varies | 10 MHz 10 MHz 100 MHz | 0.002 0.02 0.02 | $10^8$ Ω $10^8$ Ω $10^{10}$ Ω | Good Fair Good | Temp. compensating Coupling & decoupling |
| −150 p.p.m./°C | 10 MHz | 0.0002 | $10^{12}$ Ω | Excellent | Tuned circuits Filters Timing |
| 400 p.p.m./°C | 1 MHz 0.5 MHz | 0.005 0.01 | $10^{10}$ Ω $10^{11}$ Ω | Fair | General purpose Coupling & decoupling |
| −170 p.p.m./°C | 1 MHz | 0.0003 | $10^{10}$ Ω | Fair | Mains suppression Motor start & run |
| 1500 p.p.m./°C | 0.05 MHz | 0.08 | Specified by leakage current | Fair | Decoupling l.f. Reservoir & smoothing-in power supplies |
| 500 p.p.m./°C | 0.1 MHz | 0.01 | Specified by leakage current | Good | Coupling and decoupling at l.f. |
| 200 p.p.m./°C | | 0.001 | | Excellent | Timing, etc. |

Electrolytics can be divided into the following sub-classes:

| | |
|---|---|
| *Aluminium* | plain foil |
| | etched foil |
| | solid |
| *Tantalum* | solid |
| | wet sintered. |

In the ALUMINIUM ELECTROLYTICS, an aluminium foil of very high purity (99.9%) is immersed in a bath of electrolyte and moved through this bath at a constant velocity while a fixed voltage is applied. This voltage causes a forming current to flow which gradually falls in value as aluminium oxide grows on the surface of the foil. This thin covering of aluminium oxide, an insulator, is to be the dielectric of the final capacitor. The anode foil with its thin

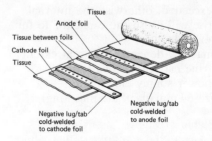

*Figure C8*  Aluminium electrolytics

coating of oxide is wound together with another foil and tissue paper spacers on a winding machine. The paper spacers are saturated with an electrolyte such as ammonium borate or ethylene glycol and this electrolyte, being a conductor, is the true cathode of the final capacitor (fig. C8). By chemically etching

the aluminium foil before the forming process, the effective surface area of the plates is increased. The "etched foil" type can achieve up to 4 times the capacitance value of a plain foil because of the effective increase in the surface area of the plates.

A few further points on aluminium electrolytics are worth note:

1) The capacitor is usually polarised and the voltage across it should not be reversed. If a reverse voltage is applied, the dielectric will be removed from the anode and a large current will flow as oxide is formed on the cathode. The gases released from the electrolyte may build up and cause damage to the capacitor, or cause the capacitor to explode and damage other components.

2) The tolerance of the capacitance value is wide, typically −20% to +50%.

3) Leakage current can be high and is usually specified at maximum rated voltage and 20°C five minutes after the voltage is applied.

4) The equivalent series resistance (ESR) is made up of the resistance of the leads, aluminium electrodes, electrolyte and the series losses of the dielectric. Typical values of ESR are in hundreds of milliohms.

Any alternating voltage across the capacitors will cause a "ripple current" to flow through this series resistance. Heat losses ($I^2R$) will result in a temperature rise inside the capacitor and this heat must be effectively removed by radiation from the surface area of the capacitor case. A maximum value of ripple current $I_R$ at 70°C is quoted for electrolytics used in smoothing applications, and this maximum value must not be exceeded, otherwise the internal temperature rise will cause the capacitor to fail and possibly explode.

Values of max. $I_R$ depend to a great extent on the size of the capacitor: for example

Computer power supply electrolytic
47 000 $\mu$F (size 64 mm dia.×115 mm long)

$$I_R \simeq 18 \text{ A rms}$$

Miniature electrolytic for coupling and decoupling (size 12.9 mm dia.×25 mm long)

$$200 \ \mu\text{F at 10 V} \qquad I_R \simeq 125 \text{ mA rms}$$

5) TANTALUM ELECTROLYTICS use tantalum oxide as the dielectric. This has a higher permittivity than aluminium oxide, and therefore a very high capacitance is available

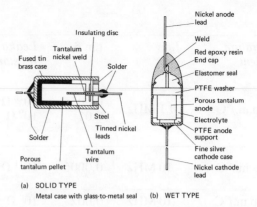

(a) SOLID TYPE
Metal case with glass-to-metal seal    (b) WET TYPE

**Figure C9**  Tantalum electrolytic capacitor

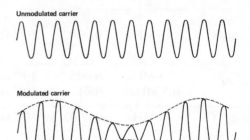

**Figure C10**

in a small size. Apart from this, tantalums have lower leakage, higher reliability, and better selection tolerance than the aluminium types. The cost is therefore higher. (See fig. C9.)

## Carrier

It is not possible to transmit and receive low-frequency signals, such as audio messages, directly over any appreciable distance except by using wires. For "wireless" transmission the low-frequency signals must be superimposed onto a high-frequency radio wave which can be radiated from an aerial. The high-frequency radio wave is called the carrier and the process by which the lower-frequency signals are impressed onto the carrier is called MODULATION (fig. C10).

A communication channel is open when the carrier wave is transmitted but no information will be transmitted or received until the carrier is modulated. Radio waves are not the only types of carrier; light beams, optical links, and ultrasonic waves can also be used.

▶ *Amplitude modulation*    ▶ *Ultrasonics*

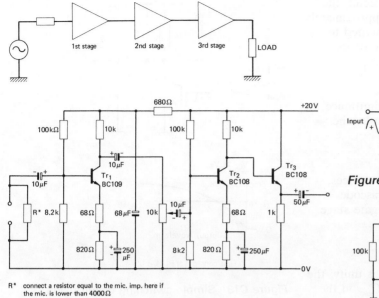

Figure C11 Cascaded amplifier; the circuit shows a microphone pre-amplifier

R* connect a resistor equal to the mic. imp. here if the mic. is lower than 4000 Ω

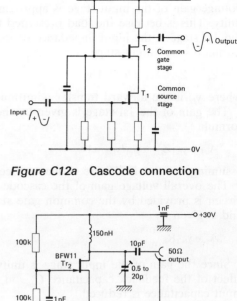

Figure C12a Cascode connection

Figure C12b Practical cascode circuit (100 MHz tuned amplifier)

## Cascaded Stage

When a high value of voltage gain is required from an amplifier, several stages may be connected as shown (fig. C11), with the output of the 1st feeding the input of the 2nd and so on. The overall amplification is then the product of the voltage gains of the individual stages. Each stage can be designed so that its load is the input of the next stage and its source the output of the preceding stage.

In the example, $Tr_1$ is a CE stage feeding an identical stage $Tr_2$ which then feeds the output emitter follower. The overall amplification is about 64 dB, the frequency response is from 18 Hz to 30 kHz, and the output is 5 V pk–pk into 600 Ω.

▶ Amplifier

## Cascode

A connection method used in amplifiers where an active device serves as the load for a similar active device. In the example (fig. C12a) the common source input stage has as its load the common gate output stage. The cascode connection is used to increase the high-frequency response of an amplifier because it markedly reduces the effect of device feedback capacitance. When a single-stage common source FET amplifier is used, the small feedback capacitance between drain and gate $C_{gd}$ is amplified by the Miller effect to appear as a much larger capacitance in parallel with $C_{gs}$ [▶ Miller effect].

The total input capacitance for the stage is

$$C_{in} = C_{gs} + C_{gd}(1 - A_v)$$

where $A_v$ is the voltage gain of the stage. For a JFET, typical values might be $C_{gs}$ 5 pF, $C_{gd}$ 1.5 pF, and $A_v$ −20. Therefore

$$C_{in} = 5 + 1.5(21) = 36.5 \text{ pF}$$

The source resistance $R_S$ acts with the input capacitance as a low pass filter and the gain of the stage will be 3 dB down when the reactance of $C_{in} = R_S$. In other words the high-frequency cut-off will be given by

$$f_h = 1/2\pi R_S C_{in}$$

If $R_S = 1 \text{ k}\Omega$, then $f_h \simeq 4.4 \text{ MHz}$.

However, by using the cascode circuit, the voltage gain of the input stage is approximately unity. This is because the load presented to the 1st stage is the input impedance of the common gate output stage

$$Z_{L1} = Z_{i2} = \delta v_{gs2}/\delta i_{d2} = 1/y_{fs2}$$

where $y_{fs}$ is the forward transfer admittance.

The gain of the 1st stage is given by the formula

$$A_{v1} \simeq y_{fs1} Z_{L1} \simeq y_{fs1}/y_{fs2} \simeq 1$$

assuming the two FETS are similar devices.

The overall voltage gain of the cascode circuit is provided by the common gate stage and is given by

$$A_v \simeq y_{fs2} R_L$$

Since the gain of the input stage is unity, the effect of the feedback capacitance $C_{gd}$ on the input capacitance is reduced.

$C_{in}$ for the cascode circuit, taking the same values previously used, is

$$C'_{in} = C_{gs} + C_{gd}(2) = 5 + 3 = 8 \text{ pF}$$

and the high-frequency cut-off caused by the input filter formed by $R_S$ and $C'_{in}$ is now

$$f_h = 1/2\pi R_S C'_{in} \simeq 20 \text{ MHz}$$

The cascode arrangement can be used with other active devices such as bipolar transistors and valves to reduce effects of feedback capacitance. A practical circuit of a 100 MHz tuned amplifier using two BFW11 FETs is shown in fig. C12b. $Tr_1$ is the common source stage and $Tr_2$ the common gate. Two 100 kΩ resistors fixed the bias voltage for $Tr_2$ gate and this is decoupled by a 1 nF capacitor. The drain current for both transistors is set by the 1 kΩ potentiometer in $Tr_1$ source.

## Catching Diode

A modification used in pulse and digital circuits to improve the rise time of an output signal. In the simple switching circuit shown (fig. C13), when the transistor switches off the output resistance of the circuit is equal to $R_L$. This load resistor has to charge the circuit capacitance plus any stray capacitance existing from output to ground. This slows the rise of the output voltage and the time taken for the output voltage to rise from $V_{CE(sat)}$ to $+V_{CC}$ will be about four time constants ($4R_L C_S$).

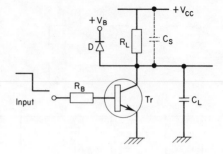

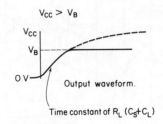

*Figure C13* Simple switching circuit using catching diode

In the catching diode circuit the diode conducts when the output exceeds $V_B$ and "catches" the output waveform. The rise time of the output waveform is improved.

## Cathode Ray Tube

The c.r.t is a thermionic tube widely used as a display device in oscilloscopes, t.v. receivers, VDUs, and radar sets. The cross-section (fig. C14) shows that it consists of three main parts: an electron gun, a deflection system, and a fluorescent screen all contained within an evacuated glass envelope. The electron gun produces a finely focussed high-velocity beam of electrons, and this beam can be deflected to strike any part of the screen by applying suitable deflection signals. On impact the beam causes a visible spot of light to be given off from the fluorescent screen. An internal graphite (aquadag) coating collects any secondary electrons given off from the screen. Deflection and focussing methods can be either magnetic or electrostatic. The description that follows is for a tube with electrostatic deflection and focussing.

In the electron gun, the metal cathode cylinder, coated with a mixture of barium and strontium oxides, is heated by a thin electrically insulated tungsten filament so that it

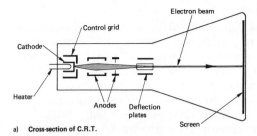

a) **Cross-section of C.R.T.**

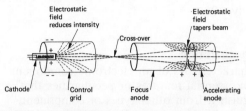

b) **Principle of electrostatic focussing**

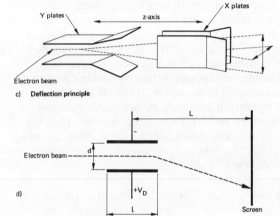

c) **Deflection principle**

d)

*Figure C14* **Cathode ray tube**

emits electrons. A metal cylinder with a small aperture surrounding the cathode is used as the control grid. When this is negative with respect to the cathode, the number of electrons in the beam is reduced. Since the intensity of the resulting spot on the screen is proportional to the number of electrons in the beam, changing the value of the grid voltage will give brightness control. Typical grid to cathode voltages are from zero to $-50\,V$. The latter voltage gives complete cut-off of the beam so that negative going pulses of this amplitude at the grid can be used for blanking.

The rest of the electron gun consists of accelerating and focussing anodes which are

specially shaped metal cylinders all at higher voltages than the cathode. The final anode voltage in a tube may be from 1 kV up to 10 kV or higher. Electrostatic focussing is achieved as shown, the lines of equal potential between the anodes forcing the electrons within the beam to come to a point at the screen. Varying $A_2$ voltage will alter the position of the potential pattern between the anodes and will therefore alter the point at which the beam is focussed. After leaving the final anode, the electron beam enters the deflection system, which for electrostatic deflection will be two sets of metal plates at right angles to each other. For the Y or vertical plates, a positive voltage applied to $Y_1$ with respect to $Y_2$ will cause the beam to be deflected upwards to hit the top of the screen; and if $Y_2$ is positive with respect to $Y_1$ the beam will be deflected towards the bottom of the screen. Voltages applied to the X or horizontal plates will cause the beam to move to the left or right. The deflection sensitivity depends upon the length of the plates and the velocity of the electrons. For tubes with low anode voltages, giving low velocity beams, deflection sensitivities of a few volts per cm are typical, but tubes with high anode voltages may have deflection sensitivities as low as 50 volts per cm. The amount of deflection achieved at the screen is given by

$$D = V_d Ll/2V_a d$$

where $V_d$ = deflection voltage
$V_a$ = accelerating voltage
$d$ = distance between plates
$l$ = length of plates
$L$ = distance from centre of plates to screen

The fluorescent screen is made up of a phosphor material deposited evenly on the internal side of the glass face plate. Various materials are used each giving a distinct colour during fluorescence and having a particular persistence. Persistence is defined as the time for the luminous spot on the screen to decay to 10% of its peak value.

Referring to Table C4, a P1 phosphor would be used in a c.r.t for a general-purpose oscilloscope, a P4 in a t.v. tube, a P11 in a tube for high-speed photography of transients, and a P33 in a radar situation display.

▶ *Measurement* (oscilloscope)

**Table C4**  *Cathode ray tube characteristics*

| Phosphor | Colour | Persistence |
|---|---|---|
| P1 | yellow/green | medium |
| P2 | yellow/green | medium short |
| P4 | white | medium |
| P7 | yellow/green | long |
| P11 | blue | medium short |
| P20 | yellow/green | medium |
| P24 | green | short |
| P32 | blue/green | long |
| P33 | orange | very long |

| Persistence | Typical time |
|---|---|
| very short | less than 1 $\mu$s |
| short | 1 to 10 $\mu$s |
| medium short | 10 $\mu$s to 1 ms |
| medium | 1 ms to 100 ms |
| long | 100 ms to 1 sec |
| very long | greater than 1 sec |

## Central Processor Unit

A CPU is the vital heart within a computer system. It carries out the arithmetical, logical and manipulative operations and controls the overall working of the computer. A very simple view of the arrangement is shown in fig. C15. The CPU operates on binary data which is presented to it over a data bus from the store or from the input/output devices. The data and address information is in sets of 8 or 16 parallel bits.

A microprocessor is an integrated circuit version of a CPU and forms the controlling i.c. in a microcomputer.

▶ *Microprocessor*

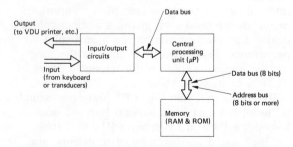

*Figure C15*  Arrangement of a central processor unit

## Cermet

A thick film, made from a mixture of ceramic and tiny particles of a metal, that is used in the manufacture of fixed and variable resistors. The proportion of metal to ceramic in the film paste determines the sheet resistance of the film deposited onto an insulating inert substrate. The assembly is fired at a high temperature and the resulting resistive track has excellent stability and a low temperature coefficient, typically $\pm 200$ ppm/°C.

## Characteristics

The characteristics of an electronic component are the essential features or peculiarities that distinguish it from other types of component. For example, one of the characteristics of thermistors is that their resistance falls rapidly with increasing temperature, whereas fixed resistors have a fairly stable temperature characteristic.

A characteristic curve is a very useful graph (fig. C16) used for active devices to show how the current through the device changes as its electrode voltages or other electrode currents are varied. These curves are used to describe the operation of active devices such as bipolar transistors and FETs and will be found on the manufacturers data sheet in the section Electrical Characteristics. The three most important are shown in Table C5.

From any characteristic curve it is possible to evaluate one of the device parameters. For example, from the slope of the transfer characteristics for a bipolar transistor one of the $h$-parameters is given:

$$h_{fe} = \frac{\delta i_c}{\delta i_b}\bigg|_{V_{ce} \text{ held constant}}$$

This is the small-signal current gain of the transistor.

From the transfer characteristic of a FET, the slope gives one of the y-parameters:

$$y_{fs} = \frac{\delta i_d}{\delta V_{gs}}\bigg|_{V_{ds} \text{ held constant}}$$

$y_{fs}$ is the forward transfer admittance. ($y_{fs}$ at low frequencies is equivalent to the $g_m$ of the device.)

▶ *Parameter*

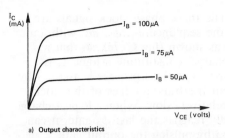

a) Output characteristics

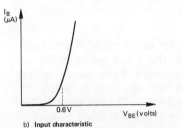

b) Input characteristic

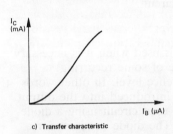

c) Transfer characteristic

Figure C16 Characteristics for a bipolar transistor in common emitter mode

## Charge Coupled Device

CCDs are MOS integrated circuits with multiple gates and are used as shift registers, delay lines, stores, and also for optical imaging. The construction usually consists of a p-type substrate, two heavily doped n regions for input and output, and a series of gates all insulated from the substrate by a thin layer of silicon dioxide (fig. C17a). The n-type region at the input together with the input gate act as the source and gate of a MOSFET, which can inject an input consisting of a packet of electrons into a "drain" area just beneath the first $\phi_1$ gate, while this gate is high (active). The charge packet can then be transferred to the next location by applying a specific sequence of

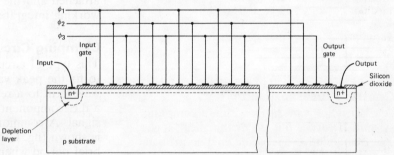

Figure C17a Basic structure of a three-phase charge coupled device (no clock pulses applied)

**Table C5**

| Characteristic | Meaning | Bipolar transistor in common emitter | Field effect transistor in common source |
|---|---|---|---|
| Output | Graphs showing how the output current varies with output electrode voltage for various values of input. | Collector current ($I_C$) against collector/emitter voltage ($V_{CE}$) for various values of base current ($I_B$). | Drain current ($I_D$) against drain/source voltage ($V_{DS}$) for various values of gate to source voltage ($V_{GS}$). |
| Input | A graph showing how the input current varies with input voltage for a fixed value of output voltage. | Base current ($I_B$) against base/emitter voltage ($V_{BE}$) for a fixed value of $V_{CE}$. | Not given because the input current to a FET is very small. |
| Transfer | A graph showing how the output current varies with input quantity at a fixed value of output voltage. | Output current ($I_C$) against base input current ($I_B$) for a fixed value of $V_{CE}$. | Output current ($I_D$) against input voltage ($V_{GS}$) for a fixed value of $V_{DS}$. |

clock pulses to the gates (fig. C17b). When a gate electrode is taken positive, the depletion region just beneath that gate increases to form the "potential well" that can receive and store electrons. The three phase clock signals are applied in the sequence $\phi_1$, $\phi_2$, $\phi_3$, with waveforms as shown (fig. C17c), so that a packet of charge, constituting a logic 1, is shifted to the right. This process is assisted by ensuring that the trailing edges of the clock pulses are relatively slow. When a logic 1 charge packet finally reaches the last $\phi_3$ gate, it can be extracted by pulsing the output gate.

## Chip

A name given to the small piece of semiconductor in which all the active and passive elements of a microcircuit have been diffused. Chips for MSI devices may only be a few square millimetres in area. Leads have to be attached and the chip encapsulated to make a working integrated circuit.

## Clamping Circuit

This type of circuit is used when it is necessary to fix the peak value of some recurring waveform to a reference level. In other words a d.c. component is introduced into the output signal. A commonly used circuit using a diode is shown (fig. C18). The diode conducts for a brief period when the input waveform is at its most positive excursion and effectively clamps the output signal to a d.c. voltage of $V_R + V_D$, where $V_D \simeq 0.6\,\text{V}$ for a silicon diode. The input time constant $CR$ should be long compared with the periodic time of the lowest signal frequency of the input.

A clamp to some negative voltage level can be made by reversing the diode and the polarity of the reference voltage.

If the reference voltage is zero, in other words the diode is connected to ground, the output waveform is clamped to nearly zero volts and the circuit is commonly referred to as a D.C. RESTORER.

## Clapp Oscillator

This type of oscillator is a modified version of the Colpitts and it gives better frequency stability. The frequency is determined by a series capacitor $C_s$ and $L_0$ instead of the parallel capacitors $C_1$ and $C_2$ as in the standard Colpitts circuit. For the Clapp circuit

$$f_0 = \frac{1}{2\pi}\sqrt{\frac{1}{L_0 C_s}}$$

and positive feedback is provided by $C_1$ and

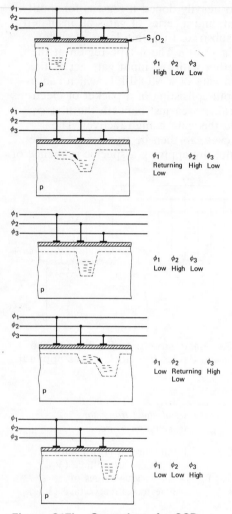

$\phi_1$
$\phi_2$
$\phi_3$

$S_1O_2$

$\phi_1$ $\phi_2$ $\phi_3$
High Low Low

p

$\phi_1$ $\phi_2$ $\phi_3$
Returning High Low
Low

p

$\phi_1$ $\phi_2$ $\phi_3$
Low High Low

p

$\phi_1$ $\phi_2$ $\phi_3$
Low Returning High
Low

p

$\phi_1$ $\phi_2$ $\phi_3$
Low Low High

p

*Figure C17b*  Operation of a CCD

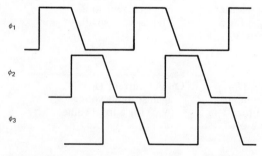

$\phi_1$

$\phi_2$

$\phi_3$

*Figure C17c*  Three phase clock waveforms

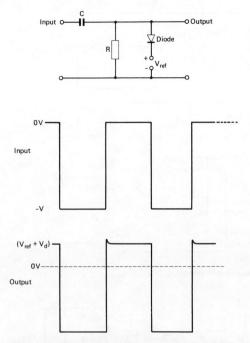

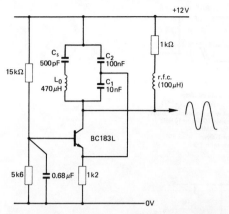

Figure C19 Clapp oscillator ($f_0 = 300$ kHz)

Figure C18 Positive clamping circuit

$C_2$. These capacitors should be much higher in value than $C_s$.

An example (fig. C19) shows a circuit designed to give sine waves at about 4 V pk–pk with a frequency of 300 kHz. A parallel path for the transistor current has to be provided through a resistor (1 kΩ) and an r.f. choke because no d.c. current can flow through the tuned circuit.

## Clipping Circuit

These circuits are used to select only a portion of an input signal that is either above or below some reference level. Simple diode circuits can be used to limit the positive or negative amplitude of a waveform or both as shown (fig. C20). If capacitive coupling is used, the input time constant $C_1 R_1$ must be long compared to the lowest frequency of the input signal.

A double clipping circuit or slicer limits both the positive and negative amplitude and can be created using two parallel diodes or two zener diodes connected in series across the signal path. In all circuits the resistor ($R_s$ or $R_p$) should be large compared to the forward resistance of the diode, a typical value for $R_s$ or $R_p$ being 1 kΩ or higher. In the waveform diagrams the volt drop across the diode has been assumed to be negligible.

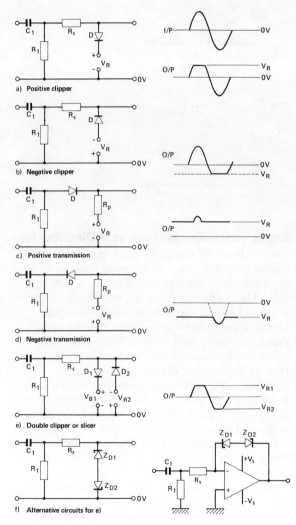

Figure C20 Clipping circuits

**59**

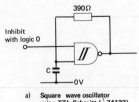

a) **Square wave oscillator using TTL Schmitt ($\frac{1}{4}$ 74132)**

simple circuit with reasonable stability.
Frequency range 100 Hz to 10 MHz
C = 200 pF for 10 MHz
C = 20 µF for 100 Hz
Mark-to-space ratio 1:3

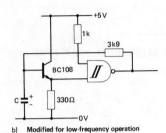

b) **Modified for low-frequency operation**

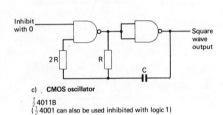

c) **CMOS oscillator**

$\frac{1}{2}$ 4011B
($\frac{1}{2}$ 4001 can also be used inhibited with logic 1)

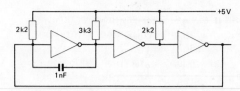

d) **TTL ring oscillator using open collector invertors**

1 MHz square waves.
Vary 3k3 to trim frequency

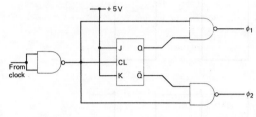

e) **Two-phase clock generator**

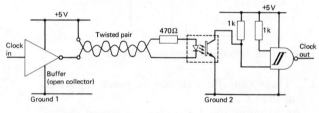

f) **Clock transmission system**

*Figure C21* **Oscillators as clock pulse generators**

## Clock Pulse Generators and Distribution

The majority of digital systems require a clock circuit that delivers a continuous train of pulses to synchronise the operation of the whole system. The main requirements of a clock circuit are good frequency stability, defined logic levels, a low output impedance to give good drive capability, and a clean waveshape. The pulse produced must be compatible with the type of logic being driven, for example standard TTL i.c.s would require the clock to have levels of less than 0.4 V for logic 0 and greater than 2.4 V for logic 1 with rise and fall times of better than 100 ns.

For the highest stability in operating frequency, a crystal-controlled oscillator should be used with a buffer drive to the output. However in many systems the specification on frequency tolerance is not tight and there are several square wave or pulse oscillators that can usefully serve as clock generators (fig. C21):

Transistor astable multivibrator
The 555 or CMOS 555 in astable mode
Blocking oscillator
Unijunction relaxation oscillator
Square wave generators using i.c. gates with *CR* networks
Schmitt trigger i.c. gates with positive feedback via *CR* network
Specialised i.c.s such as the CD 4047B.

It is important that the output from the clock can drive the required gates and i.c.s without loss of waveshape, in other words no degradation of the rise and fall times and minimum ringing. In many cases the clock has to supply a relatively large capacitive load, which means that the clock output circuit must have a low output impedance in both states to rapidly charge and discharge this load capacitance. A special buffer or drive interface may be required to do this.

The method used to distribute the clock

pulse round a system also requires careful consideration. The important factors are: the preservation of the waveshape; the prevention of noise signals or crosstalk from appearing on the clock lines; and the prevention of clock "skew". The latter effect can cause false outputs and is due to the clock signal at one point in a system being slightly out of phase with the clock at another point. The effect can be minimised by ensuring that the track lengths of all clock lines within a system are almost the same. Waveforms can be reshaped by using Schmitt triggers, and noise and crosstalk kept to low levels by careful layout, correct decoupling, and the use of ground planes. If long connections, of greater than 60 cm, have to be used then a line driver and receiver system are probably necessary. A simple arrangement for transmitting clock pulses from one system to another where the grounds are not common is shown using an opto-isolator. Interference which will be common mode to both leads will be rejected.

Several MOS i.c.s require multiphase clock signals; for example, MOS shift registers require a two-phase clock, and CCDs may require a three- or four-phase clock. The usual requirement is that the phases of the clock should not overlap. An example is given showing how TTL can be used to generate a two-phase clock with non-overlapping pulse outputs. To drive a MOS shift register, a TTL-to-MOS interface will be required for each phase.

▶ *Crystal oscillator*

## Closed Loop

Most amplifiers and control systems use negative feedback to stabilise the gain, to improve frequency response, and to reduce non-linearity and distortion. When a feedback loop is connected to feed a portion of the output back to oppose the input signal, the amplifier or system is said to be in a closed loop condition. The closed loop voltage gain of an amplifier with negative feedback is

$$A_{VCL} = \frac{A_{VOL}}{1 + A_{VOL}\beta}$$

The opposite condition to closed loop is OPEN LOOP, where no feedback signal is applied and the gain of the system is at maximum uncontrolled value. In a control system, open loop will give an output that bears

no relationship to the effect it produces, whereas a closed loop system is error-actuated and automatically adjusts the output to give a fixed effect.

An example of open loop is a motor speed controller where the input sets the power applied to the motor and no adjustment is made for loading effects. As the load varies, so will the speed of the motor. By adding a feedback loop to give closed loop control, the power applied to the motor can be altered to suit changing loads and thus keep the motor's speed relatively constant. A transducer, such as a tachogenerator, senses the speed of rotation of the motors' shaft and feeds back a voltage proportional to speed. This can be compared with the input reference to automatically adjust the final output to the required level.

▶ *Negative feedback*     ▶ *Amplifier*

## CMOS

Complementary Metal Oxide Silicon (field effect transistor logic), is a family of i.c.s that is made from combinations of p and n channel enhancement mode MOSFETS. This type of construction results in a number of particular advantages when compared with TTL and ECL. These unique features are

*a*) A very low power consumption (about 10 nW/gate static)
*b*) A wide operating supply voltage range (+3 V to +18 V)
*c*) A very high fan-out (at least 50)
*d*) Excellent noise immunity (45% of the supply voltage).

These properties show why CMOS is the logic chosen for low cost, low power consumption systems, especially those used in electrically noisy environments and where speed of operation is not the prime consideration. The basic CMOS range of devices is not as fast as TTL, a typical propagation delay being 35 ns at a supply voltage of +5 V. However by using a supply of +10 V the switching speed improves to 20 ns. One other disadvantage of CMOS is the relatively high output impedance, which means they cannot drive large capacitive loads and are more susceptible to current-injected noise.

To fully appreciate the unique features of CMOS it is useful to look at the operation of some basic gates. The structure and circuit of

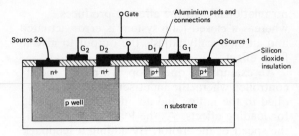

Figure C22a Typical CMOS structure (invertor)

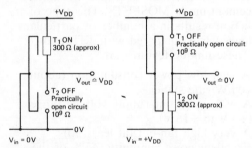

Figure C22b CMOS invertor circuit

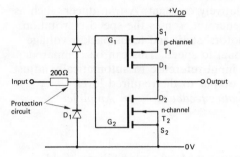

Figure C22c Equivalent circuits for two input conditions

an invertor is shown in fig. C22. Both devices are enhancement mode MOSFETs. In this type of device there is no conducting channel between source and drain until a suitable bias voltage is applied to the gate. Then the layer beneath the gate inverts (changes from p to n for an n-channel) and conduction takes place. Consider the invertor circuit with $V_{DD}$ applied and the input at 0 V. The n-channel MOSFET $T_2$ will be off, since there is zero volts between its gate and source. But the p-channel $T_1$ will be conducting and *on* because there is a large negative bias between its gate and source. The on-resistance of $T_1$ is about $300 \, \Omega$ while the off-resistance of $T_2$ is very high at $10^9 \, \Omega$. The

output is connected to $+V_{DD}$ via the low channel resistance of $T_1$.

If the input is now taken to $+V_{DD}$, the p-channel MOSFET $T_1$ is turned off, effectively disconnecting the output from $+V_{DD}$, and at the same time the n-channel device $T_2$ turns *on* connecting the output to 0 V via its conducting channel. The equivalent circuits for these two possible input states are shown in fig. C22c.

For CMOS gates   logic $0 = 0$ V to $0.3 V_{DD}$
                         logic $1 = +V_{DD}$ to $0.7 \, V_{DD}$

From the description of operation it can be seen that

1) Since the gates of the MOSFETs are insulated from the substrate, the input impedance is extremely high ($10^{12} \, \Omega$ or greater). It is this high input impedance that gives CMOS such a high fan-out capability.

2) In the static state, one device is on while the other is off. Thus the power taken from the supply is almost negligible.

3) The operation of the circuit is independent of the value of the supply voltage.

4) The output swings from $+V_{DD}$ to 0 V.

5) The input threshold is one half of the supply giving a noise margin that is typically 45% of $V_{DD}$.

A closer look at the circuit will show that, when the input is changing state, there must be a brief instant when both devices conduct. A small current pulse will be taken from the supply. Therefore power dissipation of CMOS increases with operating frequency and is typically 1 mW/MHz per gate. It is also important that inputs are not left open circuit, otherwise the small gate capacitance will slowly charge up and put the devices into their active regions. When this happens all devices in the package conduct and a very large current may be taken from the supply, causing the i.c to overheat and possibly burn out. All unused inputs, including those of unused gates in an i.c., *must be connected somewhere*. As a general rule, spare inputs to gates should be connected to another driven input or to an appropriate voltage level ($+V_{DD}$ for a NAND) and inputs to unused gates should be disabled by connecting them to 0 V or $+V_{DD}$.

Because the thin insulating region between the gate and the body of a MOSFET is very easily damaged by electrostatic discharge, CMOS circuits have built-in input protection—

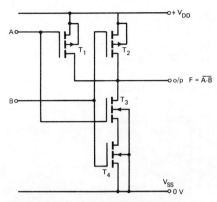

Figure C23　CMOS NAND gate

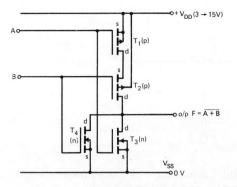

Figure C24　CMOS NOR gate

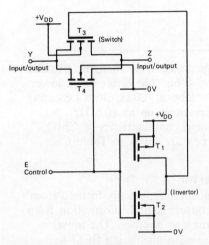

Figure C25a　CMOS transmission gate

a typical example being a $200\,\Omega$ series resistor and two diodes as shown.

The CMOS NOR and NAND gates are simply extensions of the basic invertor. With the NAND gates (fig. C23) the output can only be low when both $T_3$ and $T_4$ conduct and both $T_1$ and $T_2$ are off. This condition only occurs when both inputs $A$ and $B$ are at logic 1 $(+V_{DD})$. The operation is more easily explained using a truth table.

The truth table for CMOS NAND, with logic $1 = V_{DD}$ and logic $0 = 0\,V$, is

| Inputs | | State of MOSFETs | | | | Output |
| A | B | $T_1$ | $T_2$ | $T_3$ | $T_4$ | F |
|---|---|---|---|---|---|---|
| 0 | 0 | ON | ON | OFF | OFF | 1 |
| 0 | 1 | ON | OFF | OFF | ON | 1 |
| 1 | 0 | OFF | ON | ON | OFF | 1 |
| 1 | 1 | OFF | OFF | ON | ON | 0 |

Similarly for the NOR gate:

| Inputs | | State of MOSFETs | | | | Output |
| A | B | $T_1$ | $T_2$ | $T_3$ | $T_4$ | F |
|---|---|---|---|---|---|---|
| 0 | 0 | ON | ON | OFF | OFF | 1 |
| 0 | 1 | ON | OFF | OFF | ON | 0 |
| 1 | 0 | OFF | ON | ON | OFF | 0 |
| 1 | 1 | OFF | OFF | ON | ON | 0 |

With the NOR gate (fig. C24) the output will go to 0 if either input $A$ or $B$ is at logic 1.

Note that for clarity the protection circuits have been omitted from the diagrams. The two circuits are typical of the older A series type of CMOS. The later versions, called the B series, have two additional invertor stages to provide

buffering. The B series give a sharp transfer characteristic and better drive than the A series.

Some circuit configurations are unique to CMOS. A good example is the transmission gate (fig. C25). These switches are bilateral and can be used to switch either digital or analogue signals, the only requirement being that the input must not exceed $V_{DD}$ or go below $0\,V$. When the control input E is low at $0\,V$, both switch transistors are off and there is virtually an open circuit between Y and Z. The off-state leakage current between Y and Z is about $100\,nA$. If the control input E is high at $+V_{DD}$, both switch transistors $T_3$ and $T_4$ conduct and Y is connected to Z via a low-resistance path of about $300\,\Omega$. Transmission gates can be used in applications such as multiplexing, sample and hold (fig. C25b), digital filters, and remote control. Two typical i.c.s are

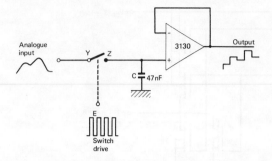

*Figure C25b*   Application of CMOS transmission gate in sample and hold circuit

the 4016, a quad digital or analogue bilateral switch, and the 4066. The latter has a lower on-resistance of about 100 Ω. Both i.c.s can operate at frequencies up to 10 MHz.

The ratings and characteristics of CMOS 4000B series i.c.s are given in Table C6.

## Code Convertor

These are logic circuits, usually formed from gates, which convert digital information from one coded form into another. The most commonly used examples being BCD to decimal and BCD to seven-segment. These code convertors are more usually referred to as DECODERS and are available as i.c.s:

| | | |
|---|---|---|
| TTL | 7442 | BCD to decimal |
| | 74141 | BCD to decimal |
| | 7447 | BCD to seven-segment |
| CMOS | 4028 | BCD to decimal |

Code convertors which convert decimal numbers or hexadecimal into some other coded form are called ENCODERS.

The i.c. types are dealt with elsewhere but in fig. C26a–d some simple examples of code convertors are shown together with the truth tables. Very simple convertors can be made using diodes, and two examples are given. The first shows how decimal can be converted into BCD and the second a method for converting from decimal to seven-segment. In the first case it is assumed that only one decimal input will be set to 0 at any one time, while with the decimal to seven-segment only one input is allowed to be high at any one time.

▶ *Decoder*   ▶ *Encoder*

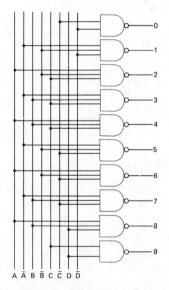

| Decimal | D | C | B | A |
|---|---|---|---|---|
| 0 | 0 | 0 | 0 | 0 |
| 1 | 0 | 0 | 0 | 1 |
| 2 | 0 | 0 | 1 | 0 |
| 3 | 0 | 0 | 1 | 1 |
| 4 | 0 | 1 | 0 | 0 |
| 5 | 0 | 1 | 0 | 1 |
| 6 | 0 | 1 | 1 | 0 |
| 7 | 0 | 1 | 1 | 1 |
| 8 | 1 | 0 | 0 | 0 |
| 9 | 1 | 0 | 0 | 1 |

*Figure C26a*   Decimal to 8421 BCD counter. (All inputs are high except the energised number.)

| D | C | B | A | Simplification | Decimal |
|---|---|---|---|---|---|
| 0 | 0 | 1 | 1 | $\bar{C}.\bar{D}$ | 0 |
| 0 | 1 | 0 | 0 | $\bar{A}.B.\bar{C}$ | 1 |
| 0 | 1 | 0 | 1 | $A.B.\bar{C}$ | 2 |
| 0 | 1 | 1 | 0 | $\bar{A}.B.C$ | 3 |
| 0 | 1 | 1 | 1 | $A.B.C$ | 4 |
| 1 | 0 | 0 | 0 | $\bar{A}.\bar{B}.C$ | 5 |
| 1 | 0 | 0 | 1 | $A.\bar{B}.\bar{C}$ | 6 |
| 1 | 0 | 1 | 0 | $\bar{A}.\bar{B}.C$ | 7 |
| 1 | 0 | 1 | 1 | $A.B.D$ | 8 |
| 1 | 1 | 0 | 0 | $C.D$ | 9 |

*Figure C26b*   Excess-3 to decimal convertor

**Table C6**  *CMOS 4000B i.c.*

**Ratings**
Limiting values in accordance with the Absolute Maximum System (IEC 134)

| | | |
|---|---|---|
| Supply voltage | $V_{DD}$ | $-0.5$ to $+18$ V |
| Voltage on any input | $V_I$ | $-0.5$ to $V_{DD}+0.5$ V* |
| Current into any input | $\pm I_I$ max. | 10 mA |
| Operating ambient temperature | $T_{amb}$ | $-40$ to $+85°C$ |
| Storage temperature | $T_{stg}$ | $-65$ to $+150°C$ |

* $V_{DD}+0.5$ V should not exceed 18 V.

**Characteristics**     d.c. at $V_{SS}=0$ V

| | Symbol | $T_{amb}$ (°C) | | | | | | | Conditions |
|---|---|---|---|---|---|---|---|---|---|
| | | $-40$ | | $+25$ | | $+85$ | | $V_{DD}$ | |
| | | min. | max. | min. | max. | min. | max. | (V) | |
| Input voltage HIGH | $V_{IH}$ | 3.5 | 5 | 3.5 | 5 | 3.5 | 5 V | 5 | |
| | | 7.0 | 10 | 7.0 | 10 | 7.0 | 10 V | 10 | |
| | | 10.5 | 15 | 10.5 | 15 | 10.5 | 15 V | 15 | |
| Input voltage LOW | $V_{IL}$ | 0 | 1.5 | 0 | 1.5 | 0 | 1.5 V | 5 | |
| | | 0 | 3.0 | 0 | 3.0 | 0 | 3.0 V | 10 | |
| | | 0 | 4.5 | 0 | 4.5 | 0 | 4.5 V | 15 | |
| Output voltage HIGH | $V_{OH}$ | 4.99 | — | 4.99 | — | 4.95 | — V | 5 | note 1 |
| | | 9.99 | — | 9.99 | — | 9.95 | — V | 10 | |
| | | 14.99 | — | 14.99 | — | 14.95 | — V | 15 | |
| Output voltage HIGH | $V_{OH}$ | 4 | — | 4 | — | 4 | — V | 5 | note 2 |
| | | 9 | — | 9 | — | 9 | — V | 10 | |
| | | 13 | — | 13 | — | 13 | — V | 15 | |
| Output voltage LOW | $V_{OL}$ | — | 0.01 | — | 0.01 | — | 0.05 V | 5 | note 1 |
| | | — | 0.01 | — | 0.01 | — | 0.05 V | 10 | |
| | | — | 0.01 | — | 0.01 | — | 0.05 V | 15 | |
| Output voltage LOW | $V_{OL}$ | — | 0.5 | — | 0.5 | — | 0.5 V | 5 | note 2 |
| | | — | 1.0 | — | 1.0 | — | 1.0 V | 10 | |
| | | — | 2.0 | — | 2.0 | — | 2.0 V | 15 | |
| Output current HIGH | $-I_{OH}$ | 0.7 | — | 0.7 | — | 0.4 | — mA | 5 | note 3 |
| | | 1.4 | — | 1.4 | — | 0.8 | — mA | 10 | |
| | | 2.2 | — | 2.2 | — | 1.4 | — mA | 15 | |
| Output current HIGH | $-I_{OH}$ | 1.5 | — | 1.5 | — | 1.0 | — mA | 5 | note 4 |
| Output current LOW | $I_{OL}$ | 1.0 | — | 0.8 | — | 0.4 | — mA | 5 | note 5 |
| | | 2.6 | — | 2.0 | — | 1.2 | — mA | 10 | |
| | | 3.6 | — | 3.6 | — | 2.0 | —mA | 15 | |
| Input leakage current | $I_{IN}$ | — | — | — | 100 | — | — nA | 5 | note 6 |
| | | — | — | — | 100 | — | — nA | 10 | |
| | | — | — | — | 1000 | — | — nA | 15 | |

*Notes*
1) $I_{OH}=0$; inputs at 0 V or $V_{DD}$.
2) $I_{OH}=0$; inputs at specified worst case conditions.
3) $V_O=V_{DD}-0.5$ V; inputs at 0 V or $V_{DD}$.
4) $V_O=2.5$ V; inputs at 0 V or $V_{DD}$.
5) $V_O=0.4$ at $V_{DD}=5$ V; $V_O=0.5$ V at $V_{DD}=10$ V and 15 V; inputs at 0 V or $V_{DD}$.
6) Pin under test at 0 V or $V_{DD}$; all other inputs simultaneously at 0 V or $V_{DD}$.

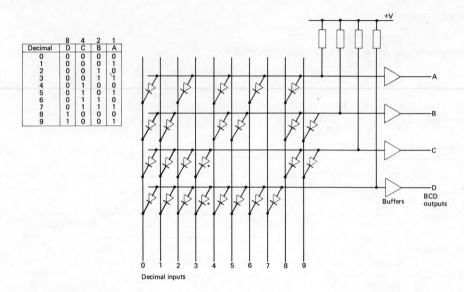

| Decimal | 8 D | 4 C | 2 B | 1 A |
|---------|---|---|---|---|
| 0 | 0 | 0 | 0 | 0 |
| 1 | 0 | 0 | 0 | 1 |
| 2 | 0 | 0 | 1 | 0 |
| 3 | 0 | 0 | 1 | 1 |
| 4 | 0 | 1 | 0 | 0 |
| 5 | 0 | 1 | 0 | 1 |
| 6 | 0 | 1 | 1 | 0 |
| 7 | 0 | 1 | 1 | 1 |
| 8 | 1 | 0 | 0 | 0 |
| 9 | 1 | 0 | 0 | 1 |

*Figure C26c* Decimal to BCD code convertor using diode matrix. (One line low at any time while all the rest are high.)

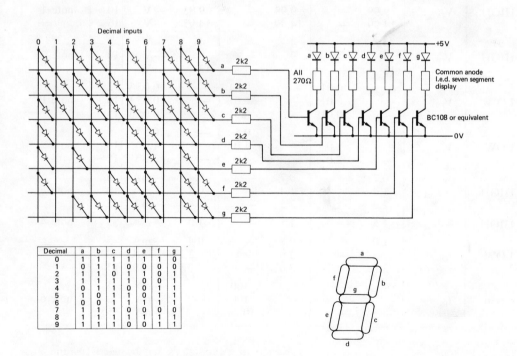

| Decimal | a | b | c | d | e | f | g |
|---------|---|---|---|---|---|---|---|
| 0 | 1 | 1 | 1 | 1 | 1 | 1 | 0 |
| 1 | 0 | 1 | 1 | 0 | 0 | 0 | 0 |
| 2 | 1 | 1 | 0 | 1 | 1 | 0 | 1 |
| 3 | 1 | 1 | 1 | 1 | 0 | 0 | 1 |
| 4 | 0 | 1 | 1 | 0 | 0 | 1 | 1 |
| 5 | 1 | 0 | 1 | 1 | 0 | 1 | 1 |
| 6 | 0 | 0 | 1 | 1 | 1 | 1 | 1 |
| 7 | 1 | 1 | 1 | 0 | 0 | 0 | 0 |
| 8 | 1 | 1 | 1 | 1 | 1 | 1 | 1 |
| 9 | 1 | 1 | 1 | 1 | 0 | 1 | 1 |

*Figure C26d* One method for converting from decimal to 7 segment. (Only one decimal input is high at any one time; all other inputs must be low.)

## Colour Codes for Components

### RESISTORS

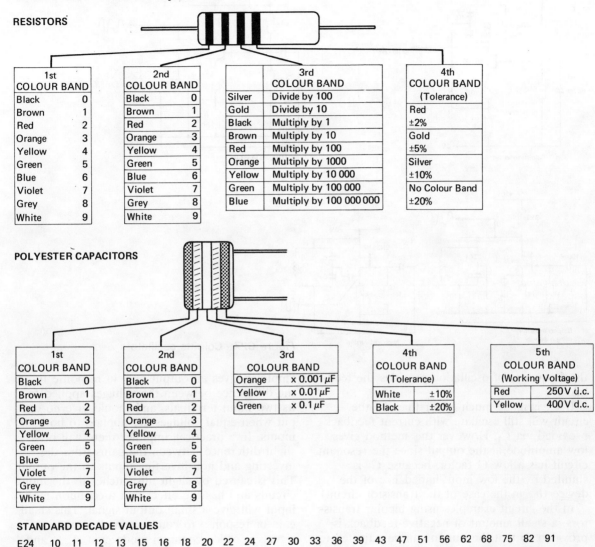

| 1st COLOUR BAND | |
|---|---|
| Black | 0 |
| Brown | 1 |
| Red | 2 |
| Orange | 3 |
| Yellow | 4 |
| Green | 5 |
| Blue | 6 |
| Violet | 7 |
| Grey | 8 |
| White | 9 |

| 2nd COLOUR BAND | |
|---|---|
| Black | 0 |
| Brown | 1 |
| Red | 2 |
| Orange | 3 |
| Yellow | 4 |
| Green | 5 |
| Blue | 6 |
| Violet | 7 |
| Grey | 8 |
| White | 9 |

| 3rd COLOUR BAND | |
|---|---|
| Silver | Divide by 100 |
| Gold | Divide by 10 |
| Black | Multiply by 1 |
| Brown | Multiply by 10 |
| Red | Multiply by 100 |
| Orange | Multiply by 1000 |
| Yellow | Multiply by 10 000 |
| Green | Multiply by 100 000 |
| Blue | Multiply by 100 000 000 |

| 4th COLOUR BAND (Tolerance) | |
|---|---|
| Red | ±2% |
| Gold | ±5% |
| Silver | ±10% |
| No Colour Band | ±20% |

### POLYESTER CAPACITORS

| 1st COLOUR BAND | |
|---|---|
| Black | 0 |
| Brown | 1 |
| Red | 2 |
| Orange | 3 |
| Yellow | 4 |
| Green | 5 |
| Blue | 6 |
| Violet | 7 |
| Grey | 8 |
| White | 9 |

| 2nd COLOUR BAND | |
|---|---|
| Black | 0 |
| Brown | 1 |
| Red | 2 |
| Orange | 3 |
| Yellow | 4 |
| Green | 5 |
| Blue | 6 |
| Violet | 7 |
| Grey | 8 |
| White | 9 |

| 3rd COLOUR BAND | |
|---|---|
| Orange | x 0.001 µF |
| Yellow | x 0.01 µF |
| Green | x 0.1 µF |

| 4th COLOUR BAND (Tolerance) | |
|---|---|
| White | ±10% |
| Black | ±20% |

| 5th COLOUR BAND (Working Voltage) | |
|---|---|
| Red | 250 V d.c. |
| Yellow | 400 V d.c. |

### STANDARD DECADE VALUES

| E24 | 10 | 11 | 12 | 13 | 15 | 16 | 18 | 20 | 22 | 24 | 27 | 30 | 33 | 36 | 39 | 43 | 47 | 51 | 56 | 62 | 68 | 75 | 82 | 91 |
|---|---|---|---|---|---|---|---|---|---|---|---|---|---|---|---|---|---|---|---|---|---|---|---|---|
| E12 | 10 | | 12 | | 15 | | 18 | | 22 | | 27 | | 33 | | 39 | | 47 | | 56 | | 68 | | 82 | |
| E6 | 10 | | | | 15 | | | | 22 | | | | 33 | | | | 47 | | | | 68 | | | |

## Colpitts Oscillator

This is a useful circuit for generating fixed frequency sine waves from about one kilohertz up to a few megahertz (fig. C27). It uses an *LC* tuned circuit and positive feedback via a capacitive pick-off from the tuned circuit. This feedback can be either series or shunt fed as shown. The frequency of oscillation is given by the formula

$$f_0 = \frac{1}{2\pi} \sqrt{\frac{1}{L}\left(\frac{1}{C_1}+\frac{1}{C_2}\right)}$$

and in the case of a transistor circuit the required value of current gain to maintain oscillations is

$$h_{fe} = -\left(\frac{-j\omega_0 L}{1/j\omega_0 C_1}+1\right)$$

Either of the capacitors can be used to tune the resonant circuit but the usual arrangement is for $C_2$ to be made much larger than $C_1$. In this case, $C_1$, the lower-value capacitor, sets the frequency and $C_2$ having a low reactance at

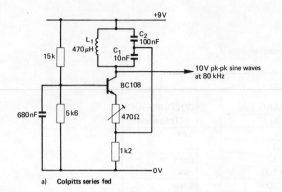

a) Colpitts series fed

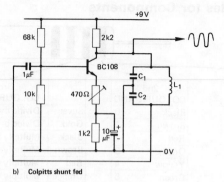

b) Colpitts shunt fed

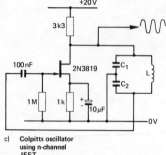

c) Colpitts oscillator using n-channel JFET

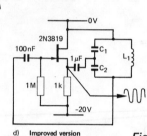

d) Improved version

*Figure C27* Colpitts oscillators

the frequency of oscillation provides the feedback.

If $C_1$ is made much larger than $C_2$ the circuit will still oscillate with current feedback provided via $C_1$. However this method gives low amplitude at the output since the resonant circuit has a low $Q$-factor, because $C_2$ is shunted by the low input impedance of the device ($h_{ib}$ in the case of the transistor circuit).

In the circuit examples using bipolar transistors, a small amount of negative feedback is provided so that the output can be adjusted to give a good sine wave. Stability of the circuit is good but the series-fed circuit which uses the common base configuration provides the best choice.

Two FET Colpitts oscillators are shown, the second allowing the drain load to be eliminated and is therefore not so dependent on spread of device parameters.

▶ *Clapp oscillator*   ▶ *Crystal oscillator*

## Common Mode Rejection Ratio

This is a term used in the specification of differential amplifiers (op-amps) and shows the quality of the amplifier in its ability to reject common mode signals. An ideal differential amplifier gives an output only in response to the difference between the voltages applied to its two input terminals, and should give no output when equal voltages are applied to both inputs. In a practical amplifier there will be a slight difference between the gains from the inverting and non-inverting inputs to the output. This is caused by slight mismatches in the input circuits and has the effect that a common mode input will give a small output signal. This small gain or response to common mode input signals is called common mode gain $A_{vcm}$, and the common mode rejection ratio (CMRR) is the ratio of the differential gain to the common mode gain:

$$\text{CMRR} = \frac{A_{vd}}{A_{vcm}}$$

This is often quoted in dB by taking $20 \log_{10}$ of the ratio.

For example, suppose an op-amp has a differential gain of 50 000 and a common mode gain of 0.5, then

$$\text{CMRR} = 20 \log 50\,000/0.5 = 100\,\text{dB}$$

Ideally the CMRR should be very large so that errors in the output are minimised.

An alternative method of specifying CMRR is

$$CMRR = \frac{\text{Common mode input voltage}}{\text{Common mode error voltage}}$$

Then the error due to common mode rejection can be represented in the equivalent circuit by a voltage generator $V_{ecm}$ in series with the input. Note that this does not apply when an amplifier is wired in the inverting mode because, as the non-inverting terminal is grounded, the common mode error must be zero. CMRR is only a problem when an amplifier is wired in the non-inverting or differential mode.

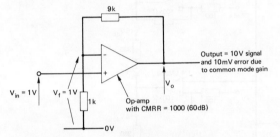

**Figure C28**

An example is shown (fig. C28) where a voltage signal is applied to a non-inverting amplifier which has a gain of 10. If the op-amp used has a CMRR of 1000 then the common mode error voltage will be 10 mV giving a measuring error of 0.1%. By selecting an amplifier with a CMRR of say 50 000 the common mode error would fall to 20 $\mu$V giving a measuring error of 0.002%.

▶ *Operational amplifier*

## Comparator

A comparator is any circuit that compares an input quantity against a reference level (or against another input) and produces a change of state at the output when one input exceeds the other. There are both analogue and digital type comparators.

**1** The basic *analogue circuit* is usually formed from a high-gain differential input op-amp; for example the 741 or 3130 can be used as comparators. The specialised i.c. comparators are designed to have very fast switching action and low values of hysteresis. A typical example of an i.c. comparator is the 710 which has a

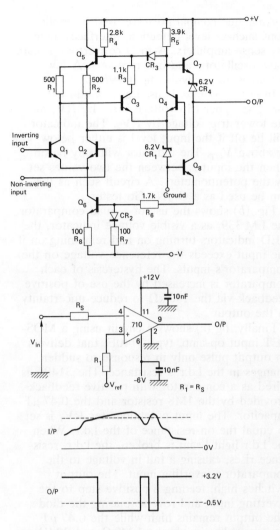

**Figure C29a** Differential voltage comparator (the 710)

response time of 4 ns and a hysteresis as low as 2 nV. The 710 circuit (fig. C29a) shows that it has a differential amplifier-type input stage and a switch-type output. If the input on the inverting terminal is less than the reference voltage, then the output remains high at +3.2 V. As the input rises and just exceeds the reference level the output switches very rapidly to a low of −0.5 V. Many other i.c. comparators are available such as the 319, a dual circuit with each comparator having similar characteristic to the 710, but which only requires a single supply rail. Its open collector output can interface directly with TTL or drive lamps and relay loads up to 25 mA.

Analogue comparators are used in applications such as level detection, interfaces, memory sense amplifiers, A-to-D convertors, square wave oscillators, and line receivers. A few examples are shown in fig. C29b.

In (a), a 319 i.c. is used as a window detector. The upper trip voltage is set by $RV_1$ and the lower trip voltage by $RV_2$. The indicator will be off if the input level is either below $V_{LT}$ or above $V_{UT}$. The indicator will only be on when the input is between the two values set by the potentiometers. A circuit such as this can be used as a Go/No-Go tester.

Fig. (b) shows the use of a quad comparator, the LM 339, as a visible voltage indicator, the LED indicators turning on and remaining on if the input exceeds the reference voltage on the comparator's inputs. The hysteresis of each comparator is increased by the use of positive feedback via the 100 kΩ to reduce uncertainty in the output.

Finally, fig (c) shows a circuit using a MOS-FET input op-amp, type 3140E, that delivers an output pulse only in response to sudden changes in the l.d.r.'s resistance. The 3140E is wired as a comparator with positive feedback provided by the 1M8 resistor and the 0.47 $\mu$F capacitor. The input potentiometer $RV_1$ is set to equal the on-resistance of the l.d.r. When the l.d.r light beam is broken, the l.d.r. resistance rises, causing a fall in voltage to the comparator's inverting input. The output switches high, feeding a positive step to the inverting input and reverse biasing the diode. The output remains high while the 0.47 $\mu$F capacitor charges via the 1 M8 Ω and 2 MΩ resistors until $D_1$ again conducts or the level on the non-inverting input falls below that on the inverting input. In this way a wide duration pulse occurs at the output every time the light beam is interrupted but slow changes in l.d.r. resistance are ignored.

**2** *Digital comparators* are logic circuits that are used to determine if two binary numbers are equal or which has the greater magnitude. The function for two single bit numbers $A$ and $B$ to be equal is

$$F = A \cdot B + \bar{A} \cdot \bar{B}$$

The logic for this can be performed by the AND-OR-NOT gate or by using wired logic, as shown (fig. C30).

To compare two binary words

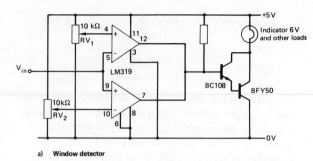

a)   Window detector

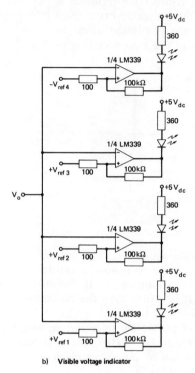

b)   Visible voltage indicator

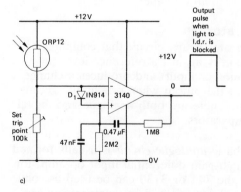

c)

*Figure C29b* Examples of comparator application

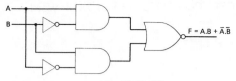

a) Single bit comparator using AND-OR-NOT gate

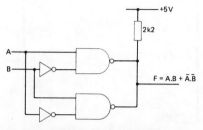

b) Single bit comparator using TTL open collector gates

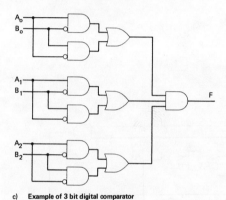

c) Example of 3 bit digital comparator

*Figure C30* Digital comparators

$A_0$ to $A_n$ compared to $B_0$ to $B_n$

the function is

$$(A_0 B_0 + \bar{A}_0 \cdot \bar{B}_0) \cdot (A_1 B_1 + \bar{A}_1 \cdot \bar{B}_1)$$
$$\cdot (A_2 B_2 + \bar{A}_2 \cdot \bar{B}_2) \cdots \cdots (A_n B_n + \bar{A}_n \cdot \bar{B}_n)$$

An example of a TTL comparator is the 7485. This 4-bit magnitude comparator gives three fully decoded decisions about two 4-bit words and can be expanded to any number of bits without extra gates by connecting comparators in cascade. In the example two 24-bit words can be compared (fig. C31). One comparator can be used with five of the 24-bit comparators illustrated to expand the word length to 120-bits. Typical comparison times for various word lengths using the 54/74LS85 are:

| Word length | Number of i.c.s | |
|---|---|---|
| 1–4 bits | 1 | 24 ns |
| 5–24 bits | 2–6 | 48 ns |
| 25–120 bits | 8–31 | 72 ns |

One application of digital comparators is in the address selection of a serial or cyclic type memory. For example a 256-bit MOS shift register, used as a 16-word 16-bit serial memory, would require a 4-bit address. The address counter records the number of the word as the data is clocked through the shift register and, when the content of the address counter equals the 4-bit number set up in the address register, the comparator output goes high enabling data to be read out or written into the store for one word length.

## Complement

Something which completes, fills up, or makes another quantity up to a whole is called a complement. For example the complement of the angle 25° required to make a right angle is 65°; the complement of number 3 required to make 10 is 7; or the complement of a colour is one which combines with the original colour to give white. A word used frequently in electronics is the adjective of complement— COMPLEMENTARY. This means supplying a mutual deficiency and therefore resulting in two parts working together to make a whole or complete function. Examples are the combination of n-p-n and p-n-p transistors in power output stages, one transistor conducting while the other is off and vice versa; and CMOS logic which uses n and p channel enhancement mode MOSFETS.

### Complement of a decimal number

The ten's complement of number $A$ of order $m$ is $(10^m - A)$.

If $A = 1046$, then $m = 4$. Therefore

Ten's complement of $1046 = 10^4 - 1046 = 8954$

The nine's complement is a value less than the ten's complement by unity. Therefore

Nine's complement of $1046 = 9999 - 1046$
$$= 8953$$

The nine's complement can be used in subtraction

**71**

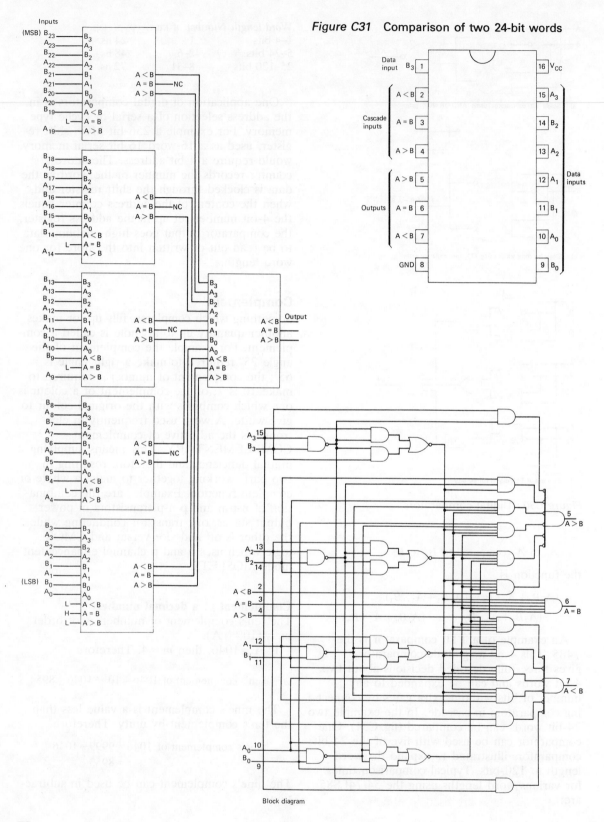

Inputs
(MSB) $B_{23}$
$A_{23}$
$B_{22}$
$A_{22}$
$B_{21}$
$A_{21}$
$B_{20}$
$A_{20}$
$B_{19}$
$A_{19}$

$B_{18}$
$A_{18}$
$B_{17}$
$A_{17}$
$B_{16}$
$A_{16}$
$B_{15}$
$A_{15}$
$B_{14}$
$A_{14}$

$B_{13}$
$A_{13}$
$B_{12}$
$A_{12}$
$B_{11}$
$A_{11}$
$B_{10}$
$A_{10}$
$B_9$
$A_9$

$B_8$
$A_8$
$B_7$
$A_7$
$B_6$
$A_6$
$B_5$
$A_5$
$B_4$
$A_4$

$B_3$
$A_3$
$B_2$
$A_2$
$B_1$
$A_1$
$B_0$
$A_0$
(LSB)

Output

### Figure C31  Comparison of two 24-bit words

Data input $B_3$ 1 — 16 $V_{CC}$

Cascade inputs
$A < B$ 2 — 15 $A_3$
$A = B$ 3 — 14 $B_2$
$A > B$ 4 — 13 $A_2$

Outputs
$A > B$ 5 — 12 $A_1$
$A = B$ 6 — 11 $B_1$
$A < B$ 7 — 10 $A_0$

GND 8 — 9 $B_0$

Data inputs

Block diagram

72

*Example*  769−237

Take the 9's complement of 237 and add it to 769.

$$
\begin{array}{r}
769 \\
762 \quad \text{9's complement of 237} \\
\hline
\text{sign digit} \quad 1\,531 \\
1 \\
\hline
532 \quad \text{Answer}
\end{array}
$$

If the sign digit is 1, the remainder is positive.

## Binary complement

The 2's complement of a binary number $A$ of order $m$ is $(2^m - A)$.

1010 has an order of 4, therefore

2's complement of $1010 = 10000 - 1010 = 00110$

1's complement of $1010 = 1111 - 1010 = 0101$

*Rules:* To obtain the 2's complement, change the 0's to 1's and the 1's to 0's and add 1.

To obtain the 1's complement change the 0's to 1's and the 1's to 0's.

*Examples*  Binary  2's comp. 1's comp.

| | | |
|---|---|---|
| 100 | 100 | 011 |
| 1011 | 0101 | 0100 |
| 001 | 111 | 110 |

The 1's complement is useful in the process of subtraction. To obtain the difference between two binary numbers the 1's complement of one of them is added to the other. This is simpler to achieve with electronic circuits than subtraction.

*Examples*

a)  1011−0101

$$
\begin{array}{r}
1011+ \\
1010 \quad \text{1's complement of 0101} \\
\hline
\text{sign digit} \quad 1\,0101 \\
1 \\
\hline
110 \quad \text{Answer}
\end{array}
$$

b)  1011.01−0101.10

$$
\begin{array}{r}
1011.01+ \\
1010.01 \quad \text{1's complement of 0101.10} \\
\hline
1\,0101.10 \\
1 \\
\hline
101.11 \quad \text{Answer}
\end{array}
$$

## Complement of a logic function

The complement of a logic signal is obtained by passing the signal through an invertor stage. De Morgan's laws are used to obtain the

complement of the logic expression:

$$\overline{A \cdot B} = \bar{A} + \bar{B}$$

and $\overline{A + B} = \bar{A} \cdot \bar{B}$

Thus if $F = A \cdot (B + C)$, the complement of $F$ is

$$\bar{F} = \overline{A \cdot (B + C)} = \bar{A} + \overline{B + C} = \bar{A} + \bar{B} \cdot \bar{C}$$

*Rule:* To obtain the complement of an expression interchange the ANDs and ORs and complement each letter. Care has to be exercised, however, with expressions containing brackets.

*Examples*

a)  $F = A + B + C$  ∴ $\bar{F} = \bar{A} \cdot \bar{B} \cdot \bar{C}$

b)  $F = A \cdot B + \bar{C}$  ∴ $\bar{F} = (\bar{A} + \bar{B}) \cdot C$

c)  $F = A \cdot (B + \bar{C})$  ∴ $\bar{F} = \bar{A} + \bar{B} \cdot C$

d) Suppose  $F = \bar{A} \cdot B \cdot C + \bar{C} \cdot (A + D) + \overline{E \cdot B}$

The complement is

$$\bar{F} = \overline{\bar{A} \cdot B \cdot C + \bar{C} \cdot (A + D) + \overline{E \cdot B}}$$

$$= (A + \bar{B} + \bar{C}) \cdot (C + \bar{A} \cdot \bar{D}) \cdot E \cdot B$$

## Contact Bounce Eliminator

Apart from the mercury-wetted types, nearly all contacts on mechanical switches and relays suffer from an effect known as "bounce". When the switch is operated, the contacts make and break several times before finally setting down in the closed position. This bounce time may last several milliseconds and, if the switch is used to operate logic circuits, all of these vibrations will be recorded as input pulses and cause false triggering. One common method of eliminating the effects of bounce is to use an R–S flip-flop circuit.

With the switch in position A (fig. C32), the set input will be low (0) and the reset input high (1). The $Q$ output will be held at 1. When the switch is operated, the contact first breaks from A but the flip-flop cannot change

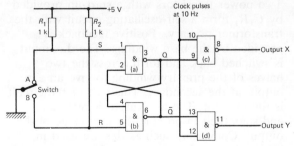

*Figure C32*  Contact bounce eliminator using 7400

its state until the contact makes to B, forcing the reset to logic 0. The first time the contact makes to B, the flip-flop will be reset and any switch bounce will be ignored. The circuit uses the ability of a flip-flop to be set or reset by a momentarily applied low signal. Bounce-less switch circuits are used in the control of nearly all logic systems. ▶ *Interface circuit*

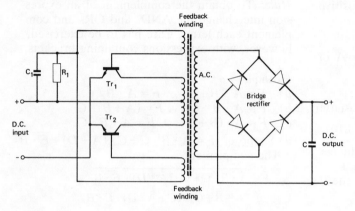

Figure C33   A typical convertor circuit

## Convertor

The operation of a unit called a convertor might seem obvious and also applicable to many varied circuits, such as code convertor, analogue-to-digital convertor, frequency convertor, and so on. However in electronics the name is usually reserved to describe a power supply circuit that converts a d.c. input voltage to another d.c. value. The circuit is an INVERTOR, that is d.c. to a.c, followed by rectification, to produce the new required value of d.c. voltage. The advantages of a convertor system are that higher values of d.c. voltage than the supply can be obtained and that the output is isolated. Fig. C33 shows a typical example. Two power transistors with a start-up provided by $C_1R_1$ form a self-oscillating circuit with the transformer primary. Positive feedback is provided by the base windings. The d.c. input is switched back and forth across the two halves of the primary winding to give an a.c. output at the secondary. This part of the circuit is the invertor. The a.c. output is rectified and smoothed to provide the new value of d.c. output. Convertors such as this are used in battery powered equipment and h.v. supplies.
   ▶ *Power supply*

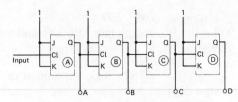

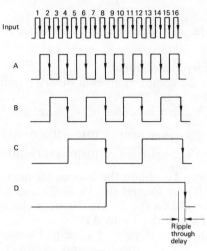

| Dec.i/p | D | C | B | A |
|---------|---|---|---|---|
| 0 | 0 | 0 | 0 | 0 |
| 1 | 0 | 0 | 0 | 1 |
| 2 | 0 | 0 | 1 | 0 |
| 3 | 0 | 0 | 1 | 1 |
| 4 | 0 | 1 | 0 | 0 |
| 5 | 0 | 1 | 0 | 1 |
| 6 | 0 | 1 | 1 | 0 |
| 7 | 0 | 1 | 1 | 1 |
| 8 | 1 | 0 | 0 | 0 |
| 9 | 1 | 0 | 0 | 1 |
| 10 | 1 | 0 | 1 | 0 |
| 11 | 1 | 0 | 1 | 1 |
| 12 | 1 | 1 | 0 | 0 |
| 13 | 1 | 1 | 0 | 1 |
| 14 | 1 | 1 | 1 | 0 |
| 15 | 1 | 1 | 1 | 1 |
| 16 | 0 | 0 | 0 | 0 |

Figure C34   Asynchronous divide-by-16 counter

## Counters and Dividers

These are created by linking bistable circuits together, so that the bistables within the counter change state in a predetermined sequence. The sequence is called the CODE and the number of different states is called the MODULO of the counter.

**1** If four bistables are connected together to make a pure binary divide-by-16 counter (fig. C34), then the modulo is 16 and the code is 1248 binary. Each bistable has its $J$ and $K$ inputs connected to logic 1 and the clock input of bistables D, C and B are connected respectively to the $Q$ output of the preceding bista-

bles. This gives ASYNCHRONOUS or "ripple-through" operation. The $Q$ output of bistable A changes state on the trailing edge of every input pulse giving a divide-by-2. Bistable B will change state just after the trailing edge of every second input pulse to give divide-by-4, and so on. After the 15th pulse the state of the $Q$ outputs of the bistables will be 1111 and, therefore on the trailing edge of the 16th pulse, all the bistables reset giving a state of 0000. As the counter overflows on the trailing edge of the 16th pulse, a negative edge appears at the output of bistable D; this edge is delayed from the input by the sum of four propagation delay periods.

**2** SYNCHRONOUS counters (fig. C35) overcome the problem of cumulative delay by ensuring that, when bistables have to change state, they all change state at the same instant in time. In a synchronous ÷16 1248 binary counter, the $Q$ output is connected to the $J$ and $K$ inputs of the next bistable and all clock inputs are connected together. To ensure that bistables change state in the correct sequence, additional AND gates are required. In this way bistable C can only change state if both bistables A and B are at logic 1; this occurs after the 4th input pulse. Similarly bistable D can only change state after the 8th input pulse because this is the first time that the $Q$ outputs of A, B and C are at logic 1. Synchronous counters allow dividers of large numbers to be created without long delays appearing and are less prone to producing "glitches" when the circuit is decoded.

**3** The count sequence of a pure binary counter can be easily altered to give counters and dividers of other numbers, such as 3, 5, 7, 9, etc. A counter that divides by 10 is referred to as a DECADE counter. This is formed by a divide-by-2 followed by a divide-by-5 and a good example is the TTL decade counter type 7490. This is an asynchronous circuit which consists of four master-slave bistables. When used as a BCD decade counter, the B input has to be externally connected to the A output. The A input receives the incoming count, which must be a TTL type signal, and a count sequence of BCD is obtained. It is also possible to get the 7490 to give a symmetrical divide-by-10 by applying the count input to input B and externally connecting output D to input A. A divide-by-10 square wave is obtained at output A. An alternative decade

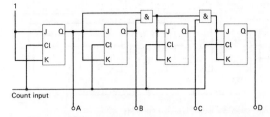

a)    Synchronous ÷16 binary counter

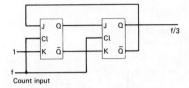

b)    Synchronous ÷3

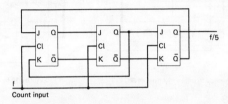

c)    Synchronous ÷5

**Figure C35   Synchronous counters**

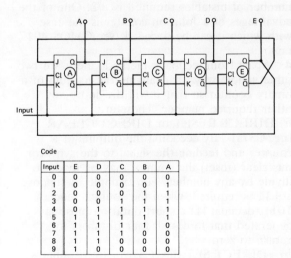

Code

| Input | E | D | C | B | A |
|-------|---|---|---|---|---|
| 0 | 0 | 0 | 0 | 0 | 0 |
| 1 | 0 | 0 | 0 | 0 | 1 |
| 2 | 0 | 0 | 0 | 1 | 1 |
| 3 | 0 | 0 | 1 | 1 | 1 |
| 4 | 0 | 1 | 1 | 1 | 1 |
| 5 | 1 | 1 | 1 | 1 | 1 |
| 6 | 1 | 1 | 1 | 1 | 0 |
| 7 | 1 | 1 | 1 | 0 | 0 |
| 8 | 1 | 1 | 0 | 0 | 0 |
| 9 | 1 | 0 | 0 | 0 | 0 |

**Figure C36   Johnson divide-by-10 counter**

counter can be made using five bistables wired as a Johnson counter (twisted ring) (fig. C36). This type of counter is synchronous in operation and generates what is called a walking code sequence. Johnson counters can be used

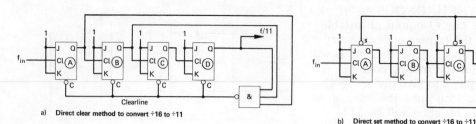

a) Direct clear method to convert ÷16 to ÷11

b) Direct set method to convert ÷16 to ÷11

**Figure C37**

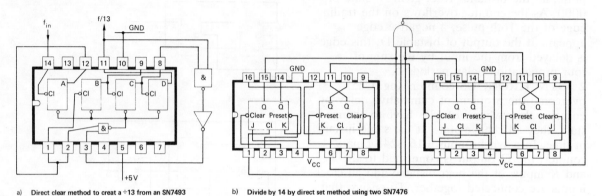

a) Direct clear method to creat a ÷13 from an SN7493

b) Divide by 14 by direct set method using two SN7476

**Figure C38**

for divide by 4, 6, 8, 10, 12, etc. and the number of bistables required is $n/2$. One of the advantages of a Johnson counter is the ease with which it can be decoded (see CMOS 4017 i.c.).

**4** Apart from feedback methods there are two other simple ways of changing the modulo of a binary counter so that it will divide by any other required number. These are

*a*) DIRECT RESET or DIRECT CLEAR (fig. C37a) By decoding the outputs of a counter and feeding the signal to the synchronous clear (reset) the counter may be made to divide by any number. For example, if suppose a ÷11 is required; when the counter reaches 1011 (decimal 11) a short duration pulse is generated that forces all four bistables to return to zero.

*b*) DIRECT SET (fig. C37b) In a similar manner it is possible to decode the output and apply a pulse to the asynchronous preset inputs of selected bistables to force the counter to skip certain states. For example to use this method for a ÷11, on the 10th pulse bistables A and C must be directly set to 1s, then on the 11th input pulse all four bistables return to zero.

These two methods are useful in creating dividers as shown in the examples in fig. C38: (i) a 7493 four-stage binary counter i.c. wired to divide by 13 (direct clear method) and (ii) a divide-by-14 using two 7476 i.c.s (direct set method) but the maximum frequency of operation is limited by the restriction that the clock must be held on for the time it takes to either clear or set the bistables; this is typically 40 ns for TTL circuits. Also invalid states occur very briefly and, if the circuit is to be used as a counter, a strobed gate has to be provided to the outputs to prevent unwanted voltage spikes.

**5** An UP/DOWN counter (fig. C39) is one which can be made to count in either the forward or reverse direction. The method used requires gates between bistables that connect either the $Q$ or the $\bar{Q}$ to the J–K inputs of the next bistable. This gives a serial-carry method. If the count control input is low, then the $\bar{Q}$ outputs are gated on and the counter direction is reversed so that counts are subtracted from the counter. Another method that gives faster operation is the parallel-carry method, or look-ahead carry, where multiple-input AND gates are used to generate the carry signal. This re-

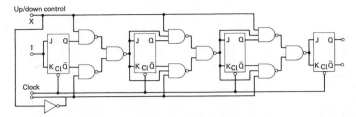

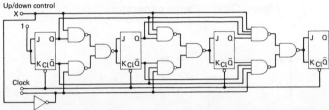

b) 4-Bit synchronous binary up/down counter (parallel carry) using SN7473s

*Figure C39* **Up/down counters**

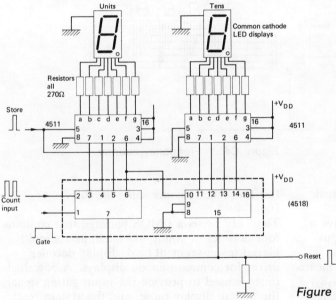

*Figure C40a* **Two-decade counter**

duces the inherent delays of the gates used in the serial-carry method. Parallel carry is always used in the fastest synchronous counters.

Several counter i.c.s are available in CMOS logic, the 4020 (14 stage), 4040 (12 stage), and 4024 (7-stage) are examples of CMOS binary counters. They are ripple-through types and are obviously useful when division by a large number is required. The 4017 is a fully de-coded Johnson counter that has a cycle length of ten. Since the circuit is synchronous the decoded outputs will be glitch-free. The 4017

and the 4022 (eight output states) are useful in making waveform generators and digital filters.

CMOS counters are also available with direct drive to seven-segment displays, typical types being the 4026 and 4033. In general CMOS counter i.c.s are about one third to one sixth as fast as TTL types.

**6 COUNTER APPLICATIONS: a few examples**

(i) *Two-decade counter with seven-segment digital readout* (fig. C40a)

This is one of the most common uses of coun-

77

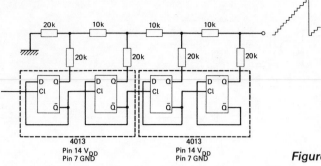

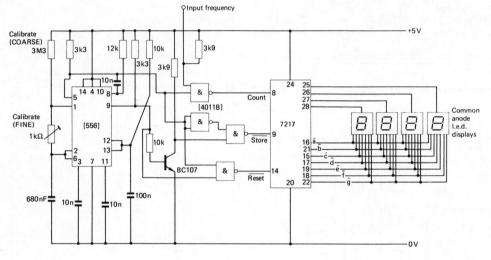

*Figure C40b*  Ramp generator

*Figure C40c*  Frequency counter

ters and in this example a CMOS 4518 dual BCD counter i.c. is used with two 4511 latch/seven-segment decoder drivers to give a divide-by-100 with display. The count input has to be conditioned correctly so that the CMOS counter does not mistrigger. This means having the correct logic levels ($0$ V to $+V_{DD}$) with reasonably fast edges (better than 15 $\mu$s) and no "bounce" such as would be obtained from an unconditioned pair of switch contacts.

The BCD outputs from the counters are stored in the 4511 latches to give a flicker-free display. Additional circuits would be required to provide the necessary timing pulses.

(ii) *Ramp generator* (fig. C40b)
Two CMOS 4013 dual D bistables are wired to give a straight ripple-through ÷16 binary counter. As the counter is advanced its binary coded outputs are applied to an R-2R D-to-A convertor. The output of the resistor ladder is a ramp of 16 levels.

(iii) *Frequency counter using the 7217*
(fig. C40c)
The 7217 i.c. is a CMOS package that contains four decades of count and latch, and a multiplexed seven-segment l.e.d. display decoder driver for common anode displays. A 556 dual timer is used to provide the input gating signal, the update display pulse, and the system reset. With the values given, the gate signal is set to 1 sec. This sets the range of the counter from 1 Hz to 10 kHz.

(iv) *Digital triangle waveform generator*
(fig. C40d)
Two CMOS 4029 up/down counter i.c.s with their outputs decoded by an R-2R ladder network are used to provide a linear triangle wave. The 4029s are set to count up in binary since pin 9 of each is connected to +12 V and the carry output from the first is fed to the second i.c. The counters advance when positive pulses are fed to the input and on the 256th

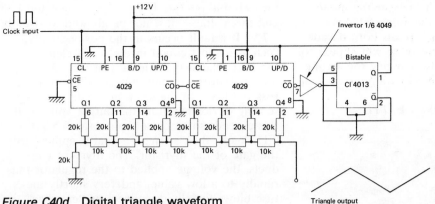

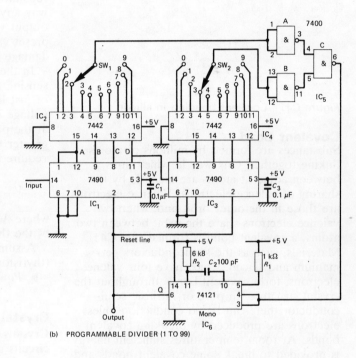

**Figure C40d** Digital triangle waveform generator

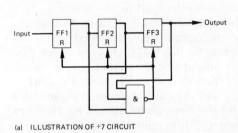

(a) ILLUSTRATION OF ÷7 CIRCUIT

**Figure C40e** Programmable divider

(b) PROGRAMMABLE DIVIDER (1 TO 99)

input pulse the carry output from $IC_2$ $\overline{CO}$ goes high to trigger the 4013 bistable. The $Q$ output changes state and alters the up/down control of the 4029s so that the counters count down from 256 to zero. At zero the carry out $\overline{CO}$ again goes high to force the 4013 to change state. Thus while clock pulses are applied a triangle waveform will be obtained at the output. This will have a frequency of 1/512th of the input. To achieve the best linearity the resistors in the R-2R ladder network should be close tolerance, at least ±1% types. The small steps in the triangle waveform can be removed by a filter circuit.

(v) *Programmable divider* (fig. C40e)
In this circuit the input frequency can be divided by any number from 1 to 99. Two 7490 TTL decade counters are used and these are both reset when the appropriate division is reached. The BCD outputs from the 7490s are decoded by 7442 BCD to decimal decoders, one output line from a decoder going low at any one time. Two switches set the required division. Suppose 32 is selected; when the input is applied after 30 pulses the 3 output of $IC_4$ goes low. Following another 2 input pulses, the 2 output of $IC_2$ also goes low. Two logic 1s appear on the inputs of NAND gate C and a

negative edge is applied to trigger the mono-stable. The $Q$ output of the monostable is a short-duration pulse which resets both decade counters. Thus one output pulse will appear for every 32 input pulses. By changing the setting of the two switches, division by any number up to 99 is possible.

▶ *Bistable*    ▶ *CMOS*    ▶ *TTL*

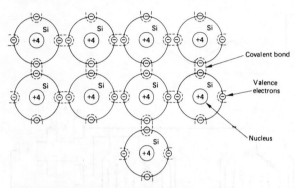

*Figure C41*   Covalent bonding in silicon

## Covalent Bond

Substances are formed by millions of atoms linking together. One way in which links between adjacent atoms are made is by the sharing of valence electrons. Valence electrons are those in the outer orbit and, when two valence electrons share the orbit between two atoms, a covalent bond is formed (fig. C41). Materials, such as the semiconductors germanium and silicon, which have four valence electrons, form covalent bonds throughout the crystal structure. In pure or intrinsic semi-conductor there can be no conduction unless electrons are produced by breaking covalent bonds. At room temperatures sufficient energy is provided to break some covalent bonds and produce hole-electron pairs which give the material its *semi*conductor status. In practical devices it is the introduction of controlled amounts of impurity which is used to radically alter the conductivity.

▶ *Semiconductor theory*

## Crowbar

This is the name given to an overvoltage protection circuit that uses a thyristor to disconnect the power being supplied to a load. Sensitive components can be burnt out or severely damaged if the voltage across them exceeds the absolute maximum rating. TTL i.c.s

for example operate from a +5 V regulated line and this voltage must not be allowed to rise above +7 V. If a fault occurs in the power supply, such as a short circuit in the series element, the unregulated input usually at least 3 V higher than the output voltage will appear across the load. In the crowbar circuit a voltage-sensing device monitors the output voltage. If this exceeds some preset value, say 6.5 V for TTL loads, then a trigger signal is applied to the gate of the thyristor. The thyristor conducts, the voltage applied to the regulator falls rapidly to a low value, and very shortly the fuse blows to fully disconnect or "crowbar" the regulator from the unregulated input. In this way any fault in the power unit that causes the output voltage to the load to rise above a preset value will be prevented from causing damage to the circuits being supplied.

In the example (fig. C42), $DZ_2$ the zener diode sensing the output voltage should have a value of just less than the normal regulated output voltage $V_o$, and the current through $R_1$ and $R_2$ at the point the circuit trips should be 10 times greater than $I_{GT}$ (the minimum gate current required by the thyristor to fire). Then

$$V_{trip} = V_Z + V_{GT}\left(\frac{R_1 + R_2}{R_2}\right)$$

where $V_{GT}$ is the minimum gate trigger voltage of the thyristor.

Assuming $V_{GT} = 0.8$ V and $I_{GT} = 0.2$ mA (thyristor C106), then $V_{trip} = 6.3$ V.

▶ *Power supply*

## Crystal Oscillator

Crystals are used in fixed-frequency oscillator circuits as the frequency-determining network when the highest frequency stability is required (fig. C43). Certain crystalline materials, in particular quartz, exhibit what is known as the piezoelectric effect. When the crystal is subjected to mechanical strain a p.d. is developed across some of the crystal faces; conversely if a p.d. is applied across the crystal it will be mechanically strained. When a thin plate of quartz is cut it will possess a natural resonant frequency, and this frequency of vibration is very stable and has a very low temperature coefficient.

If the crystal oscillator is held at a controlled temperature, in a small oven, then stabilities of 1 part in $10^{10}$ can be achieved.

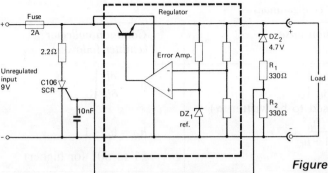

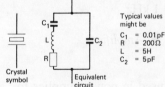

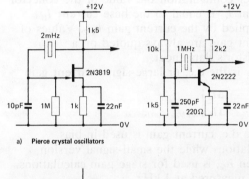

Figure C42 Crowbar circuit; values for 5 V, 1 A regulator

**Table C7** *Typical quartz crystal frequencies and temperature coefficients*

| Frequency | Temperature coefficient ($-20°C$ to $+70°C$) |
|---|---|
| 100 kHz | 200 ppm per °C |
| 1 MHz | 50 ppm per °C |
| 10 MHz | |
| 3.2768 MHz ($\div 2^{16}$ gives 50 Hz) | $\pm 12.5$ ppm per °C |
| 4.194304 MHz ($\div 2^{22}$ gives 1 Hz) | |

The equivalent circuit shows that there are two possible resonant conditions:

$$\text{Series resonance } f_s = \frac{1}{2\pi} \sqrt{\frac{1}{L_s C_s}}$$

$$\text{Parallel resonance } f_p = \frac{1}{2\pi} \sqrt{\frac{1}{L_s}\left(\frac{1}{C_s} + \frac{1}{C_p}\right)}$$

However, because $C_p \gg C_s$, these two frequencies are very close together—usually less than 5% difference.

The $Q$ of a crystal is very high—typically between $10^4$ to $10^5$.

Several circuit configurations can be used, some of the simplest being the Colpitts, Clapp and Pierce. Crystals are also used to control the frequency of square wave generators.

▶ *Clock pulse generator*

## Current Gain

The current gain of a device or circuit is the ratio of output to input current:

$$A_i = \delta i_o / \delta i_{in} = 20 \log_{10}(\delta i_o / \delta i_{in}) \text{ dB}$$

It is an important parameter of bipolar transistors. Bipolar transistors are basically current-controlled devices. For example, in common

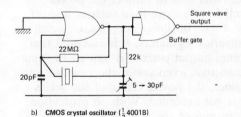

b)  **CMOS crystal oscillator** ($\frac{1}{2}$ 4001B)

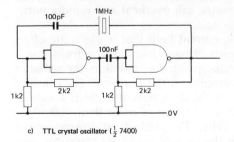

c)  **TTL crystal oscillator** ($\frac{1}{2}$ 7400)

*Figure C43* Crystal oscillators

**Table C8** *Current gains for the three transistor configurations*

| Common emitter | Common base | Common collector (emitter follower) |
|---|---|---|
| $h_{fe} = \dfrac{\delta i_c}{\delta i_b}\Big\|_{V_{ce}=0}$ <br> ($h_{fe}$ used to be written $\beta$) <br> $h_{fe} = \dfrac{h_{fb}}{1 - h_{fb}}$ <br> Value: 50 to 500 (or higher) | $h_{fb} = \dfrac{\delta i_c}{\delta i_e}\Big\|_{V_{cb}=0}$ <br> ($h_{fb}$ used to be written $\alpha$) <br> $h_{fb} = \dfrac{h_{fe}}{1 + h_{fe}}$ <br> 0.99 | $h_{fc} = \dfrac{\delta i_e}{\delta i_b}\Big\|_{V_{ce}=0}$ <br><br> $h_{fc} = h_{fe} + 1$ <br><br> 51 to 501 (or higher) |

emitter configuration the value of the collector current $I_C$ is equal to the base current $I_B$ multiplied by the current gain $h_{FE}$. Values of current gain are always quoted on a transistor's data sheet.

$h_{FE}$ is the d.c. or large-signal current gain

$$h_{FE} = \frac{\Delta I_C}{\Delta I_B}\Big|_{\text{with } V_{CE} \text{ held constant}}$$

The d.c. current gain is used in bias calculations while the small-signal version, written $h_{fe}$, is used for stage gain calculations. $h_{fe}$ is measured at 1 kHz.

▶ *Transistor*   ▶ *Amplifier*

## Current Limit

These are circuits used in power supplies to prevent damaging currents being supplied to load devices and also to protect the power supply itself from an overload such as a short circuit (fig. C44). In power supplies, especially those with series regulators, it is essential that the maximum output current is limited to some safe value so that, even under short circuit output conditions, the power rating of the series transistor is not exceeded. With no protection, if the output pins of the power unit are shorted together, a very large current will flow and the series transistor will overheat and rapidly burn out.

A simple current limit that works quite effectively consists of one resistor and a transistor. The low-value resistor $r_m$ monitors the value of the load current and the transistor $\text{Tr}_2$ is forced to conduct if the load current exceeds a value that causes a volt drop of more than 600 mV across $r_m$. When $\text{Tr}_2$ conducts, it diverts current away from the base of the series transistor and this limits the maximum current. The effect is

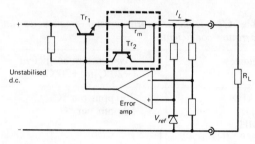

(a)  SIMPLE CURRENT LIMIT

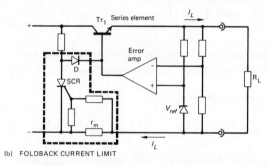

(b)  FOLDBACK CURRENT LIMIT

*Figure C44*  **Current limiting circuits**

that any increase of load current beyond the limit causes $\text{Tr}_2$ to conduct more which further reduces the base drive to the series transistor. The point at which the output current starts to limit is given approximately by the formula

$$I_{limit} \simeq V_{BE}/r_m$$

Thus if $r_m$ is 0.5 $\Omega$ and $V_{BE} = 600$ mV the current limit will operate at about 1.2 A (see fig. P7$a$).

The major drawback of this simple circuit is that under maximum overload conditions the dissipation of the series transistor is still high and it may require a large heatsink to withstand short circuit conditions. An improved limiter is one that senses the output current

and turns the series transistor off as soon as a preset value of load current is exceeded. This is called FOLDBACK CURRENT LIMITING. In the circuit the load current is monitored by a resistor $r_m$ and, if the voltage across this resistor, caused by an overload current, exceeds the gate-to-cathode trigger voltage of the thyristor, then the thyristor conducts to operate the trip. When the thyristor conducts, its anode voltage falls to just less than 1 V and this turns the series element off to reduce the load current to practically zero. The thyristor remains conducting, holding the power supply off until the fault is cleared and the circuit reset (see fig. P7b).

Both circuits can be modified to give an adjustable trip-point by wiring a potentiometer in parallel with $r_m$ and taking the wiper to the transistor base or the thyristor gate.

## Darlington

A connection method used in bipolar transistors to give a composite transistor pair with a high value of current gain and a high input impedance (fig. D1). The two transistors are connected as two cascaded emitter followers but the connection is also very useful in the common emitter mode. Many manufacturers package the circuit as a single discrete component with three leads and include a protection diode from collector to emitter and possibly an input diode to assist turn-off.

The overall current gain of the two transistors may be higher than 1000, which means that a 1 mA input current can switch 1 A at the output collector. The current gain is given by the formula

$$A_i \simeq \frac{1 + h_{fe1}h_{fe2}}{1 + h_{oe1}h_{fe2}R_E}$$

and $$R_{in} \simeq \frac{(1 + h_{fe1}h_{fe2})R_E}{1 + h_{oe1}h_{fe2}R_E}$$

If the circuit is used in the common emitter mode then $R_E$ is zero and

$$A_i \simeq 1 + h_{fe1}h_{fe2}$$

and $$R_{in} \simeq (1 + h_{fe1}h_{fe2})r_{e2}$$

where $r_{e2} = 25\ \Omega/I_{c2}(\text{mA})$.

Darlington applications include: series element in voltage regulators, audio output stages, and general interfacing from relatively high-output impedance devices to heavy loads.

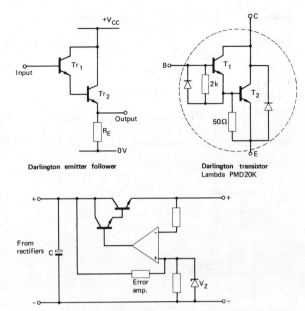

Darlington emitter follower

Darlington transistor
Lambda PMD 20K

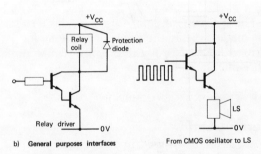

a) Series element in power supply regulator

b) General purposes interfaces

From CMOS oscillator to LS

*Figure D1* Darlington connection and some applications

One drawback to the circuit connection is the relatively high value of leakage current. A simpler component such as a VMOS power FET can replace the Darlington in many applications.

▶ *VMOS power FET*

## Data Selector (Multiplexer)

A data selector is an i.c. that can be used for multiplexing or for a circuit solution to a complex logic problem. It is simply an i.c. that acts as a large selector switch and picks one input from several and presents this input at the output pin (fig. D2). I.C.s are available giving 4-to-1, 8-to-1, and 16-to-1 line selectors. Consider the example of a 4-to-1 line data selector. If the address select code is 1 0 ($A = 0$, $B = 1$), then $X_2$, and $X_2$ only, will be selected

**Table D1**  *Typical Darlington device data*

| Device | | | $I_{C(max)}$ | $V_{CEO}$ | $h_{FE(min)}$ | $P_T$ | $f_T$ |
|---|---|---|---|---|---|---|---|
| T1S151 | npn | G.P. | 1 A | 55 V | 1000 | 625 mW | 150 MHz |
| TPSA13 | npn | high gain | 300 mA | 30 V | 10 000 | 600 mW | 200 MHz |
| T1P110 | npn⎫ | complements | 2 A | 60 V | 500 | 50 W | 1 MHz |
| T1P115 | pnp⎬ | | | | | | |
| T1P141 | npn⎫ | complements | 10 A | 80 V | 1000 | 125 W | 1 MHz |
| T1P146 | pnp⎬ | | | | | | |
| PMD20K | npn | switching | 20 A | 120 V | 3000 | 150 W | 1 μs max. turn-off time |

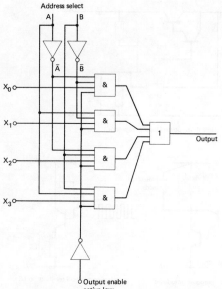

*Figure D2*  4-to-1 line data selector

| Inputs | | | | Outputs |
|---|---|---|---|---|
| D | C | B | A | F |
| 0 | 0 | 0 | 0 | 1 |
| 0 | 0 | 0 | 1 | 0 |
| 0 | 0 | 1 | 0 | 1 |
| 0 | 0 | 1 | 1 | 0 |
| 0 | 1 | 0 | 0 | 0 |
| 0 | 1 | 0 | 1 | 0 |
| 0 | 1 | 1 | 0 | 1 |
| 0 | 1 | 1 | 1 | 0 |
| 1 | 0 | 0 | 0 | 0 |
| 1 | 0 | 0 | 1 | 1 |
| 1 | 0 | 1 | 0 | 1 |
| 1 | 0 | 1 | 1 | 0 |
| 1 | 1 | 0 | 0 | 0 |
| 1 | 1 | 0 | 1 | 0 |
| 1 | 1 | 1 | 0 | 0 |
| 1 | 1 | 1 | 1 | 1 |

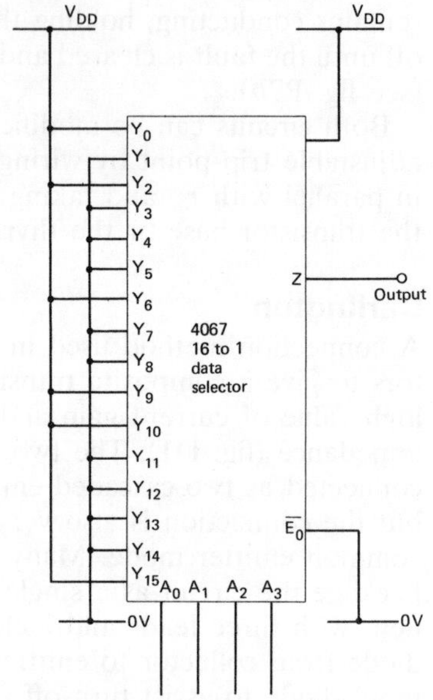

*Figure D3*  Using a data selector as the logic solution to a four-variable truth table

and presented at the output. If the address select code is changed to 0 1 ($A = 1$, $B = 0$), then only $X_1$ will be selected and presented at the output. Usually the i.c. is provided with an output enable control pin so that strobing is possible. The circuit shown is for digital signals only but CMOS data selectors, using transmission gates as the switches, can be used to multiplex analogue or digital signals.

Another use of data selectors is in an i.c. solution to logic problems. For example suppose a system required the truth table as shown in fig. D3. There are four inputs and the table has 16 states. The output is

$$F = \bar{A} \cdot \bar{B} \cdot \bar{C} \cdot \bar{D} + \bar{A} \cdot B \cdot \bar{C} \cdot \bar{D}$$
$$+ \bar{A} \cdot B \cdot C \cdot \bar{D} + A \cdot \bar{B} \cdot \bar{C} \cdot D$$
$$+ \bar{A} \cdot B \cdot \bar{C} \cdot D + A \cdot \bar{B} \cdot \bar{C} \cdot \bar{D}$$

By using simplification methods it would be possible to design a logic circuit for this. On the other hand a 16-to-1 data selector (CMOS 4067) can easily be used as shown. The inputs $Y_0$ to $Y_{15}$ are the 16 states of the table and the address is the input data $A,B,C,D$ of the table. Thus for address inputs of 0000 ($\bar{A} \cdot \bar{B} \cdot \bar{C} \cdot \bar{D}$), the input must be logic 1; for address 1000 ($A \cdot \bar{B} \cdot \bar{C} \cdot \bar{D}$), the input has to be logic 0; and so on. The data selector is simply used to represent the truth table.

▶ *Multiplexer*

## Dead Time

Usually a term used in control systems to describe the inherent time delays that occur within a system. If a step input is applied, there will be a measurable delay before the output starts

to respond; this is the dead time. In switching circuits, propagation delay time is a similar parameter.

## Dead Zone (or Dead Band)

In an electronic control system, there will be a finite range of input values, or small changes in input, that do not cause any change in the output. The band of inputs having no effect on output is called the dead zone. It is caused by several factors such as backlash in gears, control potentiometers or circuits and friction in moving parts and the inertia of the load.

## Decade Counter

This is a counting circuit that has ten states and therefore divides the input frequency by ten. This type of counter forms the basis of the majority of digital display systems, in which the content of each decade counter is first stored in a 4-bit latch and then decoded to drive the display device.

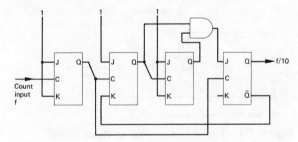

**Figure D4**   Decade counter

There are two common methods for creating decade counters:

**1**   By modifying the count sequence of a binary counter so that it resets to zero on the tenth input pulse (fig. D4). The TTL 7490 decade counter is a good example. This is asynchronous and counts in the 1248 BCD code. CMOS types such as the 4518 (dual synchronous BCD decade counter) and the 40192 (synchronous up/down BCD decade counter) are other examples.

**2**   By using a Johnson counter with five bistables. This decade counter is synchronous and relatively easy to decode. The CMOS 4017 is a typical circuit and it has a built-in decoder giving 1 out of 10 outputs.

▶ *Counter*

**Table D2**

| Decibels | Voltage ratio | Power ratio |
|---|---|---|
| 0 | 1.0 | 1.0 |
| 1 | 1.122 | 1.259 |
| 2 | 1.259 | 1.585 |
| 3 | 1.413 | 1.995 |
| 4 | 1.585 | 2.512 |
| 5 | 1.778 | 3.162 |
| 6 | 1.995 | 3.981 |
| 7 | 2.239 | 5.012 |
| 8 | 2.512 | 6.310 |
| 9 | 2.818 | 7.943 |
| 10 | 3.162 | 10.00 |
| 12 | 3.981 | 15.85 |
| 14 | 5.102 | 25.12 |
| 16 | 6.310 | 39.81 |
| 18 | 7.943 | 63.10 |
| 20 | 10.00 | 100.00 |
| 25 | 17.78 | 316.2 |
| 30 | 31.62 | 1000 |
| 35 | 56.23 | 3162 |
| 40 | 100.0 | 10 000 |
| 45 | 177.8 | 31 620 |
| 50 | 316.2 | 100 000 |
| 60 | 1000 | 1 000 000 |

## Decibel

A decibel is one tenth of a bel and is a commonly used logarithmic unit for power, voltage and current gain.

$$\text{Power gain} = 10 \log_{10} (P_o/P_i) \text{ dB}$$

where $P_o$ = output power, $P_i$ = input power.

$$\text{Voltage gain} = 20 \log_{10} (V_o/V_i) \text{ dB}$$

$$\text{Current gain} = 20 \log_{10} (i_o/i_i) \text{ dB}$$

## Decoder (Demultiplexer)

This is the name given to the group of circuits that extracts useful information from a coded signal or converts a signal from one coded form to another.

Usually the coded signal is in a digital form such as a 4-bit or 8-bit BCD word and a decoder may be required to convert this code into a suitable form to drive a 7-segment or decimal readout device. Decoders such as these are essential in any instrument that has to display the contents of a counter and typical i.c.s are

CMOS 4028 BCD to decimal
4511 BCD to 7-segment/latch/
decoder/driver
TTL 7442 BCD to decimal
7443 Excess-3 to decimal
7444 Excess-3/Gray to decimal
7445 BCD to decimal driver
7446 BCD to 7 segment
7447 BCD to 7 segment.
▶ *Code convertor*

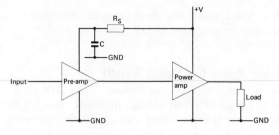

1) **Power supply decoupling**

## Decoupling

The removal or reduction of unwanted a.c. signals at some point in a circuit is called decoupling. The unwanted a.c. may be 100 Hz power supply ripple, a switching spike, or signal frequency, but the action of the circuit is the same. A bypass capacitor, with a low reactance at the frequency of the signal to be removed, is used to shunt the signal and reduce it to a low level. Typical examples of decoupling are shown (fig. D5).

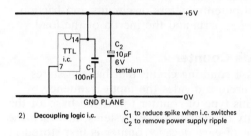

2) **Decoupling logic i.c.**    $C_1$ to reduce spike when i.c. switches
           $C_2$ to remove power supply ripple

**1** *Power supply decoupling* Removes 100 Hz ripple from the d.c. supply to a pre-amplifier and also to eliminate signal frequency feedback from the power amplifier via the supply leads to the sensitive input stages. The capacitor $C_1$ must have a low reactance at 100 Hz compared to the value of the series resistor. If $R_S$ is say 330 Ω then the decoupling capacitor $C_1$ should have a value of at least 100 $\mu$F.

The reactance of $C_1$ at 100 Hz is

$$X_{C1} = 1/2\pi fC \simeq 16 \,\Omega$$

The attenuation at 100 Hz ripple is

$$N = \frac{Z}{R_S} = \frac{\sqrt{(R_S^2 + X_{C1}^2)}}{R_S} = 20.6:1$$

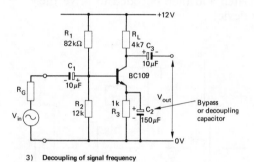

3) **Decoupling of signal frequency**

**Figure D5** Typical examples of decoupling

If there is 500 mV pk–pk ripple at the supply, then the amount of ripple present on the d.c. to the pre-amplifier is 23.4 mV pk–pk.

**2** *Decoupling logic i.c.s* When logic circuits, in particular TTL types, change state, a current pulse is demanded from the 5 V supply. This is only for a brief period but, even so, a noise pulse will appear on the power supply lead and may be propagated through the system. Since the switching spike frequency is very high, a capacitor with high resonant frequency must be used. Usually a 100 nF ceramic type is fitted at the i.c. pins. In a practical system, one decou-

pling capacitor of this type is used for about every four or five gate type i.c.s and one decoupling capacitor for every two MSI i.c.s.

**3** *Decoupling of signal frequency* Prevents reduction in gain at low frequencies. The purpose of the bypass capacitor is to prevent a.c. signals appearing at the emitter. Any a.c. signal component at the emitter will oppose the input and reduce the gain. The low-frequency gain will drop by 3 dB when the reactance of $C_2$ equals the parallel value of $R_3$ and $R_E$ where

$$R_E = r_e + R_G(1 - h_{fb})$$

$R_G$ is the parallel resistance of the bias network and the generator.

If $R_G$ is relatively low, then $R_E \simeq r_e$. $r_e$ is approximately 25 $\Omega/I_E$ (mA)

Thus $f_1 = 1/2\pi r_e C_2 \simeq 40$ Hz

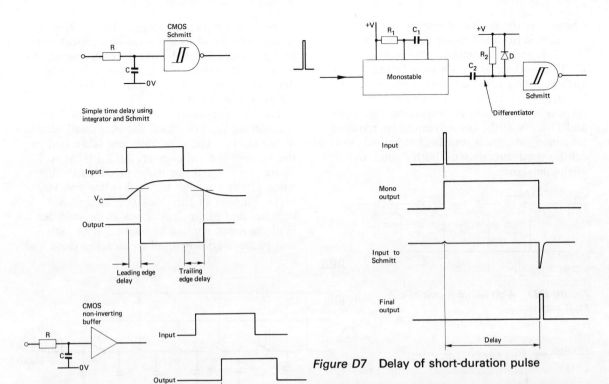

Simple time delay using integrator and Schmitt

CMOS Schmitt

Input

$V_C$

Output

Leading edge delay

Trailing edge delay

CMOS non-inverting buffer

Input

Output

Delay

**Figure D6**  Simple delay circuits

+V  $R_1$  $C_1$

Monostable

+V  $R_2$  D

$C_2$

Schmitt

Differentiator

Input

Mono output

Input to Schmitt

Final output

Delay

**Figure D7**  Delay of short-duration pulse

If the circuit is fed from a high impedance source, then the value of $C_2$ can be reduced.

## Delay Circuit

There are many situations in electronic systems where a signal, either analogue or digital, has to be delayed for a fixed time interval before being used. Circuit techniques for achieving this range from simple $RC$ networks, monostables, delay lines, to shift registers and bucket brigade devices.

**1**  Simple delay circuits can be made up using an $RC$ network with Schmitt or CMOS gates as shown in fig. D6. With TTL type gates, $R$ is limited to a maximum value of about $390\,\Omega$ but if CMOS gates are used $R$ can be a value up to several megohms. Circuits like these can be used only when the input pulse width is long compared to the required delay. As the input changes state, the capacitor takes time to charge:

$$V_C = V(1 - e^{-t/CR})$$

and the gate output will only change state when the voltage across the capacitor just exceeds the threshold level of the gate. A Schmitt circuit is preferable because of its "snap" action operation.

If a short-duration pulse has to be delayed, one solution is to use a monostable followed by a differentiator and Schmitt (fig. D7). The monostable (for example a 74121, 4047B, or 555) provides the basic delay. Its output is differentiated and the trailing edge is used as the input to the Schmitt. The shunt diode eliminates the positive edge. $C_1R_1$ sets the monostable timing and hence the delay and the differentiator $C_2R_2$ should have a reasonably short time constant. If a fixed duration pulse is required at the output, then another monostable can be used instead of the Schmitt.

A delay line is a length of low-loss cable. The signal travels along the cable with a velocity determined by the properties of the line, thus the longer the line the greater the delay. The line should be correctly terminated in its characteristic impedance $Z_0$ and should be driven from a generator of impedance $Z_0$. This prevents reflections and distortions of the signal. The delay time is given by

$$T_d = x\sqrt{(LC)}$$

where $x$ is the length of line in metres
  $L$ is series inductance per metre
  $C$ is shunt capacitance per metre.
**2** The modern approach to delaying digital or analogue data is to use shift registers and bucket brigade devices. Shift registers are digital in operation so A-to-D conversion at the input and D-to-A at the output would be required for analogue signals, but bucket brigade devices can be used directly with both digital and analogue signals.

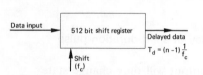

Figure D8a   4-bit serial-in/serial-out shift register

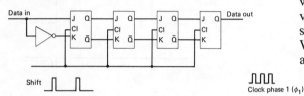

Figure D8b   n-stage shift register

A shift register is a series of bistables connected in cascade with a common clock or shift line (fig. D8a). A simple 4-bit serial-in/serial-out type is shown. When a digital signal of logic 1 is presented at the input of the first bistable, it will take four shift pulses to move this logic 1 to the output. The frequency of the clock pulse will determine the time delay from input to output.

The time delay for an $n$-stage shift register (fig. D8b) with a clock frequency $f_c$ is

$$T = (n-1)/f_c$$

Thus a 512-bit shift register driven at a clock frequency of 1 kHz will give a delay of 511 ms.
**3**  Bucket brigade delay lines, such as the TDA 1022 or the SAD 512, are ideal for use with analogue signals to produce medium delay times such as those required for reverberation units. It is not useful for really long delays (>200 ms) because the noise and distortion will increase to unacceptable levels. The name is used because the action of the i.c. is similar to water being passed along a chain of buckets, but in the i.c. the "buckets" are capacitors and

"water" is a packet of charge. By clocking the device, the signal in the form of a charge pattern is moved through from input to ouput. As with a shift register, the time delay is dependent on the number of stages and the frequency of the clock.

The basic operation can be understood by considering fig. D9 which shows a small section of the device. The capacitors are fabricated in the i.c. and the switches are MOSFETS. A 2-phase clock signal has to be provided so that when phase 1 is high phase 2 is low and vice versa. Phase 1 of the clock operates all $S_1$ switches and phase 2 operates all $S_2$ switches. With an input applied as phase 1 goes high and phase 2 goes low, all $S_1$ switches close and

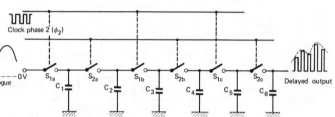

Figure D9   Basic principle of bucket brigade delay line

all $S_2$ switches open. $C_1$ charges to the instantaneous value of the analogue input; in other words, the amplitude of the analogue input is sampled for the time that the phase 1 of the clock is high. When phase 1 goes low, all $S_1$ switches open and, since phase 2 must go high, all $S_2$ switches close. The input is disconnected from $C_1$, but $C_2$ can now take some of the charge from $C_1$. On the next clock period, $C_1$ again samples the input and $C_3$ is charged from $C_2$. In this way a number of samples of the amplitude of the analogue input signal are taken and these samples are transferred through the i.c. as packets of charge. Because the signal is shifted through two stages for each clock period the delay time is given by

$$T_d = n/2f_c$$

where $n$ is the number of stages and $f_c$ is the clock frequency.

Note that the clock frequency must be at least twice the input signal frequency $f_s$ so that the maximum delay relative to signal frequency is

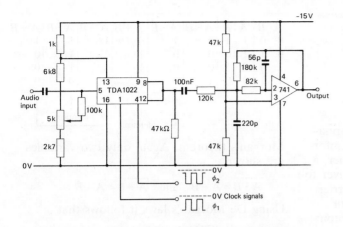

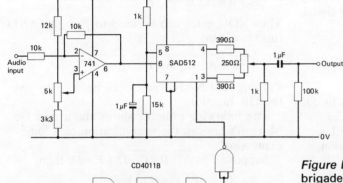

$$T_d = n/4f_{s(max)}$$

Suppose $f_s$ is 2.5 kHz, then $f_c$ must be at least 5 kHz giving a maximum delay of 51.2 ms for $n = 512$ stages.

If the bandwidth of the system is increased, then the clock frequency has to be higher and the delay time is reduced. In practice, band limiting of the analogue input is carried out to ensure that the maximum signal frequency can still be sampled by the clock.

Analogue delay lines using the TDA 1022 and the SAD 512 are shown in fig. D10. The TDA 1022 uses a negative supply and the input must be negative with respect to 0 V. This is achieved by the potential divider on the input. The SAD 512 requires a +15 V supply and positive input. In this case an op-amp wired as an inverting amplifier with positive d.c. offset is used. The TDA 1022 has to be

*Figure D10* Analogue delay lines using bucket brigade devices

provided with a two-phase clock with amplitudes of −15 V. A simple CMOS square wave oscillator running from approximately 10 kHz to 100 kHz is shown directly connected to the i.c. The SAD 512 has a built-in clock buffer and generates its own 2-phase clock signals.

The output of a bucket brigade i.c. will appear as a series of amplitude modulated pulses synchronised to the clock frequency, the variations in amplitude being the envelope of the original input frequency. The same method has to be used to remove the clock signal from the output. In the SAD 512 circuit the outputs from the last two stages are summed. Since these signals are in antiphase, the clock component will be very small in the final output. A filter circuit such as the one shown with the TDA 1022 can also be used to remove the clock signals. Circuits such as these can form the basis of reverberation units. In this a small portion of the delayed signal at the output is fed back and mixed with the new input. This gives the effect of a decaying delayed audio signal.

▶ *Monostable* ▶ *Schmitt trigger* ▶ *Timer*

| $A$ | $B$ | $\bar{A}$ | $\bar{B}$ | $A \cdot B$ | $\bar{A} \cdot \bar{B}$ | $A+B$ | $\bar{A}+\bar{B}$ | $\overline{A \cdot B}$ | $\overline{A+B}$ |
|---|---|---|---|---|---|---|---|---|---|
| 0 | 0 | 1 | 1 | 0 | 1 | 0 | 1 | 1 | 1 |
| 0 | 1 | 1 | 0 | 0 | 0 | 1 | 1 | 1 | 0 |
| 1 | 0 | 0 | 1 | 0 | 0 | 1 | 1 | 1 | 0 |
| 1 | 1 | 0 | 0 | 1 | 0 | 1 | 0 | 0 | 0 |

identical     identical

*Figure D11*    Demodulation

## Demodulator

In any communication system where information is transmitted by superimposing the information signal on a higher frequency carrier, a demodulator will be required in the receiver to extract the signal from the modulated carrier (fig. D11). The input may be amplitude or frequency modulated but the resulting output from the demodulator should be a close replica of the original low-frequency modulating signal. Demodulators are quite often called DETECTORS.

## De Morgan's Laws

These are based on a theorem published by De Morgan in *Syllabus of a proposed system of logic* (Watton & Maberly, 1860) and are laws that are extremely useful in the simplication and application of logic.

The results of his theorem written in Boolean algebra are

1) $\overline{A \cdot B \cdot C} = \bar{A} + \bar{B} + \bar{C}$

2) $\overline{A + B + C} = \bar{A} \cdot \bar{B} \cdot \bar{C}$

The proof, for two variables, is as follows:

$$A \cdot B \cdot \overline{A \cdot B} = 0 \quad \text{since } X \cdot \bar{X} = 0$$

and $\quad A \cdot B + (\overline{A \cdot B}) = 1 \quad \text{since } X + \bar{X} = 1$

$$(A \cdot B) \cdot (\bar{A} + \bar{B}) = A \cdot \bar{A} \cdot B + A \cdot B \cdot \bar{B}$$
$$= B \cdot 0 + A \cdot 0 = 0$$

and $\quad (A \cdot B) + (\bar{A} + \bar{B}) = \bar{A} + B + \bar{B}$
$$= \bar{A} + 1 = 1$$

$\therefore \quad \overline{A \cdot B} = \bar{A} + \bar{B}$

Similarly

$$(A+B) \cdot \bar{A} \cdot \bar{B} = \bar{A} \cdot \bar{B} \cdot \bar{B} + B \cdot \bar{A} \cdot \bar{B} = 0$$
and $\quad (A+B) + \bar{A} \cdot \bar{B} = A + B + \bar{B} = A + 1 = 1$

$\therefore \quad \overline{A+B} = \bar{A} \cdot \bar{B}$

A truth table can also be used to verify De Morgan's theorem. Again only two variables are shown.

$$\overline{A \cdot B} = \bar{A} + \bar{B} \quad \text{and} \quad \overline{A+B} = \bar{A} + \bar{B}$$

Using De Morgan's laws it follows that

$$\overline{\bar{A}+\bar{B}} = \overline{\overline{A \cdot B}} = A \cdot B$$

Thus NOR gates can be used to give the AND function. Also

$$\overline{\bar{A} \cdot \bar{B}} = \overline{\overline{A+B}} = A+B$$

showing that NAND gates can be used to give the OR function.

The following example shows the use of De Morgan's laws in the simplification of a logic expression.

Suppose $F = \overline{A + B \cdot C \cdot \bar{D} + E \cdot D}$, then

$$F = \bar{A} \cdot \overline{B \cdot C \cdot \bar{D}} + \bar{E} + \bar{D}$$
$$= \bar{A} \cdot (\bar{B} + \bar{C} + D) + \bar{E} + \bar{D}$$
$$= \bar{A} \cdot \bar{B} + \bar{A} \cdot \bar{C} + \bar{A} \cdot D + \bar{E} + \bar{D}$$
$$= \bar{A} \cdot \bar{B} + \bar{A} \cdot \bar{C} + \bar{A} + \bar{E} + \bar{D}$$
$$\text{(since } \bar{A} \cdot D + \bar{D} = \bar{A} + \bar{D})$$
$$= \bar{A} \cdot (\bar{B} + \bar{C} + 1) + \bar{E} + \bar{D}$$
$$= \bar{A} + \bar{E} + \bar{D}$$
$$= \overline{A \cdot E \cdot D}$$

In other words, only one NAND gate is required for the logic circuit.

## Depletion Region

A term used to describe a region in a semiconductor device that is empty of mobile charge carriers (fig. D12). Other names given to this region are "space-charge region" or "transition region". Consider a p-n junction diode that is not forward biased. The initial electron-hole recombination that took place at the junction sets up fixed charges either side which repel the mobile charge carriers (loosely bound electrons in the n material and holes in the p type). Thus a small region exists either side of

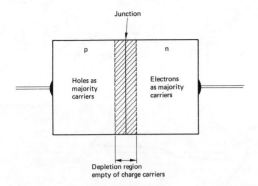

Figure D12   Depletion region in a semiconductor

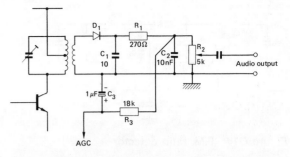

Figure D13   A.M. detector

the junction which is depleted of charge carriers. The thickness of this depletion region may be less than 0.5 $\mu$m. It is increased if reverse bias is applied to the diode.

▶ *Semiconductor theory*

## Derating

If the components that go to make up an electronic instrument or system are all operated at values of current, voltage, and power that are much lower than their rated values, then a significant improvement in the overall reliability is achieved. This is because the operating stresses acting on the components will be greatly reduced giving a lower failure rate for each component and a higher MTBF for the system.

Derating is a technique used during the design stage to improve the overall reliability of the instrument. It could mean operating all resistors at no more than 50% of their maximum power rating (using a 1 W rated resistor for $\frac{1}{2}$ W dissipation) and fitting capacitors with voltage ratings of double the expected value. For resistors the failure rate is almost halved if their dissipation is halved, and capacitors may show an even more dramatic improvement when derated. Derating has much the same effect on active components such as transistors and diodes.

▶ *Reliability*

## Detector

Like many other words used in electronics the description of detector can be applied to a wide range of devices and circuits. However "detector" is generally used for the group of circuits that act as demodulators for r.f. signals. Among these there are several types, but the two most commonly used are the following.

**1   *A.M. detector***
A typical a.m. demodulator/detector for use in a radio receiver is shown in fig. D13. The diode $D_1$ acts as a half-wave rectifier to charge $C_1$ on every positive half-cycle of the i.f. waveform. $C_1$ then discharges through the load resistors $R_1$ and $R_2$ when the diode is off. The time constant formed by $C_1$ and $R_1 + R_2$ has to be arranged to be longer than the periodic time of the i.f. but relatively short compared to the highest audio frequency. The voltage across $C_1$ then consists of the audio signal and a d.c. component proportional to the amplitude of the i.f. carrier. Further filtering to remove i.f. ripple is performed by $R_1 C_2$ and the audio signal is taken from the wiper of $R_2$ via a coupling capacitor to the input of the audio amplifier.

The d.c. component, filtered by $R_3 C_3$ to remove any a.c., is then used to provide an automatic gain control level. Negative AGC voltage can be provided by simply reversing the diode.

**2   *F.M. ratio detector*** (fig. D14)
This is the most popular demodulator circuit used in f.m. receivers being preferred to the Foster–Seeley mainly because it is fairly insensitive to amplitude variations. It gives about 30 dB of a.m. rejection.

The operation of the circuit depends upon the phase relationship between the signals developed across the tuned circuits $L_1 C_1$ and $L_2 C_2$. Both are tuned to the i.f. but the two signals are in quadrature. A third coil $L_3$ is tightly coupled to $L_1$ so that the signal in $L_3$ is in phase with that in $L_1$. This signal is injected at the mid-point of $L_2$ with the result that signals at x and y will be equal when there is no frequency deviation. When the input carrier does change in frequency, i.e. contains modulating information, the signal levels at x and y

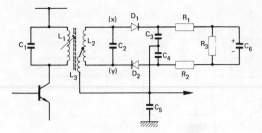

*Figure D14*   F.M. ratio detector

change. These changes are detected by the two diodes $D_1$ and $D_2$ which charge $C_3$ to the peak value of the signal at x and $C_4$ to the peak value of the signal at y respectively. The voltage that then appears across $C_5$ is proportional to the frequency deviation of the carrier, that is the a.f. signal.

## Diac

This is a bidirectional trigger diode that is used mostly in simple firing circuits for triacs in a.c. power controllers (fig. D15).

It has a p-n-p type structure which gives an equivalent circuit of two back-to-back diodes. If an applied voltage, of either polarity, is less than the breakover voltage $V_{BO}$ then the device remains in a high-resistance state. As soon as the voltage exceeds $V_{BO}$ the diac exhibits a negative resistance as the current through it rises while the voltage across it falls.

Typical data: DIAC type BR 100

| $P_{tot}$ | $I_{peak}$ | $V_{BO}$ | $I_{BO}$ | $\Delta V$ |
|-----------|------------|----------|----------|------------|
| 150 mW | 2 A | 32±4 V | 100 μA | 5 V |

This type of characteristic can be used in simple sawtooth or pulse type oscillators. As the voltage across $C$ rises towards $+V_S$ at $V_{BO}$, the diac switches and the capacitor is discharged through $R_L$. The voltage across $C$ then falls and, after a short interval depending upon the values of $C$ and $R_L$, the diac switches back to its high resistance state. The cycle then repeats.

In the a.c. power control circuit, the $RC$ network and the diac determine the phase at which the triac is fired. The value of $R$ determines the rate at which $C$ charges and, when the voltage across $C$ just exceeds the breakover voltage of the diac, it switches to a low resistance state and discharges $C$ into the triac gate. Since the diac is bidirectional, this ac-

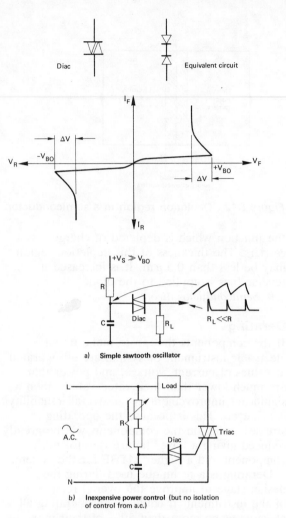

a)   **Simple sawtooth oscillator**

b)   **Inexpensive power control** (but no isolation of control from a.c.)

*Figure D15*   **Diac characteristics and applications**

tion occurs in both positive and negative half cycles of the a.c. supply. By varying the value of $R$ the triac can be made to switch on earlier or later in each half-cycle to give a wide range of control over the power dissipated in the load.

## Differential Amplifier

This is a linear amplifier which has two input leads and gives an output that is proportional to the difference in signal between these two input leads. It is the basic amplifying stage in op-amps and d.c. amplifiers, used because it has low values of voltage drift with temperature. This low drift results because of its built-in ability to reject signals that are common to both inputs. If the two transistors are closely

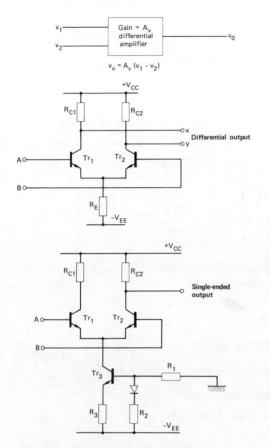

Figure D16 Differential amplifiers

gain. To achieve high common mode rejection, $R_E$ should be as large as possible and in most practical circuits $R_E$ is replaced by a constant-current stage as shown. The quiescent current can be set by the bias components to $Tr_3$ and this current then splits almost equally between $Tr_1$ and $Tr_2$ when both inputs are identical. $R_E$ is then the output resistance of $Tr_3$, typically 100 kΩ.

A single-ended output can be taken from one collector only and, provided that a constant current stage is included, the rejection of common mode signals will still be very high.
▶ CMRR    ▶ Op-amp

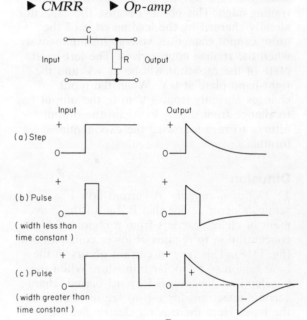

Figure D17    A differentiator

matched, relatively easy in an i.c., then the changes in $V_{BE}$ of both transistors with temperature will be almost identical and will appear as common mode inputs.

In the basic circuit (fig. D16), when input A is the same as input B, both transistors conduct equally and the differential output will be zero. When input A is made slightly more positive than B, then $Tr_1$ will conduct more than $Tr_2$ so that point x falls with respect to y to give a differential output. The output signal will be reversed if B is made more positive than A. As long as the changes in input signal are relatively small, the total current flowing through $R_E$ hardly changes at all so that there will be virtually no differential signal voltage set up at the common emitter point. However common mode signals will cause both transistor currents to change equally in the same direction causing a voltage to be set up across $R_E$. This gives negative feedback and a low common mode

## Differentiator

Basically a high pass filter circuit which allows high frequencies to pass but attenuates the low frequencies (fig. D17). It is a circuit that is commonly used to generate short-duration spikes from a pulse or square wave input when the input width is much longer than the time constant of the differentiator. The product of capacitance in farads and resistance in ohms is called the TIME CONSTANT and is the time taken for the voltage across the capacitor to change by about 63%.

When a step input is applied, and assuming that C is uncharged, then the voltage across the capacitor cannot change instantaneously. The voltage across a capacitor can only change

**93**

as it acquires charge ($Q = C \, dV/dt$), and this naturally takes time. Thus the output rises to the same value as the input. As $C$ charges, the voltage across it increases and the output voltage across $R$ falls. This fall is exponential:

$$V_r = V_1 e^{-t/CR}$$

The result is that a spike with an amplitude equal to the change of state at the input and a width of about one time constant is generated.

If the input pulse has a duration which is long compared to the differentiator's time constant, then the output will go negative on the trailing edge. This occurs because the capacitor already charged by the leading edge of the pulse cannot change its voltage instantaneously when the trailing edge arrives. The left-hand plate of the capacitor will be at $+V$ and the right-hand plate at 0 V. When the input changes abruptly from $+V$ to 0, the output has to change from 0 to $-V$. Again the output returns to zero following the exponential formula.

## Diffusion

**1** *Diffusion current*   A current flow in a semiconductor device that is due to the movement of charge carriers from a region of high concentration to regions of lower concentration (fig. D18). This type of current occurs in the base region of a bipolar transistor. When the base/emitter junction is forward biased, charge carriers (electrons for n-p-n) are injected into the base. Here there is no electric field to accelerate them to the collector until the electrons reach the collector/base depletion region. The electrons diffuse from the region of high concentration just inside the base/emitter junction towards the regions of low concentration. When they enter the collector/base depletion region, they are swept up by the collector. Because diffusion is relatively slow it is important for high-frequency performance to have a base width that is as small as possible. Typical base widths are as low as 1 $\mu$m in modern planar transistors.

**2** *Solid state diffusion*   The process by which selected regions of semiconductor material are converted into p or n type. This process forms the basis of the manufacture of many modern transistors and i.c.s.

Slices of silicon are coated with silicon dioxide which is impervious to impurities, and then

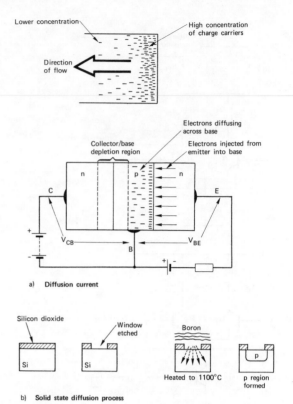

a)   Diffusion current

b)   Solid state diffusion process

**Figure D18**   Diffusion

windows are etched through the silicon dioxide to expose selected areas of the silicon slice. To dope these selected regions the slice is heated to about 1100°C in an oven with an impurity such as boron. The impurity will slowly spread and diffuse into the silicon to form a p region. Phosphorus would be used as the impurity to give an n region. The process is relatively slow (1 to 2 hours) and therefore controllable, so that areas can be well defined.

▶ *Planar process*

## Digital Circuit

Much of modern electronics utilises digital logic circuits. These are the group of circuits which only respond to input signals that are within well defined limits rather than a wide range of input values. There are many important factors that make a digital type system superior to an analogue system. Consider a simple communication link in which the analogue information is transmitted direct by wire (fig. D19a). At the receiver the signal will be degraded, since some of its energy will have been lost during

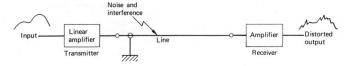

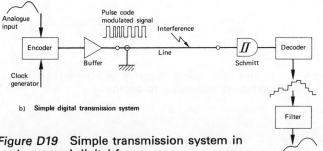

a) Analogue transmission of information

b) Simple digital transmission system

**Figure D19** Simple transmission system in analogue and digital form

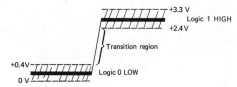

**Figure D20** Logic levels in digital circuits (positive logic)

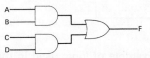

**Figure D21** Combinational logic
$F = (A$ and $B)$ or $(C$ and $D)$
In Boolean form: $F = A \cdot B + C \cdot D$

transmission, and it may be difficult to pick out the useful information among the noise and interference signals. However, if the same signal is first coded into digital form and then transmitted as a series of pulses (pulse code modulation), noise and interference will have much less effect and pulses at the receiver can be easily reshaped before being decoded. The output of the decoder will be a reproduction of the original input signal. Although more complex, the digital transmission system is vastly superior to the analogue.

Digital circuits are built using two-state binary devices, which means that devices are either ON or OFF and give either a LOW or HIGH state output. The low state for a positive logic system will be near zero volts and the high state will be a few volts positive. Typical values for TTL gates are 0 V to +0.4 V for the low state and between +2.4 V and +3.3 V for the high state (fig. D20). Other logic families have differently defined levels but what is important in digital systems is that these levels are within specified limits. The low state is referred to as Logic 0 and the high state as Logic 1. It negative logic is used, the most negative level is referred to as logic 1.

Some of the important advantages of digital systems in comparison to analogue are

1) They are less susceptible to noise and interference.

2) There should be no ambiguity about an output as it will be either a logic 1 or a logic 0.

3) Digital data can be more easily stored without degradation. This is because it is held as a group of 1 s and 0 s in a digital memory.

4) Signals are more easily transmitted, processed and manipulated when in digital form.

There are two main forms of digital logic:

*a*) COMBINATIONAL LOGIC in which a combined set of input conditions are simultaneously required to give a particular output. For example to achieve an output logic 1 from a circuit inputs of *A* and *B*, or *C* and *D*, are required. Statements such as this can be written down using Boolean Algebra

$$F = A \cdot B + C \cdot D$$

The logic circuit of two AND gates followed by an OR gate is shown in fig. D21.

*b*) SEQUENTIAL LOGIC These are logic circuits that possess a memory and give an output, in response to an input, that is dependent upon the circuit's previous state. Typical examples of sequential logic circuits are bistable multivibrators, counters, and shift registers

In general, a digital logic system contains a mixture of combinational and sequential logic and such systems can all be built using the basic logic gates of AND, OR, NOT, NAND and NOR.

Let's consider a simple example of a machine control. Suppose that to start the machine the following conditions have to be met

**95**

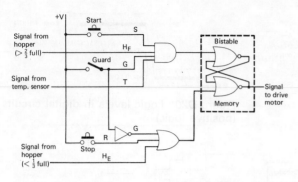

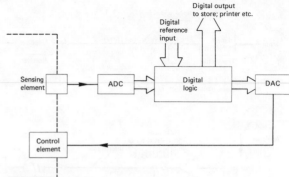

*Figure D22*  Example of logic used in machine control

*Figure D23*  Block diagram of how a digital system is interfaced to analogue

1) the start button is momentarily pressed  *S*
2) A hopper must be more than $\frac{2}{3}$ full  $H_F$
3) The safety guard must be closed  *G*
4) The temperature of a heater must be above +400°C  *T*

All these input represent an AND function and can be written in Boolean algebra as

Start pulse $= S \cdot H_F \cdot G \cdot T$  (see fig. D22)

A logic 1 pulse will result from the AND gate output if all input conditions are met. Since this is only a short-duration pulse, a memory or latch is required to remember that this pulse has been received. Any of the following conditions will stop the machine:

5) The safety guard is open, this is the same as saying not-closed, i.e. not *G* (written as $\bar{G}$).
6) The stop button is pressed momentarily  *R*
7) The hopper becomes less than $\frac{1}{3}$ full  $H_E$

These inputs represent an OR function and if any one of them is present the memory must be reset and the machine stopped.

Reset pulse $= \bar{G} + R + H_E$

The full logic diagram is shown in fig. D22.

This, of course, is an example of a digital switching system to give on/off control. In practice, inputs may be analogue and outputs may be required to give smooth control over a wide range. Again it is advantageous to convert the input analogue signal, via an analogue-to-digital convertor, into suitable digital signals and to change the digital output to an analogue control signal via a digital-to-analogue conver-

tor if this is required. Suppose the temperature inside an oven has to be controlled. The output from the temperature sensor (a thermocouple) is converted to digital information by the ADC. If an 8-bit word is used the coded temperature may be represented thus

20°C = 0000 0010

800°C = 1000 0000

870°C = 1000 0111   and so on

The digital input could be displayed, compared with a reference word, stored, and also used to generate a signal to control the heat output to the oven (fig. D23).

There are several ways of implementing digital systems. The system may use mostly NAND gates or NOR gates, and depending on the application the designer may choose from standard TTL, Schottky TTL, CMOS, or ECL. Each of these is covered in a separate section but it is useful to look at a general comparison between the types. From the table it can be seen that fast speed of operation will mean higher power consumption. Note that CMOS dissipates power while switching, typically 1 mW/MHz, and this is comparable in power consumption to low-power Schottky TTL at a frequency of about 2 MHz. The big advantage of CMOS is the fact that it has a high value of noise immunity which makes it the logic family to choose if the system has to be used in an electrically noisy environment.

▶ *Boolean algebra*  ▶ *Bistable*  ▶ *CMOS*
▶ *Gate*  ▶ *TTL*

**Table D3**  *Comparison of digital logic families*

| Logic family | Propagation delay time | Fan out | Noise immunity | Bistable clock freq. | Quiescent power dissipation |
|---|---|---|---|---|---|
| Standard TTL | 10 ns | 10 | 1 V | 35 MHz | 10 mW/gate |
| Schottky TTL | 3 ns | 10 | 1 V | 90 MHz | 20 mW/gate |
| Low-power Schottky TTL | 7 ns | 20 | 1 V | 45 MHz | 2 mW/gate |
| ECL | 2 ns | 30 | 400 mV | 140 MHz | 60 mW/gate |
| CMOS  5 V | 35 ns | >50 | 2.25 V | 8 MHz | 10 nW/gate |
| 10 V | 20 ns | >50 | 4.5 V | 16 MHz | 10 nW/gate |

## Digital-to-Analogue Convertor

If you want to drive an analogue device, say a chart recorder, from a digital system, you must first change the digital output signal into a suitable analogue voltage or current. A DAC is the essential interface circuit between digital systems and analogue-type devices.

**1**  As an example, suppose it is required to convert 3-bit digital data into an analogue voltage, and that the maximum analogue output, corresponding to all three bits being high, is 3.5 V (fig. D24). A table can be drawn up to show how the analogue output changes for various values of the 3-bit digital word:

| Most significant bit ↓ Bit 1 | | Least significant bit ↓ Bit 3 | Analogue output |
|---|---|---|---|
| Bit 1 | Bit 2 | Bit 3 | |
| C | B | A | |
| 0 | 0 | 0 | 0 V |
| 0 | 0 | 1 | 0.5 V |
| 0 | 1 | 0 | 1 V |
| 0 | 1 | 1 | 1.5 V |
| 1 | 0 | 0 | 2 V |
| 1 | 0 | 1 | 2.5 V |
| 1 | 1 | 0 | 3 V |
| 1 | 1 | 1 | 3.5 V |

Here the analogue output changes in 0.5 V steps and there are eight possible analogue output values. It is obviously not possible to resolve a signal better than 1 part in 8, showing that it is the number of bits converted that primarily determines the resolution of a DAC. If the arrangement were changed to an 8-bit convertor, then the possible resolution would be 1 part in 256 giving a much smoother

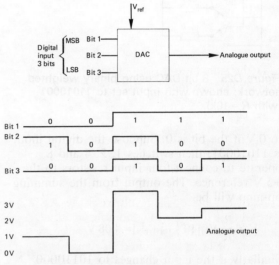

**Figure D24**  Principle of digital-to-analogue conversion (3 bits only)

analogue output with many more values. For an 8-bit conversion, with the same full-scale output of 3.5 V, each step would be about 13.72 mV. This shows that the accuracy of the various resistors and limiting factors such as offset voltages in the switching circuits and amplifiers become much more important as the number of bits being converted increases. The ability of a DAC to give an output for every possible input code is referred to as *monotonicity*.

**2**  The simplest method of building a DAC is to use a weighted-resistor network and a summing op-amp (fig. D25). Each of the resistors has to be weighted in value in a binary sequence, i.e. $R$, $2R$, $4R$, $8R$, etc. The digital word to be converted is used to drive electronic switches which connect the resistors to the +5 V reference if the bit is 1, and

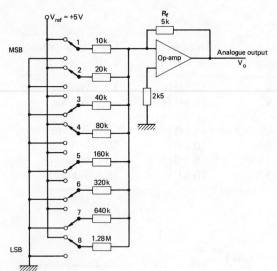

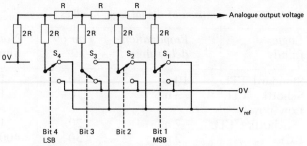

Figure D26a  R-2R ladder for D-to-A conversion to voltage output; switches (electronic types) set to convert 1101

*Figure D25*  **8-bit DAC using binary weighted network; shown with input set to 11010001 (with $R = 10k$)**

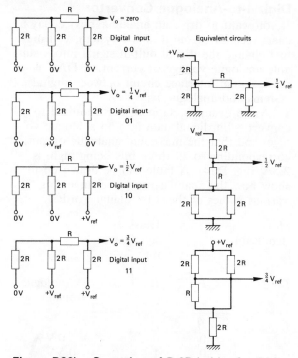

Figure D26b  Operation of R-2R ladder for 2-bit digital inputs

to 0 V if the bit is 0. Suppose the digital input is 11010001; then switches 1, 2, 4 and 8 operate to connect four input resistors to the +5 V reference. The output from the summing op-amp will be

$$V_0 = \frac{R_f}{R} V_{ref}[1 + \tfrac{1}{2} + \tfrac{1}{8} + \tfrac{1}{128}] = 4.08 \text{ V}$$

Similarly if the input changes to 10110000,

$$V_0 = \frac{R_f}{R} V_{ref}[1 + \tfrac{1}{4} + \tfrac{1}{8}] = 3.4375 \text{ V}$$

The problem with a simple circuit such as this is that the range of resistor values required is high. For a 10-bit convertor the resistor range is more than 500:1. To achieve good linearity and accuracy, the resistors must be close-tolerance and all track together with temperature. This is naturally difficult to obtain and another method of conversion is preferred. **3** The most commonly used system is based on the R-2R ladder network shown in fig. D26. The output voltage or current is generated by switching sections of the ladder to either $V_{ref}$ or to 0 V corresponding to a 1 or a 0 of the digital input. Again the switches are electronic types and are usually incorporated in the DAC i.c. The operation of the R-2R ladder can be understood by considering only a 2-bit convertor as shown and using Ohm's law.

There are several advantages of the R-2R

ladder compared to the weighted resistor network;

1) Only two values of resistor are used.

2) It can easily be extended to as many bits as desired.

3) The absolute value of the resistors is not important, only the ratio needs to be exact.

4) The resistor network can be fairly easily manufactured as a film network or in monolithic form—thus, temperature characteristics of the resistors will be all very similar.

A 4-bit DAC using discrete components and an R-2R network is shown in fig. D27 to il-

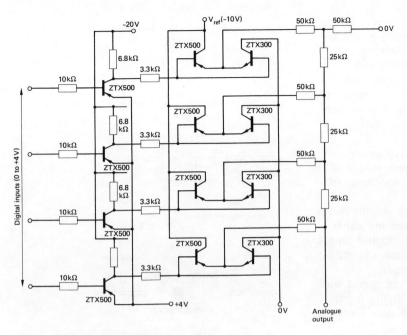

**Figure D27**  4-bit DAC using complementary transistors as switches

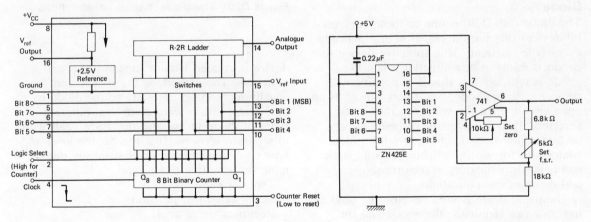

**Figure D28a**  Block diagram of ZN425E 8-bit DAC

**Figure D28b**  8-bit DAC using the ZN425E

lustrate the principle, but in most cases an i.c. can be found that will do the job better. The ZN 425E 8-bit DAC is a good example of a relatively low-cost high-performance i.c. It contains an 8-bit switch array with an R-2R ladder network plus an 8-bit binary counter and a 2.5 V precision voltage reference all on one chip. The counter makes the i.c. suitable for A-to-D conversion and ramp generator applications (fig. D28).

**4**  Yet another type of fast DAC uses what are called quad current switches. A set of transistors which are controlled by the digital input are used to conduct currents in the binary ratio 8, 4, 2, 1. A special construction technique gives the transistors similarly weighted emitter areas. This gives equal current densities for the four transistors and equal $V_{BE}$ drops with the nice result that their temperature characteristics track almost perfectly.

**5** In DACs the important parameters that have to be specified when choosing a unit for a particular application are the following:

*Resolution* is the ability to distinguish between adjacent values of the input and is a function of the number of bits being converted. However, in the limit, resolution is ultimately determined by noise, non-linearity and monotonicity.

*Linearity* is the amount of error in the analogue output between adjacent values of the input and should typically be a maximum of ±0.5 of the LSB.

*Monotonicity* means that there should be an increasing output for every increasing value of digital input—in other words all digital inputs can be decoded. Most manufacturers specify a temperature range for this parameter.

*Settling time* is a measure of the speed with which the convertor decodes a digital input into its analogue value. It is the time for the output to settle within ±1 LSB step.

▶ *Drift*

## Diode

The diode (fig. D29) is one of those very useful devices that finds application in practically every type of circuit. It is an active device that conducts easily in one direction when the anode is typically less than 1 V positive with respect to the cathode, but it acts almost as an open circuit when the voltage across it is reversed. This characteristic of being either on and conducting, or off, makes it an ideal component for use in switching circuits, logic, rectifiers, demodulators, waveform shapers, and many other applications.

The name diode is used because the device has only two terminals; the anode and the cathode. These names for the terminals are still used for modern semiconductor diodes, and were taken from the thermionic diode valve invented in 1907 by Flemming. In the valve diode the cathode is heated so that it emits electrons. These electrons are then collected by the anode plate when the anode-to-cathode voltage is sufficiently positive. This is typically a few volts. Electrons are repelled when the anode is negative with respect to the cathode. Thus anode current flows only when the diode is forward biased. In the semiconductor diode however, current flow in the forward direction is due to both holes and electrons since the

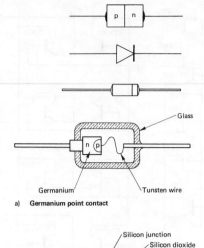

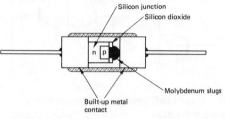

a) **Germanium point contact**

b) **Modern silicon planar**

*Figure D29* The diode : typical constructions

device is created from a junction of p and n materials. The cathode is the n-type and this is usually marked on the body of the diode by a bar. There are, of course, several different types of diode in use and these are dealt with in separate sections. It is the doping level and type of manufacture used that mostly determine the resulting diode type.

*Types of diode*
Small-signal p-n junction
Rectifiers (large area)
Backward
Tunnel
PIN
Schottky (or hot carrier)
Zener and voltage regulator
Varactor (or varicap)
Light-emitting
Photosensitive.

This section deals mainly with the small-signal junction diodes. The earliest type manufactured was germanium point contact. This consists of an n-type semiconductor against which a fine tungsten spring wire presses. During the manufacturing process a pulse of

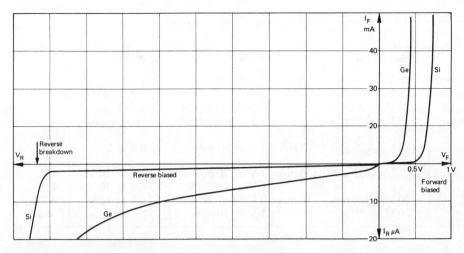

*Figure D30* Typical diode characteristics at 25°C

current is passed down the wire and a small p-region is formed at the point where the wire presses into the germanium. This give a very small junction area and therefore low capacitance. The germanium point contact diode was widely used as an efficient u.h.f. detector. Current rating is relatively low and an improved version using a gold wire was introduced. Some gold-bonded types can handle currents up to 200 mA peak.

The germanium point contact diodes are still available but are not intended for new designs. The majority of general-purpose diodes are made from silicon using the planar process. Silicon diodes have much lower reverse leakage currents than germanium and can operate at higher temperatures. The construction is shown in fig. D29. Under conditions of no bias, a depletion region exists at the junction because the fixed charges (ionised atoms) either side of the junction set up a potential barrier [▶ *Semiconductor theory*]. When a forward bias is applied making the anode positive with respect to the cathode, this potential barrier is overcome causing holes to cross from p to n and electrons from n to p. The flow of both types of charge carrier makes up the diode current $I_F$ which increases rapidly as the forward bias is increased (fig. D30). In fact the diode current can be seen to follow an exponential law:

$$I_F = I_S(e^{qV/KT} - 1)$$

where $I_S$ is the reverse or saturation current

e is the base of natural logs
q is the charge on an electron ($1.6 \times 10^{-14}$ coulombs)
V is the applied voltage
K is the Boltzmann constant ($1.38 \times 10^{-23}$ J/°K)
T is the absolute temperature in degrees Kelvin.

At room temperature $K \simeq 290°$, therefore $q/KT = 40$, and the basic diode equation can be written

$$I_F = I_S(e^{40V} - 1)$$

Most modern devices conform well to this equation.

The reverse saturation current is the tiny leakage current that flows when the anode is negative with respect to the cathode. Under these conditions the depletion region gets wider so that current flow is due only to minority charges in the material and thermally generated hole-electron pairs. In a modern silicon diode the reverse current can be as low as a few nanoamps. If the reverse voltage is made sufficiently high then the diode will go into its breakdown region in which the reverse current rapidly increases. The maximum reverse voltage rating is $V_{RRM}$.

For high-frequency detection the important parameter in a diode is its capacitance which must be as low as possible. Schottky diodes with junction capacitances less than 1 pF are commonly used for this application.

**101**

**Table D4** *Short form data for some commonly used diodes* (In all cases $T_{am} = 25°C$)

| Type no. | Description | $V_{RRM}$ | $I_{FRM}$ | $I_{AV}$ | $V_F$ at $I_F$ | $I_R$ at $V_R$ | $t_{rr}$ |
|---|---|---|---|---|---|---|---|
| 0A90 | Ge point contact | 30 V | 45 mA | 10 mA | 2 V at 30 mA | 300 µA at 30 V | — |
| 0A91 | Ge point contact | 115 V | 150 mA | 50 mA | 2.1 V at 30 mA | 75 µA at 45 V | — |
| 0A47 | Ge gold bonded | 25 V | 150 mA | — | 0.54 V at 30 mA | 10 µA $V_{RRM}$ | — |
| AAZ17 | Ge gold bonded | 75 V | 250 mA | — | 1.1 V at 250 mA | 300 µA $V_{RRM}$ | — |
| IN914 IN916 IN4148 | Si planar | 75 V | 225 mA | 75 mA | 1 V at 10 mA | 25 nA at 20 V | 4 ns |
| IN4446 | | | 450 mA | 150 mA | 1 V at 20 mA | 25 nA at 20 V | 4 ns |
| IN4448 | | | | | 1 V at 100 mA | 25 nA at 20 V | 4 ns |
| ITT700 | Si ultrafast | 30 V | 150 mA | 50 mA | 0.88 V at 10 mA | 50 nA at 15 V | 0.7 ns |
| BAV45 | Si low-leakage | 20 V | 50 mA | — | 1 V at 10 mA | 10 pA at 20 V | 350 ns |

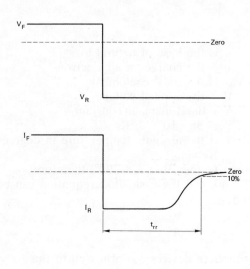

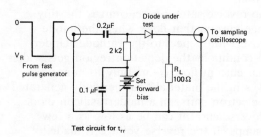

*Figure D31* Diode reverse recovery time

The capacitance of a diode arises because the thin depletion area is effectively an insulator whereas the n and p regions are conductors. By altering the reverse voltage, the capacitance of the diode can be varied and this is put to good use in varactor or varicap diodes which are used as tuning elements in v.h.f and u.h.f circuits.

When a diode has to be used in a high-speed switch application, the parameter of interest may well be the reverse recovery time $t_{rr}$. Imagine the diode forward biased and passing a moderate current ($I_F = 10$ mA). The diode will have relatively large concentrations of minority carriers near the junction; these will be holes that have just crossed from p to n and electrons that have crossed from n to p. They will exist for a short time before being neutralised by majority carriers. If a reverse voltage is suddenly applied, instead of the diode current falling rapidly to zero, the current reverses as these minority carriers are attracted back across the junction. Until all these charge carriers are depleted, the reverse current will continue to flow. Reverse recovery time (fig. D31) is the time interval from the instant the voltage across the diode is reversed to the time when the reverse current has fallen to 10% of its maximum value. Diodes such as the IN914, IN916 and IN4148 have reverse recovery time $t_{rr}$ of 4 ns. A really fast diode such as the ITT700 has a $t_{rr}$ of only 0.7 ns.

## Diode Transistor Logic (DTL)

This was one of the first successful monolithic i.c. logic families but it has now been superseded by TTL, ECL, and CMOS, all of which offer various improvements in performance compared with DTL. Diode transistor logic is relatively slow (propagation delay time about 30 ns) and is limited in fan-out to about 8. It can still be found in many working systems but

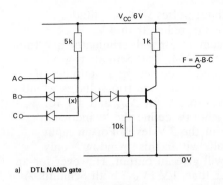

a) DTL NAND gate

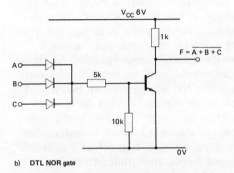

b) DTL NOR gate

*Figure D32* DTL logic gates

it is unlikely to appear in any new equipment designs.

The NOR and NAND functions were the basic gates with circuits as shown in fig. D32. It can be seen that each circuit is a diode gate followed by a transistor invertor. In the NOR gate, if any input is at logic 1 (2.3 V min.), the transistor conducts giving a logic 0 (0.8 V max.) output. The output can only be a logic 1 if all inputs are at logic 0.

In the NAND circuit, if any input is a logic 0, the associated diode conducts holding point (x) at about 0.7 V. This is insufficient to forward-bias both the series diodes and the transistor, so the output is high at logic 1. The output can only go low, 0 if all inputs are at logic 1. The series diodes, usually two, are essential to give the circuit a reasonable value of noise immunity. Noise margin is typically 1.2 V.

One useful feature of DTL gates is that gate outputs can be paralleled to give wired logic functions. This can only be achieved in TTL by using special open collector type gates. However this advantage only results from the fact that DTL has a relatively high output

impedance in the logic 1 state which restricts its drive capability especially when capacitively loaded.

*DTL reference data*

| | |
|---|---|
| Supply voltage | 6 V ± 5% |
| Operating temp. range | 0 to 75°C |
| Available d.c. fan-out | 8 |
| Fan-in | 14 |
| Noise margin: typical | 1.2 V at 25°C |
| worst case | 0.4 V |
| Logic level 1 | 2.3 V min., 6.3 V max. |
| level 0 | 0 V to 0.8 V max. |
| Propagation delay | 31 ns |

Fan out = 6

$C_L = 60$ pF

Average power dissipation 11 mW per gate

## Direct Coupling

This is a method of coupling amplifier stages that uses a direct link between the output of one stage and the input of the next, instead of a coupling capacitor. This method has several advantages when compared with *RC* and trans-former coupling since fewer components are used, resulting in better reliability. It is also important to use as much direct coupling as possible in feedback amplifiers because capacitive coupling introduces phase shifts which may cause instability or oscillations at one particular frequency.

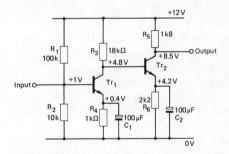

*Figure D33a* Two-stage directly coupled amp-lifier

Naturally in two directly coupled stages (fig. D33a), the d.c. operating point of the first stage is the bias level of the next and so on. This may mean that a higher value of supply voltage is required, and drift of the operating point at the final output may be excessive unless additional stabilising networks are used.

**103**

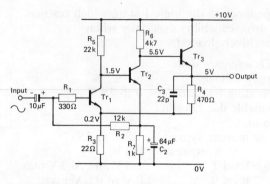

*Figure D33b* **Three-stage wideband amplifier using direct coupling between stages**

Where an amplifier is directly coupled from input through to output, its frequency response will be from d.c. and it will be referred to as a d.c. amplifier.

The circuit of the wideband amplifier shown in fig. D33b is typical of direct coupling methods. $Tr_1$ and $Tr_2$ are common emitter amplifiers and $Tr_3$, an emitter follower, gives a low output impedance. The gain is controlled by the feedback network $R_4$ and $R_3$ to be about 20 (26 dB). The d.c. operating point at $Tr_3$ emitter is held relatively stable by the two feedback networks, the second being d.c. feedback from $Tr_2$ emitter to $Tr_1$ base. The direct coupling used reduces component count and ensures that the amplifier is stable.

▶ *Amplifier*  ▶ *Negative feedback*

## Discriminator

Circuits that can distinguish between one type of input signal and another are called discriminators. The name is applied to several different circuits, namely

*Frequency discriminator* gives an output for a narrow range of input frequencies and rejects signals at other frequences.

*Pulse width discriminator* gives a pulse output when the width of an input pulse lies within a specified value.

*Pulse height discriminator* produces an output, usually a pulse, when the amplitude of the input lies within a specified value.

What is important is that the circuit gives an output only when the input is within the specified limits. For example, in a pulse height discriminator it could be arranged that an output pulse only occurs when an input is within the value of $3\,V \pm 1\,V$. There should be no

output for input pulses with amplitudes of less than 2 V or greater than 4 V.

An arrangement for a discriminator like this is shown in fig. D34. Two Schmitt trigger circuits are used to detect the amplitudes of pulses and in our case the trip voltages would have to be set to 2 V for $V_{R1}$ and 4 V for $V_{R2}$. These two Schmitts define the "window", or channel, about the 3 V level. For an input pulse that falls within this "window", only Schmitt A will give an output. However for an input greater than 4 V ($V_{R2}$) both Schmitts A and B will produce outputs. To avoid the possibility of false outputs caused by the slightly different times at which the Schmitts produce output pulses, a strobe pulse is generated from the leading edge of the input pulse using a short delay circuit and a monostable. As shown in the waveform diagram an output pulse only occurs from the NAND gate when Schmitt A is triggered by an input, i.e. only for pulses that are within the range 2 V to 4 V. Circuits such as this are commonly used in instruments where it may be necessary to select a narrow range of input signal amplitudes from a detector. In this way noise pulses and other pulses at higher levels are rejected.

A pulse width discriminator (fig. D35) can be created using a monostable and a few gates. The input signal is first shaped by a Schmitt to give fast leading and trailing edges, and the leading edge is used to trigger the reference monostable. If the monostable output pulse has a width $p$ that is greater than the width $t$ of the input pulse, then an output pulse will be generated from the lower NAND gate and there will be no output from the upper gate. The reverse occurs if $t$ is greater than p.

## Dissipation

Dissipation is the loss of electrical energy as heat.

Any component in a working circuit that has a voltage across it and a measurable current flowing through it will develop a power loss. This power is the product of the voltage and the current:

$P = VI$ watts for d.c. conditions

$P = VI \cos \phi$ for a.c. conditions (r.m.s. values)

where $\phi$ is the phase angle between the voltage and the current.

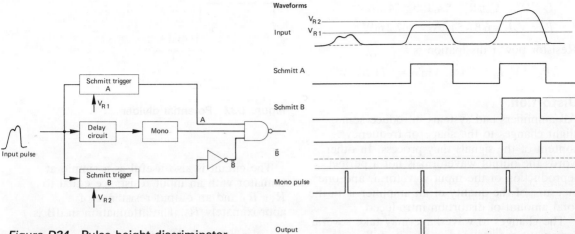

Input

$V_{R2}$
$V_{R1}$

Schmitt A

Schmitt B

$\bar{B}$

Mono pulse

Output

Figure D34  Pulse height discriminator

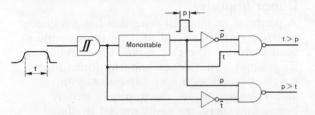

Figure D35  Pulse width discriminator

This power within the component raises its temperature and the excess heat has to be dissipated by radiation or convection. Some components require heat sinks so that the temperature of the component is held at a safe value, and instrument-enclosures will require vents or possibly forced cooling.

Components that exhibit some of the biggest power losses are resistors and semiconductors, especially those used in power output stages. It is obviously important to keep the power wasted to as low a value as possible since this will keep efficiency high, reduce the power required from the power supply, and reduce the temperature of components, thereby allowing a greater packing density. By reducing the working temperature of components the failure

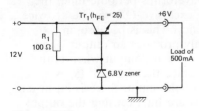

Figure D36

Waveforms

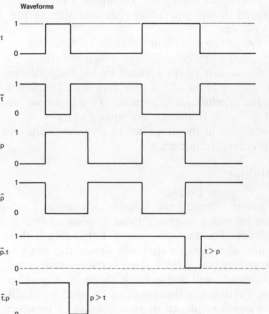

rate will be reduced and the overall equipment reliability enhanced.

▶ *Derating*  ▶ *Heat sink*

**Simple d.c. stabiliser** (fig. D36)
Transistor power dissipation is

$$P_C + P_B = V_{CE}I_C + V_{BE}I_B$$
$$= 6 \times 0.5 + 0.8 \times 0.02$$
$$= 3.016 \, \text{W}$$

Zener power dissipation is maximum when the load is disconnected.

$$I_Z = (V_S - V_Z)/R_1 = 5.4/100 = 54\,\text{mA}$$

$$P_Z = V_Z I_Z = 6.8 \times 54\,\text{mW} = 367.2\,\text{mW}$$

Resistor power dissipation is

$$P_R = V_R I_R = 5.4 \times 54\,\text{mW} = 271.6\,\text{mW}$$

## Distortion

All amplifiers and systems introduce some slight changes to the shape or frequency content of the signals they process. In other words the output waveform is not a perfect reproduction of the input waveform, and one measure of the quality of an amplifier is in the total amount of distortion introduced.

The changes to waveshape may take several forms but are grouped under the following general headings:

*Amplitude distortion*   Harmonic distortion caused by the non-linear characteristics of the amplifier.

*Frequency distortion*   Signals of differing frequency are given different amplifications.

*Phase distortion*   Caused by the amplifier introducing phase shifts that vary with frequency.

*Intermodulation distortion*   Two input signals of differing frequency are mixed by non-linearities in the amplifier to give new sum and difference frequencies.

## Divider

**1** *Frequency divider*

Circuits formed from bistable or counter i.c.s can be wired to give a fixed division of frequency. For example, one J–K bistable with its J and K inputs set to 1 will divide the clock input frequency by 2. Two bistables connected in cascade will divide by 4, three by 8, and so on. Feedback techniques can be used to modify the count sequence to give division of numbers other than pure binary. Examples are given in
▶ *Counters and Dividers.*

**2** *Potential divider*

This is one of the most commonly used networks for biasing transistors and for simple attenuators. It is formed by two resistors as shown (fig. D37). The voltage at the ouput is dependent on the relative values of the two resistors. For the unloaded circuit $R_L \rightarrow \infty$,

$$V_o = V_{in} R_2/(R_1 + R_2)$$

In practice, if $R_L \geq 10R_2$ the error in $V_o$ will be less than 10%.

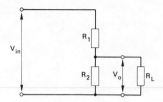

*Figure D37*   Potential divider

The circuit is also useful as a simple attenuator with an input resistance equal to $R_1 + R_2$ and an output resistance of approximately $R_2$. The attenuation in dB is

$$N = 20\log(R_1 + R_2)/R_2$$

## Donor Impurity

A pentavalent material (one that has 5 valence electrons) which is used to dope intrinsic (pure) semiconductor to create n-type semiconductor. Each atom of the donor impurity forms four covalent bonds with four adjacent silicon atoms, leaving a fifth electron only loosely bound and therefore easily moved by the application of an electric field. The electrons introduced by doping pure silicon with a donor impurity give the n-type its low resistivity.

Typical doping levels in practical semiconductors are as low as 1 part in $10^8$. Commonly used donor impurities are: Phosphorus (P), Arsenic (As), Antimony (Sb).
▶ *Semiconductor theory*

## Doubler

**1** *Frequency doubler*

An amplifier with a tuned circuit in the output is set to resonate at the 2nd harmonic frequency of the input signal. In this way the output will be at twice the frequency of the input.

A digital frequency doubler can be created using a CMOS exclusive-OR gate type 4070 and an *RC* network (fig. D38). When the *RC* time constant is set approximately equal to one half the input waveform's periodic time, the output from the exclusive-OR will be at twice the input frequency. This is because an exclusive-OR only provides an output when one input is high and no output when both inputs are high. Since the signal at (x) is a delayed version of the input, there is an instant when both inputs are high causing the output to go low, followed by a short time when the

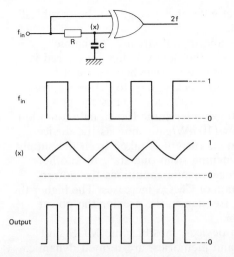

**Figure D38** Frequency doubler using 4070 CMOS Exclusive-OR

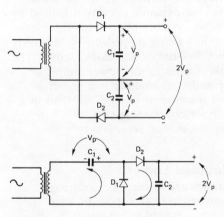

**Figure D39** Voltage doubler

input is low, but the level at (x) is high causing a second output pulse to be generated. If the *RC* time constant is made shorter, short-duration pulses are generated at the gate output for every edge of the input waveform.

**2** *Voltage doubler*
In some applications, such as instruments using c.r.t.s or photomultiplier tubes, relatively high values of d.c. voltages are required. To reduce the size of transformers, voltage doublers (or multipliers) are used to produce the required d.c. voltages. However the load regulation is poor and voltage doublers can only be used to supply relatively low values of load current.

The conventional circuit (fig. D39) is formed from two half-wave rectifier circuits. $D_1$ conducts on the positive half-cycle to charge

$C_1$ to very nearly the peak value of the a.c. secondary voltage ($V_p$). Then on the negative half-cycle, $D_2$ conducts to charge $C_2$, also to nearly $V_p$. The d.c. output voltage is therefore almost $2V_p$.

An alternative voltage doubler is shown which has the advantage that one end of the transformer secondary can be taken to ground. $C_2$ is charged to $V_p$ on the negative half-cycle and this then acts as a "battery", so that $C_1$ is charged to nearly $2V_p$ as $D_2$ conducts on the positive half-cycle of the secondary a.c. voltage.

## Down-time
When a system fails and is inoperative, there is going to be a time delay while the fault is diagnosed, located, and repaired, and before the system is again working. This time interval is often referred to as the down-time and is similar to the Mean Time To Repair (MTTR). Down-time will affect the overall availability of the system which after all is one of the main considerations. Faults may occur but if they can be rectified quickly the availability of the system will be high.
▶ *Availability*    ▶ *Maintenance*

## Drift
**1** Any slow change in some characteristic or a parameter of a device or circuit is referred to as drift. Typically it is used to describe the slow variation of the output of a d.c. amplifier. Such drift is usually caused by changes in temperature and makes the amplification of very small d.c. signals extremely difficult.
**2** *Drift current:* the movement of charge carriers in a semiconductor device under the influence of an applied voltage.

## Duty Cycle
This is the ratio of the working time of a device, when it is on, to the total time in a cycle. When applied to a pulse waveform it is the ratio of pulse width to the periodic time. Duty cycle is quoted either as a ratio or a percentage. [▶ *Mark-to-space ratio*]

*Example:* A switching device is on for 20 $\mu$s, and off for 50 $\mu$s. The duty cycle is

20:70  i.e.  0.286:1  or  28.6%

## Dynamic
This word is often used in a descriptive sense

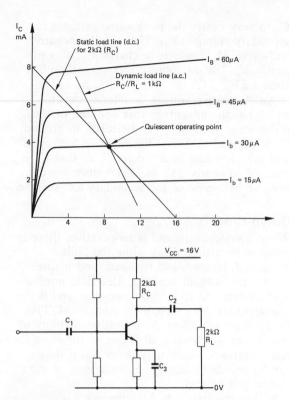

$I_C$ mA

Static load line (d.c.) for 2kΩ ($R_C$)

$I_B = 60\mu A$

Dynamic load line (a.c.) $R_C//R_L = 1k\Omega$

$I_B = 45\mu A$

Quiescent operating point

$I_b = 30\mu A$

$I_b = 15\mu A$

$V_{CC} = 16$ V

2kΩ $R_C$

$C_2$

$C_1$

2kΩ $R_L$

$C_3$

0V

*Figure D40* Dynamic load line for common emitter amplifier

to distinguish a characteristic, a parameter, or a type of device from those which are static. Dynamic, in electronics, means operating or under operating conditions, usually a.c. or pulsed, whereas static is usually taken to indicate d.c. or no-signal conditions. For example, the dynamic resistance of a tuned circuit is the effective resistance of the circuit when it is actually resonating. The inductive and capacitive reactive elements cancel leaving the tuned circuit to appear as a pure resistance. For a parallel tuned circuit the dynamic resistance is

$$r_d = L/CR$$

where $R$ is the series resistance of the coil.

A DYNAMIC CHARACTERISTIC is one which takes into account the working load of the device. In the simple common emitter amplifier stage shown in fig. D40, the d.c. (static) load line is constructed for the collector load of 2 kΩ. However the true gain of the circuit has to include the effect of the additional load $R_L$. The dynamic load line has a slope of $R_C//R_L$ (in this case 1 kΩ) and by using this load line

the correct gain of the stage can be graphically evaluated.

In logic and digital circuits, "dynamic" indicates that the device being considered is operating and being switched, or must be supplied with clock signals to achieve correct operation. In CMOS logic gates the power consumption under static (no-signal) condition is very low (10 nW/gate) but, as the devices are operated, current is taken from the supply during switching from one state to another. Thus under dynamic conditions the power consumption of CMOS increases. The higher the switching frequency, the greater the power dissipation.

Certain devices, mostly some MOS shift registers and memories, are termed dynamic. A DYNAMIC MEMORY is one which must be continuously supplied with clock pulses in order to refresh the contents of its store locations. In this case dynamic operation is used to reduce the power consumption since the memory cells are only being topped-up by relatively low-frequency low-duty-cycle clock signals. The clocks signals constitute the power supply of the memory and consequently less power is used, since it is switched on and off, than in a static memory of comparable size.

▶ *Memory* ▶ *Shift register*

## Emitter Coupled Logic

ECL, sometimes called current mode logic (CML), is one of the fastest types of logic available. It has a typical propagation delay time of only 2 ns. The standard family is the 10 000 series and the basic gate is the OR/NOR shown in fig. E1. The operation of the gate depends on the fact that $Tr_1$, $Tr_2$ and $Tr_3$ form a differential switch and are not allowed to go into saturation. A reference voltage of $-1.29$ V is developed in the i.c. and applied to the bias input of the differential switch circuit ($Tr_3$ base). If the two inputs are at logic 0 ($-1.75$ V) then the current through $R_3$ is supplied by $Tr_3$. This follows because the voltage on $Tr_3$ base at $-1.29$ V is more positive than the base voltages of $Tr_1$ and $Tr_2$. There will be a voltage drop of 0.85 V across $R_2$, and the OR output from emitter follower $Tr_5$ will be $-1.75$ V (logic 0). The emitter voltage of $Tr_6$, the NOR output, will be at $-0.9$ V (logic 1).

If a logic 1 level ($-0.9$ V) is applied to either A or B input. Then $Tr_1$ or $Tr_2$ will conduct,

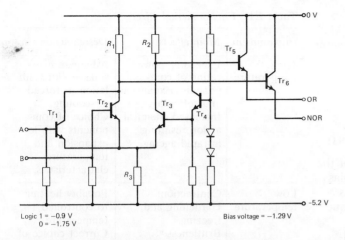

**Figure E1** ECL logic gate

Logic 1 = −0.9 V
0 = −1.75 V          Bias voltage = −1.29 V

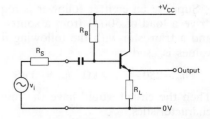

a) Emitter follower circuit

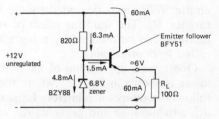

b) Emitter follower in a simple stabilisor circuit

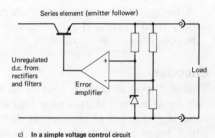

c) In a simple voltage control circuit

**Figure E2** Emitter follower and applications

diverting current from $Tr_3$. A voltage of 0.85 V will be dropped across $R_1$. The NOR output will be −1.75 V (logic 0) and the OR output −0.9 V (logic 1).

The noise margin of ECL is obviously low at about 400 mV but the circuit has the advantage of high fan-out. This is because of the emitter followers. Another advantage over other types of logic is the fact that power supply noise generation is virtually eliminated since the current taken from the supply remains almost constant even when switching takes place.

*Brief ECL data*

| | |
|---|---|
| Propagation delay time | 2 ns |
| Fan-out | 30 |
| Noise margin | 400 mV |
| Power consumption | 60 mW/gate |
| Power supply | −5.2 V |

## Emitter Follower

This is the more usual name given to the common collector connection of a bipolar transistor. It is a circuit (Fig. E2) that has near unity voltage gain (0.98), a moderately high input impedance, and a low output impedance. These features make it a good circuit for matching a high impedance source to a lower impedance load [▶ *Buffer*].

There is no signal inversion between the output on the emitter and the input on the base, and since the voltage gain is close to unity, the emitter "follows" the input, hence the name emitter follower.

Current gain $A_i = \dfrac{1 + h_{fe}}{1 + h_{oe}R_L} \simeq 1 + h_{fe}$

when $h_{oe}R_L \leqslant 0.1$.

Input resistance of transistor, $R_i = h_{ie} + A_iR_L$

Voltage gain $A_v = 1 - \dfrac{h_{ie}}{R_i}$

Output resistance $R_o = \dfrac{1}{Y_o}$

where $Y_o = h_{oe} + \dfrac{1 + h_{fe}}{h_{ie} + R_s}$

$\therefore \quad R_0 \simeq \dfrac{h_{ie} + R_s}{1 + h_{fe}}$

**109**

Suppose an emitter follower is required to drive a load of $500\,\Omega$ from a source of $10\,k\Omega$ and a transistor with the following $h$-parameter values is used:

$$h_{fe} = 250 \quad h_{ie} = 2\,k\Omega \quad h_{oe} = 20\,\mu S$$

Then the circuit would have the following characteristics:

$$A_i = 251 \quad R_i = 127\,k\Omega \quad A_v = 0.984 \quad R_o = 48\,\Omega$$

Note that the overall input impedance of the emitter follower is degraded by the base bias resistor $R_B$. The shunting effect of any bias resistors can be reduced by using a technique called ▶ *Bootstrapping*.

One typical application of an emitter follower is shown in fig. E2b where the emitter follower produces a relatively large current drive to the load and prevents the zener diode from being excessively loaded. The control element in a series regulator of a power supply is also an emitter follower (fig. E2c).

A complementary emitter follower circuit is widely used in class B power output stages.
    ▶ *Amplifier*    ▶ *Darlington*

## Encoder

This is a type of code converter in which only one input is excited at a time and an $n$-bit code is generated at the output depending on which input is activated. A good example of an encoder is a converter that changes decimal to binary or decimal to other coded from.
    ▶ *Code convertor*

## Environmental Stresses

The environment in which an electronic instrument or system operates will have a profound effect on its resulting reliability. The major environmental hazards (fig. E3) are
    a) Extremes of temperature
    b) Rapid changes in temperature
    c) High levels of humidity
    d) Mechanical vibrations and shocks.
In addition the following will also have effect on reliability:
    e) Variations in atmospheric pressure
    f) Corrosive elements in the atmosphere
    g) High levels of radiation ($\gamma$ rays, X-rays, etc)
    h) Attack by fungi and/or insects.
The various effects of temperature and methods to minimise these are shown in Table E1.

**Table E1**  *Environmental effects*

| Environment | Main effects | Design action |
|---|---|---|
| High temperature | Exceeding power rating of components. Expansion and softening. Increased chemical action resulting in rapid ageing. | Adequate heat sinks and/or ventilation or forced air-cooling. Choice of components with low expansion and temperature characteristics. |
| Low temperature | Contraction. Hardening and freezing. Brittleness. Loss of gain and efficiency. | Possibly heating to controlled temperature. Correct choice of materials and components. |
| Temperature cycling | Severe stressing. Fatigue failure. | Introduction of large thermal delays to prevent rapid changes affecting internal components. |

### Humidity
The main effects of high levels of humidity are
    1) To reduce values of insulation resistance leading to possible electrical breakdown.
    2) Corrosion resulting from moisture forming an electrolyte between dissimilar metals.
    3) Promotion of fungi growths which reduce insulation.
The worst sort of conditions are those of high temperatures together with high levels of humidity as is common in the tropics.

To minimise the effects, insulating materials must be used that do not absorb moisture or support a water film and if possible resist growth of fungi. Typical materials that do not absorb moisture are silicones and polystyrene.

Obviously the best and most effective method is to hermetically seal (totally close) any sensitive components or in some cases whole assemblies or instruments.

### Mechanical vibrations and shocks
All instruments when transported and handled (even when moved from one bench to another) will experience some degree of mechanical shock and vibration. The effects are to weaken supports, loosen wires and connections, bend and possibly fracture components, and to set up stresses that lead to fatigue failure. All of

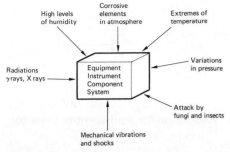

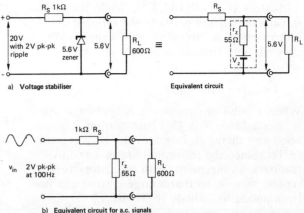

a) **Voltage stabiliser**     **Equivalent circuit**

b) **Equivalent circuit for a.c. signals**

*Figure E4*   Equivalent circuits

**Figure E3**   Environmental hazards

these can be minimised by careful design techniques such as using anti-vibration mountings, shake-proof washers, locking nuts and varnishes, and by the encapsulation of sensitive components in some protective material. Silicone rubber compounds are often used to "pot" assemblies to provide an absorber for the mechanical energy caused by vibrations and shocks.

### Variations in pressure

Atmospheric pressure variations are important at high altitudes where the pressure falls to a low value, and can cause leaking of seals in components and sub-assemblies. This can occur if equipment is transported by air in an unpressurised hold. In working instruments, low pressure causes a decrease of electrical breakdown voltage between contacts using air insulation. Therefore for equipment subjected to low pressure the distances between electrical conductors have to be increased, and care must be taken in maintaining a dust and dirt-free path.

### Epitaxial

Expitaxial means "grown out from" or "grown on top of" and is the thin layer of lightly doped silicon (usually n-type) that is deposited onto the silicon slice as one of the first process steps in making both discrete and integrated circuit semiconductor devices. The slice is usually heavily doped n-type and is typically 0.3 mm thick. The epitaxial layer, between 7 $\mu$m to 12 $\mu$m thick, is deposited onto the top surface by exposing the slice to an atmosphere of silicon-tetrachloride at a temperature of about 1200°C. It is in the epitaxial layer that all the active and passive elements in an i.c. are created.

    ▶ *Planar process*

## Equivalent Circuit

Analysis of many circuits, particularly linear amplifiers, is made easier by redrawing the circuit without bias components and replacing any active devices by suitable networks of impedances and/or admittances and generators. This gives an equivalent circuit which behaves in almost the same way as the original for a limited frequency range. By using equivalent circuits the operation and limitations of a circuit or a device can be understood and proper evaluation of a circuit's performance can be made.

Before considering an amplifier, take the relatively simple example of the voltage stabiliser shown in fig. E4. The zener diode can be replaced by its equivalent circuit of a battery with a voltage of $V_z$ and an internal resistance $r_z$. The resistance $r_z$ is the zener's characteristic resistance when it is operated in its breakdown region. To calculate the a.c. component at the output, the equivalent circuit of the potential divider formed by $R_S$ and $r_z$ can be used since the battery can be assumed short circuit to a.c. signals. Note that the effect of $R_L$ has been neglected since $R_L \gg r_z$.

    Output ripple = $V_{in}r_z/(R_S + r_z)$

where $V_{in}$ is 2 V pk-pk at 100 Hz.

For a BZY88 5.6 V zener, $r_z$ is 55 $\Omega$ at 5 mA. Therefore

    Output ripple = 104 mV pk–pk

Another use of equivalents is in describing the way in which a more complex device operates. A thyristor (SCR), for example, can be more easily understood by considering it as two

transistors as shown (fig. E5). While the anode to cathode voltage is below the forward breakover value, and if no gate signal is applied, both transistors remain in a high resistance state.

$$I = I_{co}/[1 - (h_{FB1} + h_{FB2})]$$

When a pulse of gate power is applied, current is injected into $Tr_2$. This base current is amplified and therefore increases the base current to $Tr_1$. Since the transistors are in a positive feedback arrangement, both transistors very rapidly turn on so that a large current can flow from anode to cathode [▶ *Thyristor*].

One of the main uses of equivalent circuits is in the analysis of amplifiers. Take the basic common emitter amplifier shown in fig. E6a. At medium frequencies, say 1 kHz, the following points are valid:

*a*) The three capacitors will have very low values of reactance and can be considered as short circuits.

*b*) The positive supply line $+V_{CC}$ will have a low impedance to the 0 V line and can be considered as the same point as 0 V.

*c*) The bias resistor $R_B$ can be neglected since it has a high value compared to the input impedance of the transistor.

Thus as a first step the circuit can be redrawn as in (*b*) to be equivalent to the original at a signal frequency of 1 kHz. Resistor $R_B$ has been omitted, $R_E$ is shorted out by $C_2$, and resistors $R_C$ and $R'_L$ appear in parallel to give the true a.c. load of $R_L$.

$$R_L = R_C R'_L/(R_C + R'_L)$$

The next step is to replace the transistor itself with an equivalent network of impedances and generators. There are several networks that can be chosen:

T equivalent circuit
Hybrid-$\pi$ equivalent circuit
$h$-parameter equivalent circuit
Y-parameter equivalent circuit
Z-parameter equivalent circuit

The one most commonly used for small signal analysis at medium frequencies of a BJT is the $h$-parameter circuit. One of the main reasons for this is the fact that the $h$-parameters are fairly easily measured at 1 kHz and values are specified on most transistors' data sheets. The complete equivalent circuit of

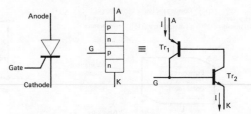

Figure E5    Two-transistor equivalent of thyristor

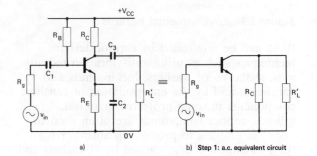

a)                                b) Step 1: a.c. equivalent circuit

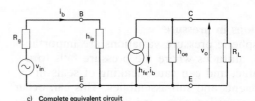

c)   Complete equivalent circuit

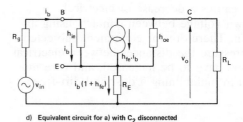

d)   Equivalent circuit for a) with $C_2$ disconnected

Figure E6    Equivalent circuit of basic common emitter amplifier

the CE amplifier is shown in (*c*). The input impedance of the transistor, between its base and emitter, is represented by the parameter $h_{ie}$; the a.c. collector current is shown as a current generator $h_{fe}i_b$; and the output admittance $h_{oe}$ is shown connected between collector and emitter. The reverse feedback voltage ratio $h_{re}$ is usually very small and has been left out of this circuit.

Using the final equivalent circuit we can obtain formulae for the input and output impedance and the current and voltage gains

of the amplifier. Suppose we wish to get an expression for voltage gain:

$$A_v = V_o/V_{in}$$

Now $\quad V_{in} = i_b\,(R_g + h_{ie})$

and $\quad V_o = V_{ce} = \dfrac{-h_{fe}i_bR_L}{1 + h_{oe}R_L}$

$$\therefore \quad A_v = \dfrac{-h_{fe}i_bR_L}{i_b(1 + h_{oe}R_L)(R_g + h_{ie})}$$

$$= \dfrac{-h_{fe}R_L}{(1 + h_{oe}R_L)(R_g + h_{ie})}$$

If $h_{oe}R_L \leqslant 0.1$ and $h_{ie} \gg R_g$, then

$$A_v = -h_{fe}R_L/h_{ie} \quad \text{(this is the same as } -g_mR_L\text{)}$$

The original circuit could be modified to give series negative feedback by disconnecting $C_2$. This would give an a.c. equivalent circuit as shown in $(d)$. $R_E$ must now be included since there is no longer a short circuit a.c. condition from emitter to 0 V. The current flowing through $R_E$ is $i_b(1 + h_{fe})$ and this sets up the negative feedback voltage that opposes the input signal.

$$V_{in} = i_b[R_g + h_{ie} + R_E(1 + h_{fe})]$$

If $h_{fe} \gg 1$ and $R_g$ is small, then

$$V_{in} \simeq i_b(h_{ie} + h_{fe}R_E)$$

The input impedance is now

$$Z_{in} = h_{ie} + h_{fe}R_E$$

which is much higher than in the original circuit.

$$A_v = \dfrac{V_o}{V_{in}} = \dfrac{-h_{fe}i_bR_L}{i_b(h_{ie} + h_{fe}R_E)}$$

$$\therefore \quad \text{Voltage gain} \simeq \dfrac{-h_{fe}R_L}{h_{ie} + h_{fe}R_E}$$

If $h_{fe}R_E \gg h_{ie}$, then $A_v = -R_L/R_E$.
In other words the voltage gain with sufficient negative feedback is dependent only on the values of the resistors [▶ *Negative feedback*].

Amplifiers using FETs can also be analysed by using suitable equivalent circuits. A common source amplifier will have an equivalent circuit for small signals at medium frequencies as shown in fig. E7. Here the FET has been replaced by its y-parameter equivalent circuit. The input admittance is $y_{is}$; the a.c. drain current is represented by a current generator

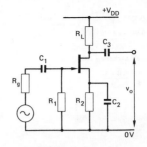

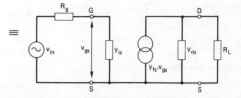

**Figure E7** Equivalent circuit of common source amplifier

$y_{fs}V_{gs}$; and the output admittance by $y_{os}$. The reverse feedback admittance $y_{rs}$ is usually very small and has been omitted. A similar analysis will give the voltage gain of the FET stage as

$$A_v = \dfrac{-y_{fs}R_L}{(1 + y_{os}R_L)}$$

There are several other useful equivalent circuits. The hybrid-$\pi$ for example is particularly useful for the analysis of transistor circuits at high frequencies.
▶ *Amplifier* ▶ *Parameter* ▶ *Transistor*

## Exclusive-OR Gate

This is one of the logic gates commonly used in digital circuits (fig. E8). It is usually a 2-input gate and it gives a logic 1 at its output if either input is at logic 1 but not when both inputs are logic 1 simultaneously. The Boolean expression for this is

$$F = A \cdot \bar{B} + \bar{A} \cdot B$$

*Truth table*

| A | B | F |
|---|---|---|
| 0 | 0 | 0 |
| 0 | 1 | 1 |
| 1 | 0 | 1 |
| 1 | 1 | 0 |

It can be seen from the truth table that the

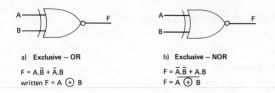

a) Exclusive — OR

$F = A.\bar{B} + \bar{A}.B$

written $F = A \oplus B$

b) Exclusive — NOR

$F = \bar{A}.\bar{B} + A.B$

$F = A \overline{\oplus} B$

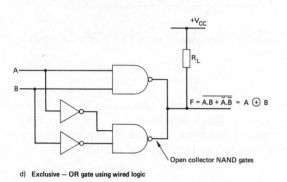

c) Exclusive — OR gate using NAND gates

$F = \overline{\overline{A.B} \cdot \overline{B.A}} = A \oplus B$

d) Exclusive — OR gate using wired logic

$F = \overline{A.B} + \overline{A.B} = A \oplus B$

Open collector NAND gates

**Figure E8**   Exclusive-OR gate

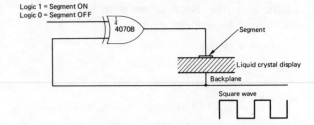

Logic 1 = Segment ON
Logic 0 = Segment OFF

¼ 4070B

Segment

Liquid crystal display

Backplane

Square wave

*Figure E9*   Use of an Exclusive-OR gate to give a.c. drive to a liquid crystal display

output is logic 1 if one and only one input is in the logic 1 state.

The exclusive-OR function can be created from other logic gates; two examples are shown using NAND gates and also wired logic. However it is also readily available in an i.c. package, the most commonly used types being the CMOS 4070 and the TTL 7486. Each of these i.c.s contain four independent exclusive-OR gates.

The special features of the gate allow it to be used in a number of interesting applications. It can be wired as a frequency doubler or used as the drive to liquid crystal displays (fig. E9). These displays require an a.c. drive for long life. A square wave is applied to the back plane of the display and also to one input of the exclusive-OR gate. When the other input to the gate is held at logic 0, the ouput of the gate will follow this square wave with the result that both segment and back plane signals

are the same, giving no drive to the display. When a logic 1 is applied to the other gate input, the gate output becomes a complement of the square wave drive, in other words it is high while the back plane is low and vice versa. This gives a.c. drive to the display to switch on the segment. In this application the gate is being used as a controllable invertor; a low on one input allows data or information on the other input to appear unchanged at the output, while a logic 1 complements the information on the other input. This is useful for DATA SCRAMBLERS when information can be transmitted down a line in coded form (fig. E10a).

One of the main uses of exclusive-OR gates is in comparator circuits (fig. E10b). With an exclusive-OR, identical inputs give a logic 0 output and different inputs a logic 1. Several gates can be wired to compare two binary words.

By following an exclusive-OR gate by an invertor we get the exclusive-NOR:

$$F = \bar{A} \cdot \bar{B} + A \cdot B$$

| A | B | F |
|---|---|---|
| 0 | 0 | 1 |
| 0 | 1 | 0 |
| 1 | 0 | 0 |
| 1 | 1 | 1 |

The CMOS 4077 is an i.c. that has four exclusive-NOR gates. This type of gate is also used as a comparator. It give a logic 1 output if both inputs are the same and a logic 0 if both inputs are different.

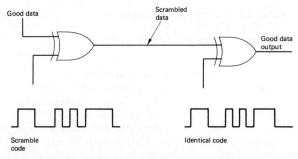

Figure E10a   Exclusive-OR gates as data scramblers

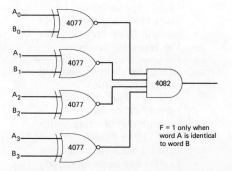

F = 1 only when word A is identical to word B

Figure E10b   Using a quad Exclusive-NOR i.c. to compare two 4-bit binary words

## Fault Diagnosis

Electronic systems are made up from many varied interconnected component parts. Each of these components has a vital part to play in the operation of the whole and, if any one of them should fail, then probably the whole system fails.

Fault diagnosis is the skill in being able to rapidly locate a faulty component within a system and it involves a good understanding and knowledge of
*a*) The system operation
*b*) The way in which the components operate and their most likely failure mode
*c*) The reliability of various components
*d*) Systematic fault location methods
*e*) Testing methods.

Table F1 indicates the more probable types of failure for various electronic components and Tables F2 and F3 show in more detail the type and causes of failure for resistors and capacitors.

## Failures in semiconductors

Semiconductor devices can be broadly classified into two areas:

1) *Bipolar*  Transistors
   Diodes, Unijunctions
   Thyristors
   Logic i.c.s such as TTL, ECl
   Linear i.c.s
2) *Unipolar*  Junction FETs
   MOSFETS
   VMOS power FETs
   CMOS logic
   Some linear i.c.s

What the two groups have in common is that they can easily be destroyed if subjected to an overload, and also the fact that the manufacturing processes are very similar.

Failure mechanisms inherent in the manufacturing process include:

*Diffusion and epitaxial growth processes*
*a*) Flaws and imperfections in basic crystal slice
*b*) Incorrect resistivity
*c*) Incorrect diffusion
*d*) Contamination
*e*) Epitaxial "washout" of pattern or "spike" formation.

*Photo-resist processes*
*a*) Mask imperfections, scratches or pinholes
*b*) Poor mask alignment
*c*) Poor mask definition
*d*) Contamination of photo-resist
*e*) Insufficient cleaning or etching.

*Metallisation processes*
*a*) Poor ohmic contact
*b*) Metal adhering badly or too thin
*c*) Microcracks or voids over oxide steps.

*Mechanical processes*
*a*) Chipping, cracking or fracture during scribe and break
*b*) Weak bonds, sagging leads, microcracks in leads during wire bonding
*c*) Poor seal.

Many of these defects will be picked up during manufacture or by "burn-in". However, some units, whether discrete or i.c., may possess inherent weaknesses from any of these sources when delivered and may fail prematurely in equipment.

Of much greater importance are failures caused by misuse, by bad handling during assembly or test, or by exceeding the maximum rated values of voltage current and power.

**Table F1**  *The more probable types of failure for various types of electronic component.*

| Component | Common type of fault |
|---|---|
| Resistors | High in value or open circuit. |
| Variable resistors | Open circuit or intermittent contact resulting from mechanical wear. |
| Capacitors | Open or short circuit. |
| Inductors (including transformers) | Open circuit. Shorted turns. Short circuit coil to frame (iron cored types). |
| Thermionic valves | Filament open circuit. Short circuited electrodes (i.e. grid to cathode). Low emission. |
| Semiconductor devices Diodes, Transistors, FETs, SCRs etc. | Open or short circuit at any junction. |

**Table F2**  *Failures in fixed resistors*

| Resistor type | Failure | Possible cause |
|---|---|---|
| Carbon composition | High in value | Movement of carbon or binder under influence of heat, voltage or moisture. Absorption of moisture causes swelling, forcing the carbon particles to separate. |
| | Open circuit | Excessive heat burning out resistor centre. Mechanical stress fracturing resistor. End caps pulled off by bad mounting on a circuit board. Wire breaking due to repeated flexing. |
| Film resistors (Carbon, metal oxide, metal film, metal glaze) | Open circuit | Disintegration of film with high temperature or voltage. Film may be scratched or chipped during manufacture. With the higher values (greater than 1 M$\Omega$) the resistance spiral has to be thin and therefore open circuit failure is more likely. |
| | High noise | Bad contact of end connectors. Usually the result of mechanical stress caused by poor assembly on a circuit. |
| Wire-wound | Open circuit | Fracture of wire, especially if fine wire is used. Progressive crystallisation of wire because of impurities—leads to fracture. Corrosion of wire due to the electrolytic action set up by absorbed moisture. Failure of welded end connection. |

Most i.c.s for example will be permanently damaged if the maximum supply voltage is exceeded or if the device is removed from or inserted into a test socket while the power is applied.

Apart from the main environmental hazards, electrical interference is another cause of premature failure in semiconductors. Voltage surges carried along the mains leads, caused by heavy machines or relays being switched, can easily cause breakdown of semiconductor junctions.

Failures are mostly those of an open or short circuit at a junction. In other words a bipolar may fail short or open between base and emitter or collector and base. In addition it is possible for a short circuit to occur between collector and emitter. With digital i.c.s it is usually only possible to identify which gate in the package has failed, but accuracy is important since good fault reporting to the design authority can help to reduce future failures.

The section on ▶ *Maintenance* deals with the location of faults in systems.

## Feedback

Feedback exists in an electronic circuit when a portion of the output signal is returned to the circuit's input. If the resulting feedback signal is 180° out of phase with the original input, so that it opposes this input, then the effective signal to the circuit is reduced and we have

**Table F3** *Failures in capacitors*

| Capacitor type | Failure | Possible cause |
|---|---|---|
| Paper foil | a) Loss of impregnant leading to eventual short circuit. b) Intermittent or open circuit. | a) Seal leak. Mechanical, thermal shock or variations in pressure. b) Damaged during assembly or mechanical/thermal shock. |
| Ceramic | a) Short circuit. b) Open circuit. c) Fluctuations in capacitance value | a) Fracture of dielectric from shock or vibration. b) Fracture of connection. c) Silver electrode not completely adhering to ceramic. |
| Plastic film | Open circuit. | Either damage to end spray during manufacture or poor assembly. |
| Aluminium electrolytic | a) Leaky or short b) Fall in capacitance value c) Open circuit. | a) Loss of dielectric. High temperature. b) Loss of electrolyte via leaking seal caused by pressure, thermal or mechanical shock. c) Fracture of internal connections. |
| Mica | a) Short circuit. b) Intermittent or open circuit. | a) Silver migration caused by high humidity. b) Silver not adhering to mica. |

**Table F4** *Feedback characteristics*

| Negative feedback | Positive feedback |
|---|---|
| $A_c = A_0/(1 + A_o\beta)$ If $A_o\beta \gg 1$ then $A_c \simeq 1/\beta$ *Main effects* Gain reduced and stabilised. Improved frequency response with wider bandwidth. Noise and distortion (internally generated) reduced. Method of applying feedback can modify input and output impedances. | $A_c = A_o/(1 - A_o\beta)$ If $A_o\beta \to 1$ then $A_c \to \infty$ *Main effects* Gain increased with less stability. If $A_o\beta \to 1$, oscillations likely at one particular frequency. |

negative feedback. The majority of amplifiers and control systems use negative feedback to provide stabilisation of the overall gain and to give good frequency response.

Positive feedback occurs if the feedback signal is in phase with the original input. It therefore enhances the effective input and causes an increase in gain. Positive feedback is used in many oscillator circuits.

The general formula for feedback is

$$A_c = A_o/(1 \pm A_o\beta)$$

where $A_c$ is referred to as the "closed loop gain" and is the resulting gain when feedback is applied,

$A_o$ is the "open loop gain", i.e. the gain before feedback is applied,

$\beta$ is the fractional gain of the feedback network.

The product $A_o\beta$ is called the *loop gain*.
▶ *Negative feedback*  ▶ *Oscillator*
▶ *Amplifier*  ▶ *Bode plot*

## Field Effect Transistor (FET)

FETs are an important family of active devices that are now used extensively in linear and digital circuits. The name "field effect" is used because the current flowing through a FET device is controlled by the electric field across its input terminals. Very little input current is required and therefore all types of FET have a very high input resistance. They are also sometimes referred to as unipolar devices because the current flow consists of only one type of charge carrier, that is only holes for a p-channel and only electrons for an n-channel device.

The various types of FET are

1) Junction field effect transistors (abbrev. JFET, JUGFET, or simply FET).

2) Metal-oxide-silicon field effect transistor (abbrev. MOSFET or MOST). Another name commonly used for a MOSFET is insulated-gate field-effect transistor (IGFET).

3) Power FETs such as the VMOS. This is a vertical metal-oxide-silicon power field effect transistor.

Each type is available as a p or n channel and the MOSFETS can be constructed as either *Depletion* or *Enhancement* mode devices.

## 1 *JFET operation*

The construction and typical characteristics for an n-channel JFET are shown in fig. F1. Modern devices are nearly all silicon and are manufactured using the planar technique. The source and drain connections to the n-channel are made via heavily doped low resistance $n^+$ regions and the gate is connected to the top p-channel and to the substrate. Since the gate is p material, the arrow on the symbol for an n-channel JFET points inward.

During operation, the source is at or very near zero volts, the drain is positive, and the gate is negative with respect to the source. It is the reverse voltage between gate and source that controls the flow of electrons from source to drain. Maximum drain current $I_D$ will flow when $V_{GS}$ is zero and $V_{DS}$ is above what is called pinch-off voltage $V_p$. This value of current is quoted in most FET short form data as $I_{DSS}$, which stands for the value of drain current in common source mode when the gate is short circuited to the source. A typical value is only a few milliamps (BFW10: $I_{DSS} = 8\,\text{mA}$). The reason the current is limited to a relatively low value is partially due to the fairly high resistance of the lightly doped n-channel and also because, even with $V_{GS}$ at zero, a depletion region exists between the p and n regions. As $V_{DS}$ is increased positively, the depletion regions nearest the drain enlarge and this accounts for the so called pinch-off effect. Below pinch-off ($V_p$) the channel acts as a resistor and the characteristics here are referred to as the ohmic region since drain current increases linearly with drain to source voltage. However when $V_{DS}$ reaches a value equivalent to $V_p$, the characteristics flatten out and very little increase in drain current takes place. Pinch-off

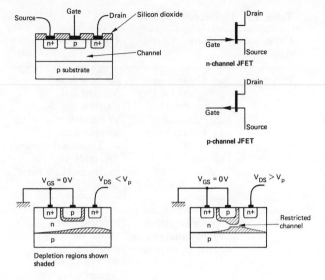

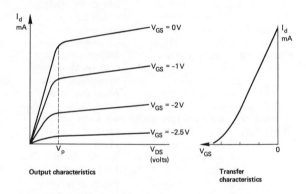

Operation to illustrate pinch-off effect

**Figure F1** JFET construction and typical characteristics (common source)

has the effect of restricting the width of the channel nearest the drain connection as the depletion regions increase and therefore prevents any large increase of drain current as drain to source voltage is increased. Naturally if $V_{DS}$ is increased beyond a certain value, breakdown will occur so all FETs have a specified maximum value of $V_{DS}$.

If $V_{GS}$ is made slightly negative, say $-1\,\text{V}$, then the depletion regions are increased further and the channel carrying the current gets narrower. The drain current therefore levels off at a lower value than when $V_{GS}$ is zero. A family of characteristics for various values of $V_{GS}$ are shown. Note that when $V_{GS}$ is made sufficiently

negative the depletion regions meet and the drain current is cut off completely.

The important parameters of a FET in common source mode are:

$y_{fs}$ the forward transfer admittance
$y_{os}$ the output admittance.

$$y_{fs} = \frac{\delta i_d}{\delta V_{gs}}\bigg|_{\text{with } V_{ds} \text{ constant}} \quad \begin{array}{l} \text{units: milliSiemens} \\ \text{(mA/volt)} \end{array}$$

$$y_{os} = \frac{\delta i_d}{\delta V_{ds}}\bigg|_{\text{with } V_{gs} \text{ constant}} \quad \text{units: microSiemens}$$

At low frequencies

$y_{fs} \equiv g_{fs}$ the forward transconductance $g_m$

and $\quad y_{os} \equiv g_{os}$ the output conductance

$$\therefore \quad i_d = g_{fs} V_{gs} + \frac{1}{r_d} V_{ds}$$

where $r_d = 1/g_{os}$ is the drain output resistance.

In amplifier applications the JFET is normally operated in the region beyond pinch-off. Here the value of drain current flowing depends upon $V_{GS}$ and is given by

$$I_{DS} = I_{DSS}\left(1 - \frac{V_{GS}}{V_p}\right)^2$$

Thus $\quad g_{fs} = g_{fso}\left(1 - \frac{V_{GS}}{V_p}\right)$

It also follows that

$$g_{fso} = 2I_{DSS}/V_p$$

Here $g_{fso}$ is the value of forward transfer conductance when $V_{GS}$ is zero.

At low frequencies the equivalent circuit of a JFET is as shown (fig. F2). Note that $y_{is}$ the input admittance and the current generator $y_{rs}V_{ds}$ have been omitted since their effects can be neglected.

The voltage gain for a resistive load is

$$A_v = -\frac{y_{fs}R_L}{1 + y_{os}R_L}$$

and if $y_{os}R_L \leqslant 0.1$, then $A_v = -y_{fs}R_L$ or $g_m R_L$.
▶ *Amplifiers*  ▶ *Equivalent circuit*

At higher frequencies the effects of device capacitance become noticeable and $C_{gs}$, $C_{gd}$ and $C_{ds}$ must be added to the equivalent circuit. It is $C_{gd}$, the feedback capacitance between drain and gate, that can have the most limiting effect on high-frequency

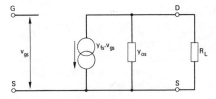

a) **Small-signal equivalent circuit (common source)**

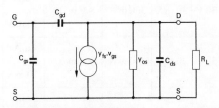

b) **High frequency equivalent circuit**

*Figure F2*  **JFET equivalent circuits**

response, since it is amplified by the gain of the circuit and appears in parallel with $C_{gs}$ as a much larger value:

$$C_{in} = C_{gs} + C_{gd}(1 - A_v)$$

On the data sheet, gate input capacitance is specified as $C_{iss}$ and the feedback capacitance as $C_{rss}$. Table F5 shows data for a few popular JFETs, which shows the spread to be expected on important parameters such as $I_{DSS}$ and $Y_{fso}$.

Applications of JFETs are those that utilise either of the following properties:

a) Its behaviour as a voltage-controlled resistor when $V_{DS}$ is below pinch-off. This makes the device useful as a gain controlling element.

b) Its high input resistance compared to the bipolar transistor. This makes it a useful amplifying device from high-resistance sources.

FETs are also ideal switching devices for use in chopper-type amplifiers, analogue switches, and sample and hold circuits. The incremental channel resistance $r_{ds}$ depends upon both $V_{GS}$ and $V_{DS}$. When $V_{GS}$ is large, the JFET is off and the channel resistance is very high ($>10^9 \, \Omega$), but when $V_{GS}$ is at zero the JFET is on with a channel resistance of only tens of ohms. The switching speed of a JFET is determined primarily by the external circuit elements and the device capacitance since there is no minority charge storage effect as in bipolar transistors.
▶ *Bias*

**Table F5**  *JFET data*

| Type | p/n | $V_{DS}$ max. | $I_D$ max. | $P_{TOT}$ | $V_p$ | $I_{DSS}$ | $Y_{FSO}$ | $C_{iss}$ | $C_{rss}$ | $I_{GSS}$ max. | Case | Application |
|---|---|---|---|---|---|---|---|---|---|---|---|---|
| BF244A | n | 30 V | 10 mA | 300 mW | 0.5 V min. to 4.8 V max. | 2 mA to 6.5 mA | 3 mS to 6.5 mS | 8 pF | 2 pF | 5 nA | T092 | R.F. amplifiers |
| BFW10 | n | 30 V | 20 mA | 300 mW | 8 V max. | 8 mA to 20 mA | 3.5 mS to 6.5 mS | 5 pF | 0.8 pF | 0.5 nA | T072 | Wideband amplifiers |
| BFW11 | n | 30 V | 20 mA | 300 mW | 6 V max. | 4 mA to 10 mA | 3 mS to 6.5 mS | 5 pF | 0.8 pF | 0.5 nA | T072 | Wideband amplifiers |
| 2N3819 | n | 25 V | 20 mA | 360 mW | 8 V max. | 2 mA to 20 mA | 2 mS to 6.5 mS | 8 pF | 4 pF | 2 nA | T092 | General-purpose amplifiers |
| 2N3820 | p | 20 V | 15 mA | 360 mW | 8 V max. | 0.3 mA to 15 mA | 0.8 mS to 5 mS | 32 pF | 16 pF | 20 nA | T092 | General-purpose amplifiers |
| 2N5460 | p | 40 V | 16 mA | 310 mW | 0.75 V to 6 V | 1 mA to 5 mA | 1 mS to 4 mS | 7 pF | 2 pF | 5 nA | T092 | General-purpose amplifiers |

$I_{GSS}$ is gate reverse current with drain shorted to source and with $V_{GS} = V_p$.

## 2  *MOSFET operation*

The MOSFET (fig. F3$a$) is constructed with the gate insulated from the main body of the device by a thin region of silicon dioxide. This is typically less than 1 $\mu$m thick and serves to isolate the gate from the substrate giving a very high input resistance. For the n-channel as shown, the substrate is p-type and the source and drain connections are made to the low-resistance n$^+$ regions. The substrate is connected to the source. The operation of the ENHANCEMENT MODE MOSFET differs completely from a JFET in that a bias voltage has to be applied between gate and source to allow a conducting channel to be set up between source and drain. With $V_{GS}$ at zero, there is no current flow and the resistance between drain and source is very high ($>10^{10}\,\Omega$). This is because the material between the source and drain is the p-type substrate. However if, with $V_{DS}$ positive, a positive gate to source voltage is applied, then the free electrons in the substrate are attracted to the area just beneath the gate. A thin n-type conducting channel, called the inversion layer, is set up between source and drain allowing drain current to flow. To enable a reasonable

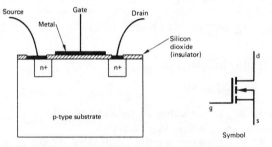

*Figure F3a*   n-channel enhancement-mode MOSFET

conduction to take place, $V_{GS}$ must be above $V_T$, the gate threshold voltage.

The characteristics of a MOSFET also show a pinch-off effect, again due to higher values of $V_{DS}$, causing a restriction of induced channel width of the drain end of the channel. A family of characteristics can be drawn showing that more current flows for a given value of $V_{DS}$ as $V_{GS}$ is increased. In other words drain current is enhanced by gate to source voltage. Enhancement MOSFETs are used extensively in digital and linear i.c.s.

▶ *CMOS*      ▶ *Charge coupled device*
▶ *Memory*

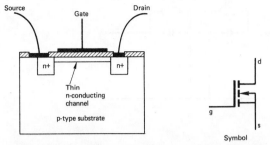

**Figure F3b** n-channel depletion-mode MOSFET

The depletion mode MOSFET (fig. F3*b*) is more usually made as a discrete component. During manufacture a thin conducting channel is diffused in beneath the gate so that current flow will take place with $V_{GS}$ equal to zero. The gate is insulated from the body of the device by a thin layer of silicon dioxide so that very high input resistance is achieved. For the n-channel shown, a negative gate bias will reduce the conducting channel so that drain current falls, giving depletion mode operation. Also by making $V_{GS}$ positive the channel width can be increased to give more drain current, i.e. enhancement mode operation. One of the more useful depletion mode MOSFETs is the dual gate type which enables relatively simple r.f. amplifiers to be constructed since the feedback capacitance is very low. Effectively the circuit is a cascode. Dual gate MOSFETs are also used for mixer stages.

*Brief data on one dual gate depletion mode MOSFET*

| Type | | $V_{GS}$ max. | $V_{DS}$ max. | $I_{D(on)}$ | $P_{tot}$ | $Y_{fso}$ | $C_{rss}$ |
|------|---|-----|------|-------|--------|------|--------|
| 3N201 | n | 5 V | 25 V | 6 mA to 30 mA | 360 mW | 8 mS to 20 mS | 0.03 pF |

VMOS and HEXFET devices are the latest additions to the FET range and are basically power semiconductors with similar action to enhancement mode MOSFETs. Devices are available that will switch several amps in fractions of a microsecond with working voltages of several hundred volts. Their operation is explained fully in the section on VMOS.

## Filter

A filter is a network of impedances or a circuit that allows signals within a certain specified range of frequencies to pass unaltered while

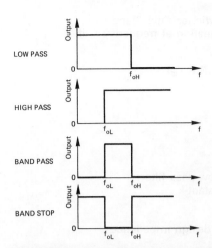

**Figure F4a** Ideal filter characteristics

highly attenuating signals at all other frequencies.

There are four main types of filter with the following "ideal" characteristics (fig. F4):

**1** LOW PASS   Passes all frequencies below cut-off with zero attenuation but provides infinite attenuation to signals with frequencies above cut-off.

**2** HIGH PASS   Passes all frequencies above cut-off with zero attenuation but provides infinite attenuation to signals with frequencies below cut-off.

**3** BAND PASS   Passes all frequencies within a limited band with zero attenuation but provides infinite attenuation to signals with frequencies above and below the pass band.

**4** BAND STOP   Gives infinite attenuation to signals with frequencies within a certain band but allows signals with frequencies above and below this band to pass with zero attenuation.

Simple *RC* networks can provide basic filters and some circuits such as the twin-tee, used for a band stop, can produce excellent results. Since op-amps are so readily available an ACTIVE FILTER is often a better alternative to the sections using *L* and *C* shown in the diagrams.

## Flip-Flop

This is simply another commonly used name for a bistable. In a circuit diagram, bistables will often be marked with a reference such as $FF_1$, $FF_2$, and so on. Flip-flops form the basic memory elements used in counters, shift registers and stores.

▶ *Bistable*

*Figure F4b* Twin-tee filter (band stop)
Infinite attenuation at frequency $f_c$ if

$C_1R_1 = 4C_2R_2$

$f_c = 1/2\pi\sqrt{(R_1R_2C_1C_2)}$

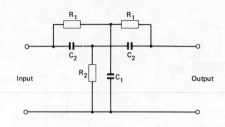

*Figure F4c* Low-pass filter;
prototype constant-k

$L = R_0/\pi f_c$

$R_0 = \sqrt{(L/C)}$, design impedance

$C = 1/\pi R_0 f_c$

$f_c = 1/\pi\sqrt{(LC)}$, cut-off frequency

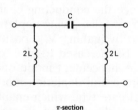

*Figure F4d* High-pass filter;
prototype constant-k

$L = R_0/4\pi f_c$

$R_0 = \sqrt{(L/C)}$

$C = 1/4\pi R_0 f_c$

$f_c = 1/4\pi\sqrt{(LC)}$

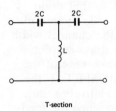

*Figure F4e* Band-pass filter;
prototype constant-k

$f_2$ = upper cut-off frequency

$f_1$ = lower cut-off frequency

$L_1 = R_0/\pi(f_2 - f_1)$

$L_2 = R_0(f_2 - f_1)/4\pi f_1 f_2$

$C_1 = (f_2 - f_1)/4\pi f_1 f_2 R_0$

$C_2 = 1/\pi R_0(f_2 - f_1)$

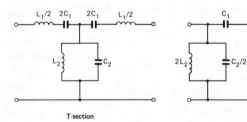

*Figure F4f* Band-stop filter;
prototype constant-k

$L_1 = R_0(f_2 - f_1)/\pi f_1 f_2$

$L_2 = R_0/4\pi(f_2 - f_1)$

$C_1 = 1/4\pi R_0(f_2 - f_1)$

$C_2 = (f_2 - f_1)/\pi R_0 f_1 f_2$

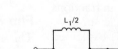

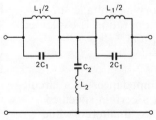

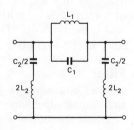

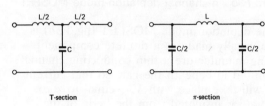

## Fourier Analysis

Any complex waveform can be shown to be made up of a number of pure sine waves consisting of a fundamental sine wave plus particular harmonics of that fundamental. For example, by taking a pure sine wave and adding to it only odd harmonics (i.e. $3f$, $5f$, $7f$, etc.) we get a square wave. Fourier analysis is the mathematical process used to solve complex waveform problems by resolving the waveform into its various sinusoidal components. A square wave with an amplitude of $\pm 1$ V can then be written as

$$f(t) = \frac{4}{\pi} \{\sin \omega t + \tfrac{1}{3} \sin 3\omega t + \tfrac{1}{5} \sin 5\omega t + \cdots\}$$

The analysis of the harmonic content of signals is very important because it gives essential information on the highest frequencies present in the signal. This then tells us the bandwidth required in systems through which the signal may be transmitted.

The general Fourier series which can be used to represent any periodic function is given by

$$f(t) = a_0 + \sum_{n=1}^{\infty} a_n \cos n\omega t$$
$$+ \sum_{n=1}^{\infty} b_n \sin n\omega t$$

where $a_n$ and $b_n$ are the coefficients to be evaluated for the various harmonics.

$$a_n = \frac{2}{T} \int_{-T/2}^{+T/2} f(t) \cos n\omega t \, dt$$
$$b_n = \frac{2}{T} \int_{-T/2}^{+T/2} f(t) \sin n\omega t \, dt$$

where $\omega = 2\pi/T$ and $T$ is the periodic time.
The d.c. term is $a_0$

$$a_0 = \frac{1}{T} \int_{-T/2}^{+T/2} f(t) \, dt$$

Note that if $f(t) = f(-t)$ the function is even, giving symmetry about the axis and then only cosine terms will be present. Conversely if $f(t) = -f(-t)$ the function is odd and only sine terms will be present.

### Example of Fourier Analysis
Since transmission of digital data is increasingly used the example is of a pulse waveform:
Width $\delta$    Amplitude $+1$    Periodic time $T$ (see fig. F5).

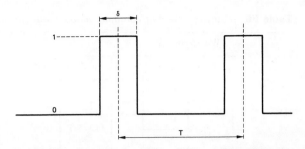

**Figure F5a**  Pulse waveform for Fourier Analysis

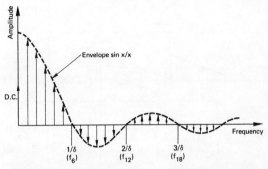

**Figure F5b**  Plot of the line spectra for $T = 6\delta$

Since the waveform is symmetrical about the origin, only cosine terms will be present. Therefore only the $a_n$ coefficients have to be evaluated.

$a_0$ is the d.c. term given by the average value of $f(t)$ in a period of $T$ seconds.

$$a_0 = \frac{1}{T} \int_{-\delta/2}^{+\delta/2} f(t) \, dt$$
$$= \frac{1}{T} [t]_{-\delta/2}^{+\delta/2} = \frac{1}{T} \left[ \frac{\delta}{2} - \left( -\frac{\delta}{2} \right) \right] = \frac{\delta}{T}$$

The $a_n$ coefficients are given by

$$a_n = \frac{2}{T} \int_{-\delta/2}^{+\delta/2} f(t) \cos n\omega t \, dt$$

In this case $f(t) = +1$. Therefore

$$a_n = \frac{2}{T} \left[ \frac{\sin n\omega t}{n\omega} \right]_{-\delta/2}^{+\delta/2}$$
$$= \frac{2}{T} [\sin \tfrac{1}{2} n\omega\delta - \sin(-\tfrac{1}{2} n\omega\delta)] = \frac{4}{n\omega T} \sin \tfrac{1}{2} n\omega\delta$$

The completed expression for $f(t)$ then becomes

$$f(t) = \frac{\delta}{T} + \frac{2\delta}{T} \sum_{n=1}^{\infty} \frac{\sin \tfrac{1}{2} n\omega\delta}{\tfrac{1}{2} n\omega\delta} \cos n\omega t$$

**Table F6**  *Fourier content of common periodic waveforms (For energy content, square the values shown)*

| Waveform | Fund | 2nd | 3rd | 4th | 5th | 6th | 7th |
|---|---|---|---|---|---|---|---|
| Square | $\dfrac{4}{\pi}E$ (127%) | 0 (0%) | $-\dfrac{4}{3\pi}E$ (42.5%) | 0 (0%) | $+\dfrac{4}{5\pi}E$ (25.5%) | 0 (0%) | $-\dfrac{4}{7\pi}E$ (18.2%) |
| Triangular | $\dfrac{8}{\pi^2}E$ (81%) | 0 (0%) | $+\dfrac{8}{9\pi^2}E$ (9%) | 0 (0%) | $+\dfrac{8}{25\pi^2}E$ (3.2%) | 0 (0%) | $+\dfrac{8}{49\pi^2}E$ (1.6%) |
| Sawtooth | $\dfrac{2}{\pi}E$ (63.6%) | $-\dfrac{1}{\pi}E$ (31.8%) | $+\dfrac{2}{3\pi}E$ (21.2%) | $-\dfrac{1}{2\pi}E$ (15.9%) | $+\dfrac{2}{5\pi}E$ (12.7%) | $-\dfrac{1}{3\pi}E$ (10.6%) | $+\dfrac{2}{7\pi}E$ (9.1%) |
| Half-wave Rectifier* | $\dfrac{E}{2}$ (50%) | $+\dfrac{2}{3\pi}E$ (21.2%) | 0 (0%) | $-\dfrac{2}{15\pi}E$ (4.2%) | 0 (0%) | $+\dfrac{2}{35\pi}E$ (1.8%) | 0 (0%) |
| Full-wave Rectifier † | 0 (0%) | $+\dfrac{4}{3\pi}E$ (42.3%) | 0 (0%) | $-\dfrac{4}{15\pi}E$ (8.5%) | 0 (0%) | $+\dfrac{4}{35\pi}E$ (3.6%) | 0 (0%) |

Example: from chart, Fourier expression for square wave is: $Y = \dfrac{4}{\pi}E(\cos \omega t - \tfrac{1}{3}\cos 3\omega t + \tfrac{1}{5}\cos 5\omega t - \tfrac{1}{7}\cos 7\omega t \cdots)$

\* D.C. component: $E/\pi$
† D.C. component: $2E/\pi$

The amplitudes of the various discrete frequencies that make up the pulse waveform can be evaluated by putting a value for $n$, corresponding to the frequency, into this expression.
The amplitudes of the discrete frequency components are seen to be contained within an envelope of

$$\dfrac{\sin x}{x} \quad \text{where } x = n\omega\delta/2$$

This envelope shows that there will be zeros at frequencies corresponding to $1/\delta$, $2/\delta$, $3/\delta$, etc.
Evaluation of amplitudes of $f_1$, $f_2$, $f_3$, for a pulse train where $T = 6\delta$:

$$f_1 \text{ amplitude} = \dfrac{2\delta}{6\delta}\left\{\dfrac{\sin 2\pi\delta/12\delta}{2\pi\delta/12\delta}\cos\dfrac{2\pi}{6\delta}6\delta\right\}$$

$$= \dfrac{1}{3}\left\{\dfrac{\sin \pi/6}{\pi/6}\cos 2\pi\right\} = 0.318$$

Similarly the amplitude of

$$f_2 = \dfrac{1}{3}\left\{\dfrac{\sin 2\pi/6}{2\pi/6}\cos 4\pi\right\} = 0.276$$

$$f_3 = \dfrac{1}{3}\left\{\dfrac{\sin 3\pi/6}{3\pi/6}\cos 6\pi\right\} = 0.2122$$

The first frequency component that has zero amplitude will be $f_6$. The plot of the line spectra is shown in fig. F5b.
Table F6 lists the harmonic composition of some other common periodic waveforms. In each case the waveforms are shown as symmetrical about the origin so that only cosine terms are present.

## Frequency

Any signal or quality that varies regularly with time as indicated in fig. F6 will have a

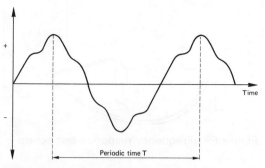

**Figure F6** Frequency

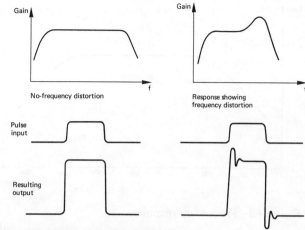

**Figure F7** Frequency distortion

frequency which is defined as the number of complete variations it makes in unit time—in other words, the number of cycles per second. The correct unit for frequency is Hertz.

Frequency is related to the periodic time of the waveform by the formula:

$$f = 1/T \text{ Hz} \quad \text{where } T \text{ is measured in seconds}$$

Angular velocity $\omega = 2\pi f$ radians per sec. For a propagated wave, the velocity $v$ is given by

$$v = f\lambda$$

where $\lambda$ is the wavelength in metres and the velocity will be in metres per sec.

Thus for a radio signal at 30 MHz the wavelength is

$$\frac{v}{f} = \frac{300 \times 10^6}{30 \times 10^6} = 10 \text{ m}$$

▶ *Measurement* (*frequency meters*)

## Frequency Band

A range of frequencies, within specified limits, that are used for particular purposes.

## Frequency Distortion

This type of distortion in amplifiers is caused

**Table F7** *Internationally agreed telecommunication bands*

| Band f | Wavelength metres | Type | Typical use |
|---|---|---|---|
| Below 30 kHz | $10^5$ to $10^4$ | V.L.F. Very low freq. | VF telegraphy. Radio telegraphy. |
| 30 kHz to 300 kHz | $10^4$ to $10^3$ | L.F. Low freq. | Carrier telephony. A.M. radio. |
| 300 kHz to 3 MHz | $10^3$ to $10^2$ | M.F. Medium freq. | A.M. radio. |
| 3 MHz to 30 MHz | $10^2$ to 10 | H.F. High freq. | Long-distance radio. |
| 30 MHz to 300 MHz | 10 to 1 | V.H.F. Very high freq. | Mobile radio. F.M. radio. Television. |
| 300 MHz to 3 GHz | 1 to 0.1 | U.H.F. Ultra high freq. | Radar. T.V. and communications. |
| 3 GHZ to 30 GHz | 0.1 to 0.01 | S.H.F. Super high freq. | Radar and communications. |

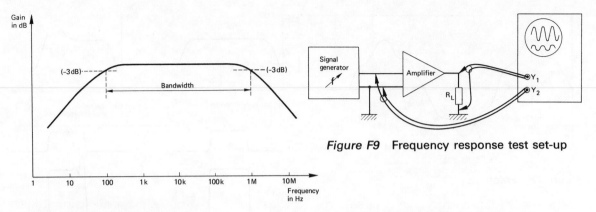

Figure F8   Frequency response

*Figure F9*   **Frequency response test set-up**

by variations in gain with frequency over the range of frequencies for which the gain should be flat. The signal components of different frequencies of a complex input signal are then amplified differently with the result that the output waveform will be a distorted version of the input (fig. F7).

## Frequency Modulation

A method of transmitting lower-frequency signals by modulating the frequency of a higher-frequency carrier wave. The angular velocity of the carrier $\omega$ is made to vary with the amplitude of the modulating signal.

$$\omega = \omega_0 + K_f V_m \cos \omega_m t$$

where
$f_0 = \omega_0/2\pi$ is the centre frequency of the modulated carrier
$f_m = \omega_m/2\pi$ is the frequency of the modulating signal
$K_f V_m$ is the degree of frequency variation proportional to the modulating amplitude.
By integrating:

$$v = V_0 \cos \left( \omega_0 t + \frac{k_f V_m}{\omega_m} \sin \omega_m t \right)$$

The maximum value of $k_f V_m/2\pi$ is called the FREQUENCY DEVIATION and is written as $\Delta f$.

$\Delta f$ is the maximum deviation in hertz from the centre frequency.

Deviation ratio $m_f = \Delta f_{max}/f_{m(max)}$

$\therefore \quad v = V_0 \cos \left( \omega_0 t + m_f \sin \omega_m t \right)$

Advantages of f.m. compared to a.m. are

*a*) The constant amplitude of the f.m. wave allows high transmitter efficiency.

*b*) Received noise present with an f.m. signal is only one third of that in an a.m. signal occupying the same bandwidth.

The disadvantage of f.m. is that it requires a wider bandwidth:

$$BW \simeq 2m_f f_{m(max)} = 2\Delta f_{max}$$

For f.m. broadcasting, the value of $\Delta f_{m(max)}$ is usually ±75 kHz allowing for a maximum modulating frequency of 15 kHz.

## Frequency Response

The gain of any system will vary in some way with signal frequency. At low frequencies the gain of an amplifier may be low because of the high reactance of coupling and decoupling capacitors, and at high frequencies the limitations of devices and stray circuit capacitance will cause the gain to fall. The frequency response is a graph showing exactly how the gain of a system varies with signal frequency (fig. F8). It is usually plotted using a logarithmic scale for frequency on the horizontal axis, so that several decades of frequency can be conveniently fitted onto one sheet of graph paper.

A frequency response test on an amplifier or system can be made using the test set-up shown in fig. F9. In this, the input amplitude is held constant and the signal frequency is varied in convenient steps. At each step the output amplitude (and phase if required) is measured to give the gain at that particular frequency. The response should be plotted directly onto the graph paper to enable smaller changes in frequency to be made when the gain is changing rapidly. A plot such as this will give

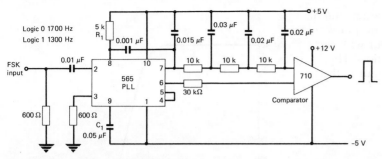

*Figure F10*　FSK receiver/decoder

the bandwidth of the system and indicate any frequency distortion that is present. An alternative is to use a swept frequency oscillator at the input. The c.r.o. will then display the frequency response of the amplifier under test.

    ▶ *Amplifier*　▶ *Bandwidth*　▶ *Bode plot*

## Frequency Shift Modulation (or FSK)

FSK stands for Frequency Shift Keying and is a method used for transmitting digital data using frequency modulation of a carrier. Logic 0 level will be one frequency, say 1700 Hz, while logic 1 will be represented by a frequency of 1300 Hz. At the transmitter the logic levels are applied to a voltage controlled oscillator (VCO) to force the output to assume either of the two frequencies. The receiver can be a phase locked loop i.c. which locks to the input frequency and then produces a level shift at its output as the received frequency shifts. Such a system has the advantage of being less affected by noise and interference.

A typical FSK receiver using a 565 PLL i.c. is shown in fig. F10. This is intended to receive and decode FSK signals of 1700 Hz and 1300 Hz. The output from the 565 will be a voltage level dependent on the received frequency and this is passed through a 3-stage *RC* filter which removes the carrier frequency to a 710 comparator. This comparator gives a high-state output for a 1300 Hz signal and a low-state output for a 1700 Hz signal. The bit rate at which the transmission can be made is 600 bits/sec.

## Fuse

A fuse is one of the simplest equipment-protection devices and consists basically of a wire link that melts and breaks the circuit when the current flowing through it exceeds its

rated value. The fuse rating is the current that the fuse will pass normally. The time taken for the fuse to blow depends on its construction and the relative size of the overload.

In many circuits, especially those with reactive loads, very high surge currents flow when equipment is switched on or off. Anti-surge, or "slow-blow", fuses are those that are designed to withstand these short-duration surges but to blow if the overload persists.

Fuses may or may not be provided to internal power lines, but must be fitted between the a.c. main line and the primary winding of the transformer of the equipment's power supply. This "mains" fuse should be an HRC (high rupture capacity) type which will safely blow if a high fault current occurs. A glass fuse may shatter violently under these conditions. The mains fuse to the primary should blow if any secondary winding is short circuited so that the possibility of the transformer overheating is prevented. However in some designs a short circuit on a secondary winding may not cause sufficient primary current to flow and to blow the fuse. Then the secondary winding itself must be fused.

Fuses are basically designed to disconnect equipment from the power source when a fault occurs and cannot be relied upon to protect semiconductor devices. This is because most semiconductor devices operate very rapidly and would probably burn out before the fuse when a fault occurs. Other methods of protection such as current limits and crowbars would have to be used.

## Gain

This is one of the basic requirements of most electronic systems and active components. A circuit that possesses gain is able to amplify a

weak input signal and produce a larger output. In the case of a linear circuit, the output would be an amplified version of the input with the smallest possible amount of distortion. The ratio of the output signal to the input is the gain of the circuit.

Voltage gain $A_v = V_o/V_i$

Current gain $A_i = I_o/I_i$

Power gain $A_p = P_o/P_i$

The above are all dimensionless ratios. In other words, if a circuit gives an output of 2 V amplitude when its input is 0.2 V, the voltage gain is 10. For convenience gain is often expressed in logarithmic units called decibels (dB):

Voltage gain $= 20 \log_{10} (V_o/V_i)$ dB

Current gain $= 20 \log_{10} (I_o/I_i)$ dB

Power gain $= 10 \log_{10} (P_o/P_i)$ dB

High values of gain can be achieved by cascading amplifier stages but in most cases it is the stability of gain and frequency response in an amplifier that is most important. In most amplifiers negative feedback is used to stabilise the gain and give wide bandwidth.

▶ *Amplifier*

## Gain-Bandwidth Product

The gain-bandwidth product is very useful in the specification of amplifying devices and amplifiers. For a transistor the frequency at which the short circuit common emitter current gain falls to unity is called the transition frequency, denoted by $f_T$. This is shown graphically on the plot of $h_{fe}$ against frequency (fig. G1). Note that at a frequency marked $f_{hfe}$ the common emitter current gain has fallen by 3 dB, i.e. the corner frequency.

If we assume a straight line of 20 dB/decade (6 dB/octave) from the corner frequency to $f_T$ then we can write

$$f_T = h_{fe}f_{hfe}$$

In other words $f_T$ is the product of gain and bandwidth.

If, for a typical transistor, $f_T$ is 300 MHz and $h_{fe}$ is 200 then the bandwidth in CE will be 1.5 MHz. Another transistor with identical $f_T$ but lower value of $h_{fe}$ will give a correspondingly larger bandwidth.

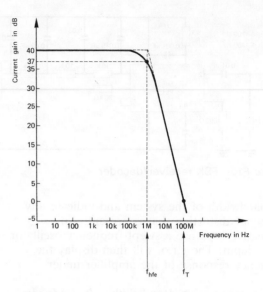

**Figure G1** Gain-bandwidth product for a transistor

## Gate

A gate can be rightly taken to indicate any circuit that allows the transmission of a signal when an appropriate switching input is applied. In CMOS, for example, special analogue transmission gates are available that have this facility. More particularly the word "gate" is assumed to mean a digital logic element, one that gives a certain logical output as long as a particular set or combination of states exists on its inputs. Such gates are the building blocks of digital logic systems. In the majority of logic circuits the logic levels have two states and the most positive level is called logic 1 (positive logic).

|  | TTL | CMOS at 10 V | ECL |
|---|---|---|---|
| Logic 1  High most-positive level | 2.4 V | +9.5 V | −0.9 V |
| Logic 0  Low most-negative level | 0.4 V | +0.5 V | −1.75 V |

The types of logic gates used in combinational logic are (fig. G2)

AND, OR, NOT, NAND, NOR, Exclusive-OR

and their operation can be described by Boolean algebra. The following assumes positive logic.

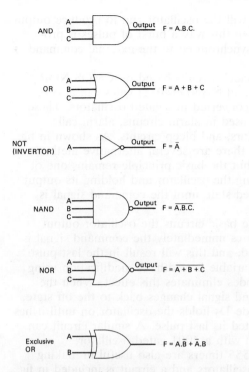

**Figure G2** Logic gates

The Exclusive-OR is a special circuit which only gives a logic 1 output if either of its inputs are at logic 1 but not when they are both 1 or both 0.

The operation of the logic gates can also be followed from a truth table. Only two input gates are assumed for simplicity.

| Inputs | AND | OR | NAND | NOR | Exclusive-OR |
|--------|-----|-----|------|-----|--------------|
| A  B | $F = A \cdot B$ | $F = A + B$ | $F = \overline{A \cdot B}$ | $F = \overline{A + B}$ | $F = A \oplus B$ |
| 0  0 | 0 | 0 | 1 | 1 | 0 |
| 0  1 | 0 | 1 | 1 | 0 | 1 |
| 1  0 | 0 | 1 | 1 | 0 | 1 |
| 1  1 | 1 | 1 | 0 | 0 | 0 |

Simple circuits using discrete components can be used to make up any of the gates but this would be totally impractical for most digital systems where hundreds of gates may be required. The integrated circuit versions in such families as TTL, CMOS and ECL are the types that are most commonly used.

The common properties and parameters used in specifying gates are:

FAN-IN   The number of inputs that can be accommodated on one gate.

FAN-OUT   The ability of a gate to drive several inputs of similar gates simultaneously.

PROPAGATION DELAY TIME   The speed at which the logic switches.

NOISE MARGIN   The measure (in volts) of a noise signal that can be accepted without causing a change of state at the output.

POWER CONSUMPTION   The amount of power taken by one gate during static, i.e. steady state, and dynamic (switched) conditions.

Groups of gates are connected together usually of the same family (i.e. all TTL or all CMOS) to create a logic system. A simple example is given in ▶ *Digital circuit.*

The AND gate gives a logic 1 output only when all its inputs are at logic 1. In other words, for a three-input gate, the output will be high only when input A *and* input B *and* input C are all high. The output will be logic 0 and low for all other input combinations.

The OR gate will give a logic 1 output if there is a logic 1 output on any one of its inputs. For a three-input OR gate, the output will be high if input A *or* input B *or* input C are high.

The NOT gate is simply a circuit that always inverts and is therefore often referred to as an invertor. If its input is logic 1 then its output will be logic 0 and vice versa.

By following an AND gate by a NOT gate we get the NAND function (NOT-AND). A NAND gate will give a logic 0 output only when all of its inputs are at logic 1. For a three-input gate the output will be logic 0 only if input A *and* input B *and* input C are logic 1.

A NOR gate is the result of following an OR gate by a NOT gate (NOT-OR). A NOR gate will give a logic 0 output if any input is at logic 1. For a three-input gate, the output will be logic 0 if A *or* B *or* C is at logic 1.

## Gated Oscillator

Being able to control the operation of an oscillator so that it only produces an output when a command signal is present is a very useful feature. Gated oscillators are employed in many applications where it is necessary to start and stop a clock signal on command. Many of the oscillator circuits that can be built using CMOS gates can be easily converted to accept a gate input. A very simple example is the gated oscillator using a 4093 Schmitt (fig. G3). While

the command input at the gate is "low" at 0 V, the output will be held in the "high" state at nearly $+V_{dd}$. Only when the command input is

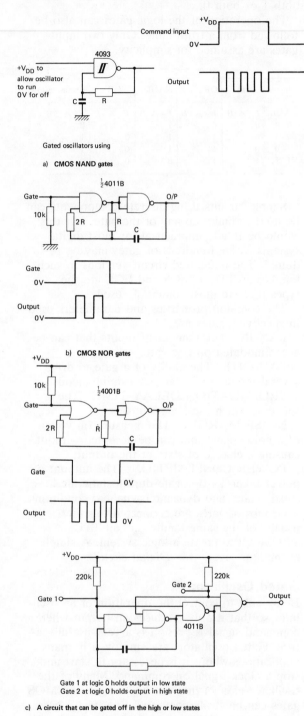

Gated oscillators using

a)  CMOS NAND gates

Gate

Output

b)  CMOS NOR gates

Gate

Output

Gate 1 at logic 0 holds output in low state
Gate 2 at logic 0 holds output in high state

c)  A circuit that can be gated off in the high or low states

*Figure G3*  Gated oscillators

"high" will the oscillator run to produce output pulses. In this way a burst of pulses is produced synchronised to the external command signal.

The clock circuits based on CMOS NOR (4001) and NAND (4011) gates can also be simply converted into gated oscillators. These can be used in alarm circuits, alarm call generators, and bleep circuits. As shown in fig. G4*a*–*c* there are several alternative methods of gating but the basic principle remains one of inhibiting the oscillator and holding its output in a fixed state until the command signal is present.

In the basic circuits the oscillator output terminates immediately the command signal is removed, and this will result in the last pulse being variable in width. A modification using two diodes eliminates this effect. When the command signal changes back to the off state, the diode $D_2$ holds the oscillator on until it has completed its last pulse. A similar circuit can be used with a 4001 gated oscillator.

The 555 timers are also useful in making gated oscillators and a circuit is included in fig. G5 of a Tone Burst Generator using a 556 (dual 555). The first half is wired up as a monostable with timing set by $R_1$ and $C_1$ and the output from this monostable "gates" the other half of the 556 which is wired as an astable multivibrator. The burst of output pulses has a frequency set by $R_2$, $R_3$ and $C_2$.

▶ *Clock pulse generator*    ▶ *Oscillator*

## Generator

In electronics, a generator is taken to indicate an instrument (or circuit) that, when provided with the necessary power, produces a signal that can be used for test or control purposes. This means that the output of a generator of whatever type has relatively high drive capability and a low output impedance.

▶ *Waveform*

## GO/NO-GO Testing

These are relatively simple test circuits used to verify that some parameter or characteristic of a device is within specification limits. Generally, circuits of this kind have limited flexibility but can be very useful for rapid checking of components.

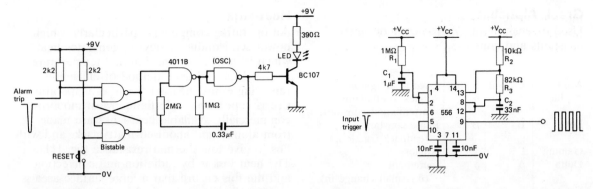

Figure G5 Tone burst generator using a 556 (dual 555)

*Figure G4a* **Alarm circuit using a CMOS 4011B gated oscillator**

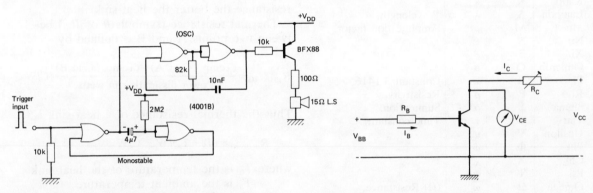

*Figure G4b* **An audible alarm circuit using a 4001 gated oscillator with auto turn-off**

*Figure G6* **Go/no-go test for $h_{FE}$**

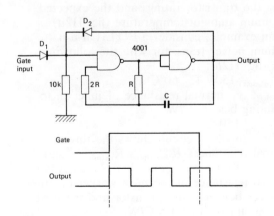

*Figure G4c* **Modification to give fixed output width**

An example in fig. G6 shows an $h_{FE}$ check on a transistor. $I_B$ is adjusted to give the base current that would be necessary for the required $h_{FE}$ value and $I_C$ is set up to the value given by

$$I_C = I_B h_{FE}$$

If the voltage $V_{CE}$ indicated on the meter is less than a certain value, then the $h_{FE}$ is greater than that required, and if $V_{CE}$ is greater than a value specified then $h_{FE}$ is less than required. Limits can be marked on the voltmeter.

With $V_{BB}$ and $V_{CC} = 10\,\text{V}$, $R_B = 180\,\text{k}\Omega$ and $R_C = 1\,\text{k}\Omega$, the value of $I_B$ is $50\,\mu\text{A}$. If $h_{FE}$ is 100, then the meter will indicate 5 V, i.e. mid-scale. Another transistor with $h_{FE}$ of 50 will give an indication of 7.5 V, while if $h_{FE}$ is 150 the voltmeter will indicate 2.5 V.

## Greek Alphabet

Used extensively in electronics to designate constants and units.

| Name | Capital | Small | Used to designate |
|------|---------|-------|-------------------|
| Alpha | A | $\alpha$ | Angles, coefficients, attenuation constant. |
| Beta | B | $\beta$ | Phase constant, feedback fraction. |
| Gamma | $\Gamma$ | $\gamma$ | Angles. |
| Delta | $\Delta$ | $\delta$ | Increment ($\delta$ = small change in). |
| Epsilon | E | $\varepsilon$ | |
| Zeta | Z | $\zeta$ | |
| Eta | H | $\eta$ | Efficiency. |
| Theta | $\Theta$ | $\theta$ | Angles. |
| Iota | I | $\iota$ | |
| Kappa | K | $\kappa$ | |
| Lambda | $\Lambda$ | $\lambda$ | Wavelength. |
| Mu | M | $\mu$ | Amplification factor. |
| Nu | N | $\nu$ | |
| Xi | $\Xi$ | $\xi$ | |
| Omicron | O | $o$ | |
| Pi | $\Pi$ | $\pi$ | Constant 3.1416 |
| Rho | P | $\rho$ | Resistivity. |
| Sigma | $\Sigma$ | $\sigma$ | Summation. |
| Tau | T | $\tau$ | Time constant. |
| Upsilon | V | $\upsilon$ | |
| Phi | $\Phi$ | $\phi$ | |
| Chi | X | $\chi$ | |
| Psi | $\Psi$ | $\psi$ | |
| Omega | $\Omega$ | $\omega$ | ($\Omega$) Resistance, ($\omega$) Angular velocity |

## Ground

Indicates a conducting path to earth or to a reference conducting plane which may be earth but can also be an array of interconnected conductors at the return point of the power supply.

A *ground plane* is a term used to describe the use of a large area of copper on a printed circuit board that is used to give a low impedance return path to the supply. Any circuit component connected to it sees a low inductance path to the supply so that current surges returning to the supply do not generate noise pulses.

A grounded connection on an active component is used to describe the electrode that is electrically neutral as in grounded base or grounded gate. The grounded electrode is common to both input and output circuits.

## Heat-sink

An operating component, particularly a high-power semiconductor device, generates heat which in turn causes an internal temperature rise. A heat-sink is a method of efficiently removing excess heat so that the temperature rise is kept within safe limits. Most modern commercially available heat-sinks are made from aluminium, anodised matt black, and with fins to give low thermal resistance (fig. H1). The heat loss is by radiation and convection and the fins ensure that a large surface area is obtained in a comparatively small space.

In the same way that electrical resistance is used to compare different types of wire, thermal resistance is the yardstick for comparing heat-sinks. The lower the value of thermal resistance the better the heat-sink.

Thermal resistance (symbols $\theta$ or $R_{th}$) between two points A and B is defined by

$$R_{th(A-B)} = \frac{(\text{Temp. °C at A}) - (\text{Temp. °C at B})}{\text{Power dissipation in watts}} \, °C/W$$

Thus the thermal resistance of a heat-sink is

$$R_{th(h-a)} = (T_h - T_a)/W \quad °C/W$$

where $T_h$ is the temperature of the heat-sink
$T_a$ is the ambient temperature
$W$ is the power required to be dissipated from the device mounted on the heat-sink.

Equivalent circuits can be drawn to show the thermal resistances of a transistor, the mounting, and its heat-sink. Calculations of the required heat-sink size can then be made using the transistor ratings and the expected maximum ambient temperature (fig. H2*a*). As an example, consider a BD 131 n-p-n medium-power transistor with the following data (fig. H2*b*):

$P_{tot(max)}$ 15 W $T_{mb}$ 60°C
$R_{th(j-mb)}$ thermal resistance of junction to mounting base 6° C/W
$T_{j(max)}$ 150°C

The equivalent circuit shows the three thermal resistances $R_{th(j-mb)}$, $R_{th(mb-h)}$, and $R_{th(h-a)}$ all in series. Note that $R_{th(mb-h)}$ is the thermal resistance of the mica washer or other interface between the transistor and the heat-sink (assumed to be 1.5°C/W).

$$T_j - T_a = P_{tot}[R_{th(j-mb)} + R_{th(mb-h)} + R_{th(h-a)}]$$

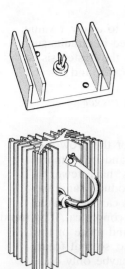

Figure H1  Heat-sinks

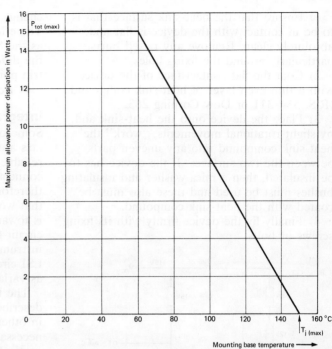

Figure H3  Power dissipation derating graph (for a BD131 transistor)

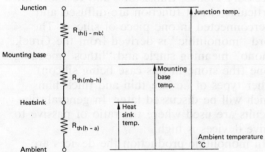

Figure H2a  Equivalent circuit showing thermal resistances between semiconductor junction and ambient

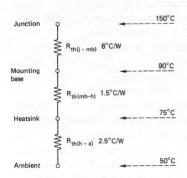

Figure H2b  Example of thermal circuit for a BD131 operating at 10 W in an ambient of 50°C

Rearranging this formula gives

$$R_{th(h-a)} = [(T_j - T_a)/P_{tot}] - [R_{th(j-mb)} + R_{th(mb-h)}]$$

Suppose a BD131 is required to dissipate 10 W to a maximum ambient temperature $T_a$ of 50°C. Then

$$R_{th(h-a)} = \frac{150-50}{10} - (6+1.5) = 2.5°C/W$$

Thus a heat-sink with a thermal resistance of better than 2.5°C/W is required. A check on the power dissipation derating graph, a plot of maximum allowable power dissipation against mounting base temperature (fig. H3), shows that with this heat-sink the mounting base temperature will be 90°C allowing a dissipation of 10 W. However it would not be good practice to design without some margin for error, so in the example either the maximum power would be limited to less than 10 W, say 8 W max, or a larger heat-sink with a thermal resistance lower than 2.5°C/W, say 2°C/W, should be used.

Mounting a power device on a heat-sink requires care to ensure that the interface between the device case and the heat-sink has the lowest possible thermal resistance.

1) Ensure that the heat-sink surface that is to be in contact with the device is flat and absolutely clean. Remove any raised burrs, particularly around the fixing holes.

2) Coat the flat contact face of the device with a thin even layer of heat-sink compound (R.S. 554-311 or Dow Corning 2633).

3) Place the device onto the heat-sink and, by slight rotational movements, "work" the heat-sink compound into any uneven parts between the two surfaces. If the device has to be insulated, then a mica washer and insulating bushes must be used and these also must be coated with the heat-sink compound.

4) Finally fix the device firmly with its fixing screws or bolts.

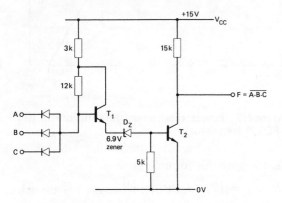

*Figure H4* High-threshold logic NAND gate

## High-threshold Logic

(Abbrev: HTL.) In many industrial environments the electrical noise level can be very high. The noise is caused by motors and machines being operated and heavy inductive loads being switched. Any digital system operated in such an environment must have a high threshold so that noise spikes are ignored and do not cause false operation. The HTL gate was developed for this requirement and is a modified form of DTL circuit with the series diode replaced by a zener (fig. H4). The power supply is increased to 15 V and the zener has a breakdown voltage of 6.9 V. This gives a noise margin of approximately 7 V.

## Hysteresis

An effect in electronic circuits when the output of a trigger circuit remains in a changed state even when the input which caused the original change of state has returned to a lower value. The effect is most pronounced in circuits such as the Schmitt trigger where the hysteresis is the difference between the upper and lower trip points.

▶ *Backlash*   ▶ *Schmitt trigger*

## Integrated Circuit

Some of the main reasons for the growth of i.c.s are that they are small and light; more reliable than discrete circuits; have fewer connections; can be mass produced and are therefore cheap; and circuits can be created that would otherwise be uneconomical. The trend is towards packaging more and more of the circuit functions inside the i.c. so that future instruments will consist of maybe only one LSI circuit plus a few discrete components such as variable resistors and switches.

The term i.c. has now come to be used to describe the monolithic integrated circuit, or in other words an element in which all the necessary diodes, transistors, and resistors for a particular circuit function are diffused and interconnected in one piece of silicon. The word "monolithic" is derived from the Greek, "mono" meaning single and "lithos" meaning stone (the stone in this case being silicon). Other types of i.c. are thin and thick films which will be discussed later. In general, film circuits are used where the ratio of passive to active devices is high.

In monolithic production the devices are produced several hundred at a time on one 50 mm or 75 mm diameter silicon slice using the planar process. There are two important differences between bipolar i.c.s and discrete semiconductors. For the i.c.,

*a*) Each transistor, diode, resistor, etc. must be isolated from adjacent elements.

*b*) All connections must be made to the top surface; this includes collector leads.

Three methods of isolation exist:
junction isolation
dielectric or oxide isolation
collector diffusion isolation.

An n-p-n i.c. transistor construction is shown in fig. I1. The collector as well as base and emitter connections must be brought out to the top surface. To reduce the collector resistance, a highly doped n-region is diffused into the p-substrate before the epitaxial growth, and this is called the buried n-region. To create the n-p-n i.c. transistor the process steps required are

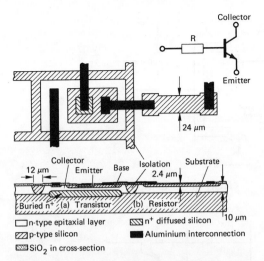

*Figure I1* I.C. transistor with resistor load

1) Preparation of silicon slice p-type.
2) Oxidisation of slice.
3) First photo-resist to define buried collector n-regions.
4) Buried collector diffusion n-type. Arsenic 5 hours at 1300°C.
5) Removal of oxide.
6) Epitaxial layer grown, lightly doped.
7) Oxidisation of slice.
8) Photo-resist to define isolation.
9) Isolation diffusion p-type. Boron 16 hours at 1180°C.
10) Oxidisation.
11) Photo-resist to define base regions.
12) Base diffusion.
13) Oxidisation.
14) Photo-resist to define emitter regions.
15) Emitter diffusion.
16) Oxidisation.
17) Photo-resist to define contact areas,
18) Metallisation of aluminium onto contacts.
19) Slice probe test.
20) Scribe and break into chips.
21) Mounting and wire bonding.
22) Encapsulated and final test.

This is a continual repeat of the planar process, and naturally during the process other elements such as diodes and resistors have to be created to make a complete microcircuit. A resistor is usually a p-type area as shown diffused into the epitaxial region.

MOS i.c.s are manufactured using fewer process steps than bipolar and the resulting MOS-FET structure can be created in an area about 60 $\mu$m by 40 $\mu$m. In addition, MOS transistors can be used as load resistors and are self-isolating. This is why the packing densities of MOS i.c.s can be so large.

Film or hybrid i.c.s fall into two main classes:

**1** THIN FILMS A film of appropriate conducting material is deposited onto an inert substrate such as borosilicate glass or glazed alumina. Film materials, such as nickel-chromium, gold, tantalum, etc., are either evaporated or sputtered from a cathode, and the required pattern is obtained either by masking the substrate or by etching after the film has been applied.

**2** THICK FILMS A silk screen process is used, where the film material is deposited in paste form onto a substrate via the screen. The unit is then fired in a furnace at about 1000°C. Cermets and resistive links are used as the film materials, and the substrates are commonly alumina or beryllium.

For both types of film circuit, adjustments of resistor values can be made to very close tolerance which is not possible in monolithic circuits. Trimming to better than ±0.5% can be achieved with spark-erosion, and with laser beams to within ±0.1%.

Low-value capacitors and inductors (spiralling film) can be made but more usually these components and active devices such as "flip-chip" transistors and diodes are added to the film network. Connections are made by ultrasonic bonding, and finally the unit is encapsulated.

### Integrated Injection Logic

Usually written as I²L, integrated injection logic is a type of logic circuitry that is used in large-scale i.c.s. Its main advantages are a high packing density of up to 200 gates per square millimetre, and very low-power dissipation of about 100 $\mu$W per gate at 1 MHz. The basic circuit is an extension of Direct Coupled Transistor Logic (DCTL) but without problems of "current hogging" which cause variations in the available drive to gate inputs, one transistor with slightly different input characteristics taking more current than others (fig. I2). The I²L gate avoids this because of the p-n-p transistor. This delivers a constant current which is either switched into or away from the input base of a multi-collector transistor. In the structure (fig. I3) the collector of the p-n-p is

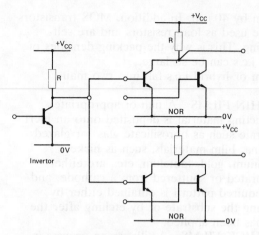

*Figure 12* DCTL gates

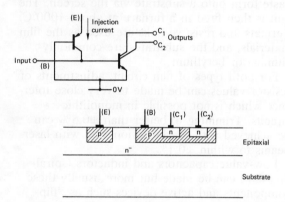

*Figure 13* Basic I²L gate and structure of I²L

common to the base (p) of the n-p-n transistor, and similarly the base of the p-n-p is common to the emitter of the n-p-n. This can be seen from the diagram. Because of these shared areas of silicon and the fact that no resistors are used, the area taken by one I²L gate is very small.

With a high value of 750 mV applied to the input, the current is switched into the base of the multi-collector transistor giving a low output of 50 mV. Conversely with a low input of 50 mV the current is steered away from the base of the multi-collector transistor. This is switched off and the output will rise to a high level determined by the load or pull-up resistor used in the design. For internal circuits where the output of one I²L gate is driving the input of another, the output will be limited to 750 mV, this being the $V_{BE}$ of the next gate. The power supply rail is typically 850 mV and

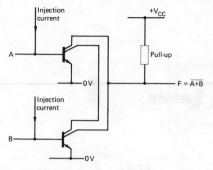

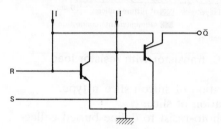

*Figure 14* NOR gates using two I²L gates

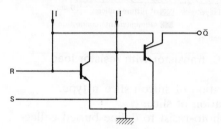

*Figure 15* R-S flip-flop from two I²L gates

the logic swing 700 mV, giving a noise margin of only 350 mV.

Logic functions such as the OR/NOR can be easily created by connecting the collector outputs of more than one gate together just as in wired-logic. Thus a NOR gate can be made as shown in fig. 14 and a R–S bistable as in fig. 15.

## Integrator

The integrator, often called a low pass filter, is shown in fig. 16 together with output signals for typical inputs. Since the reactance of the capacitor falls with increasing frequency, this circuit removes the high-frequency components from a pulse wave-form. When a step input is applied the voltage across the capacitor cannot change instantaneously. It rises exponentially according to the formula

$$v_C = V(1 - e^{-t/CR})$$

Now $CR$, the product of capacitance in farads and resistance in ohms, is called the time constant of the circuit. In one time constant the voltage across the capacitor changes by about 63%. Note that it takes nearly 4.5 time constants for the voltage across the capacitor to equal $V$.

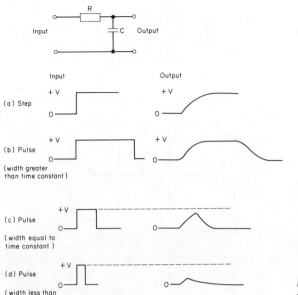

Figure l6   An integrator

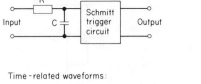

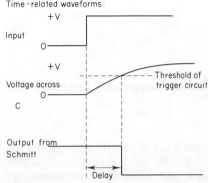

Figure l7   Use of an integrator to give a time delay

The effect of an integrator on pulses which have a long width in comparison to the integrator's time constant is to degrade the rise and fall times. If, however, the pulse is short in comparison to the integrator's time constant, then the capacitor will not have sufficient time to charge completely, and the output will appear triangular. Circuits such as these are often used to provide short time delays. An example is shown in fig. l7.

## Interface Circuit

An interface circuit (fig. l8) is one that matches a signal from one unit or device to another. The interface takes the signal from unit A and modifies it to a form that can be accepted by unit B. Logic i.c.s of one type of family, for example TTL, require an interface if they are to drive logic i.c.s of another family such as CMOS or ECL. In addition to this, most input devices, such as switches and transducers, must be interfaced to enable them to provide inputs that are compatible to the logic of a digital system. At the output the logic also has to be fitted with output interface circuits to enable load devices to be driven correctly. The range

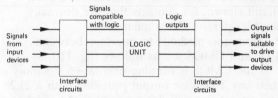

Figure l8   Use of interface circuits

of circuits and circuit techniques for interfaces is therefore very large and only some of the more standard methods can be described here.

**1**   One of the common inputs to any system will be from switch contacts. These inevitably suffer from contact bounce since they make and break several times before coming to rest in a closed or open position. These changes of state will be seen as valid input signals as far as digital logic circuits are concerned and an interface circuit that eliminates contact bounce and gives one clean change of state every time the switch is operated is essential. Two examples using CMOS Schmitt invertors are shown (fig. l9) and in each case the *RC* time constant should be at least three times the worst expected contact bounce time. In (*a*) the output of the Schmitt gate remains low until the input

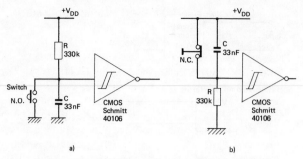

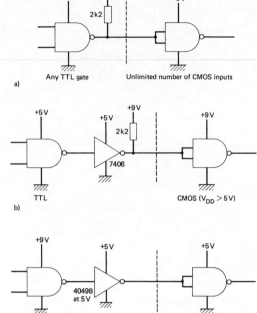

*Figure I9* Interface circuits from switch contacts to CMOS

voltage has fallen below the negative going threshold (2.3 V with a +5 V supply). The time taken for this is just under one time constant of $R_1$ and $C_1$. Any bounce of the contacts at make is ignored since this only partially discharges the capacitor, and the output only switches high after a delay of approximately 10 ms. When the push switch is released, the voltage across the capacitor has to rise above the positive going threshold of the Schmitt (2.9 V with a +5 V supply) so that contact bounce on release is also ignored. The other circuit shown, for contacts that are normally closed, operates in a similar way.

**2** Common methods to interface between TTL and CMOS are as follows (fig. I10):

If the CMOS is running from a +5 V supply then any TTL gate output provided with a 2k2 pull-up resistor will drive an unlimited number of CMOS gate inputs.

If the CMOS has a higher supply voltage, say +9 V, then a TTL open collector output gate such as the 7406 or 7407 must be used.

Any CMOS gate output, with a supply of +5 V, will drive *one* low power Schottky TTL input. Otherwise the CMOS (at +5 V) must be interfaced to the TTL via a 4049 or 4050 buffer. These will drive any two TTL inputs.

If the CMOS has a higher supply voltage, say +9 V, then a 4049 or 4050 run from +5 V can be used as the interface to any two TTL inputs.

**3** Some examples of interface circuits from logic i.c.s to load devices are shown (fig. I11). TTL gate outputs are best used in the current sinking mode. A standard TTL gate output at logic 0 will sink up to 16 mA of load current. This is sufficient to drive an LED but a current-limiting resistor must be fitted. A

*Figure I10* Interface methods from TTL to CMOS and vice versa

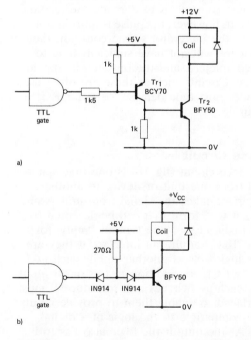

*Figure I11* TTL interface to 250 mA loads, with alternative circuit in (*b*) for loads up to 150 mA

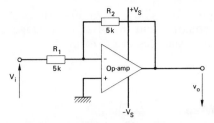

*Figure I12* Analogue invertor: $v_o = -v_i$

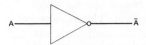

*Figure I13* Digital invertor: $F = \bar{A}$

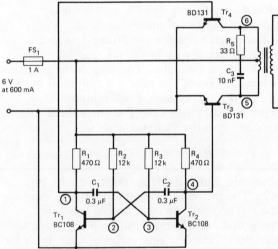
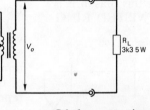

*Figure I14* A power invertor

an 8 Ω speaker to CMOS to provide an audio output.

**4** Specialised i.c.s are available for line driving and receiving. These are essential for interfacing over long distances and in electrically noisy environments. Usually a differential system is used so that noise, which will be a common signal, is rejected. Optical couplers can also be used to interface between systems that have different power supply arrangements since they provide excellent isolation. An example in ▶ *Clock pulse generator* shows how data can be interfaced and transmitted over a reasonable distance using a twisted pair of wires and an opto-coupler.

Of the many interface i.c.s available the 75492 MOS LED driver i.c. is perhaps one of the most useful for interfacing from a microcomputer system to load devices. It has six invertors each of which can sink up to 200 mA on each of its six outputs. This makes it useful for driving indicators and relays and for interfacing the computers outputs ports to a.c. loads. Again an opto-isolator should be used between the 75492 output and the a.c. control element.

**Invertor**

This has a number of possible meanings depending on the application. The three most common are:

**1** *Analogue invertor* (fig. I12) When an op-amp is wired to give an inverting amplifier and if $R_1 = R_2$ the voltage gain is −1.

The output is an inverted version of the input. In analogue computers it is also called a "sign changer" [▶ *Op-amps*].

**2** *Digital invertor* (fig. I13) This is a single input logic gate that gives an output level that is always the complement of its input. It is also referred to as a NOT gate [▶ *Gates*].

**3** *Power invertor* (fig. I14) An invertor is a power unit that produces an a.c. power output from a d.c. input. The frequency of the a.c. output may be 50 Hz or 60 Hz, but is more

typical value is 270 Ω. CMOS buffers can also be used to drive LEDs as long as a limiting resistor is used, a typical value being 1 kΩ. Any loads heavier than an LED, i.e. of more than about ten milliamps, require a transistor drive. The circuit in fig. I11a illustrates a technique for driving a relay or solenoid from TTL. When the output of the TTL gate goes low, $Tr_1$ and $Tr_2$ conduct to operate the load. Note that in both cases the inductive load is prevented from generating a large back e.m.f. by a parallel diode. Even simpler output interface circuits from CMOS can be obtained by using VMOS power FETS since these devices can be driven directly from CMOS gate outputs. If the CMOS is running from a power supply of 10 V then load currents of 1 A or greater can be controlled. An example in the section *Alarms* shows the ease of interfacing

often 400 Hz or higher. The d.c. source is typically a battery system, and one good example of an invertor is in standby power units which, in the event of a mains failure, provide a short-term emergency supply of 240 V r.m.s at 50 Hz (or 110 V r.m.s. at 60 Hz) from say a 24 V battery. The battery is trickle-charged when the mains is present. The basic circuit is a transformer with the primary switched by electronic devices.

## Johnson Counter

A synchronous type of counter formed from a basic serial-in/serial-out shift register by feedback from the $Q$ and $\bar{Q}$ outputs of the final bistable to the K and J inputs respectively of the first bistable (fig. J1). For this reason it is sometimes referred to as a TWISTED RING counter.

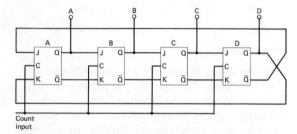

*Figure J1*   Divide-by-8 Johnson counter

Also called a WALKING CODE counter because the count sequence proceeds from all $Q$ outputs at zero (0) up to all outputs at logic 1 and then back to all zeros, i.e. the counter fills up with 1's and then empties to zeros. A table of the count sequence for a $\div 8$ Johnson counter looks like this:

| Input | D | C | B | A |
|---|---|---|---|---|
| 0 | 0 | 0 | 0 | 0 |
| 1 | 0 | 0 | 0 | 1 |
| 2 | 0 | 0 | 1 | 1 |
| 3 | 0 | 1 | 1 | 1 |
| 4 | 1 | 1 | 1 | 1 |
| 5 | 1 | 1 | 1 | 0 |
| 6 | 1 | 1 | 0 | 0 |
| 7 | 1 | 0 | 0 | 0 |

The Johnson counter is particularly useful for creating low modulo counters. The circuit has the advantage of being synchronous and rela-

tively easy to decode. The CMOS 4017 (five-stage decode counter) and the 4022 (four-stage $\div 8$ counter) are examples in i.c. form.
▶ *Counters*

## Karnaugh Map

Used as an aid in the simplification of logic. A Karnaugh map is a graphical representation of all the unique combinations of the variables involved in a logic expression. The advantages over truth tables are that the map makes it possible to see the general form of the problem and allows it to be simplified quickly. Minimization of combinational logic by mapping can yield a result more quickly than by using Boolean Algebra alone.

Karnaugh maps for two, three and four variables are shown in fig. K1 together with some simple examples. The map is the graphic representation of the minterm canonical form of the logic expression. Each minterm is represented by one cell. Thus in a three variable map the expression

$$F_1 = A \cdot B \cdot \bar{C} + A \cdot B \cdot C + A \cdot \bar{B} \cdot \bar{C} + A \cdot \bar{B} \cdot C$$

results in four cells being filled with 1's (fig. K2). The process of simplification is to examine the marked cells and to "couple" together those that are adjacent. Adjacent cells are those that differ by one variable. Cells must be grouped in binary combinations, i.e. 2, 4, 8. Therefore from the map for $F_1$ the four marked cells, being adjacent, can be coupled together giving the simplification that $F_1 = A$.

Cells on the edge of the map can also be considered as adjacent. For example (fig. K3),

$$F_2 = A \cdot \bar{C} \cdot \bar{B} + A \cdot B \cdot C + \bar{A} \cdot \bar{C} \cdot \bar{B}$$

giving the map as shown. Cell (x) cannot be combined with any other so that $A \cdot B \cdot C$ must appear in the simplified function, but cells (y) and (z) are adjacent since they differ by one variable and can be coupled. Therefore

$$F_2 = \bar{C} \cdot \bar{B} + A \cdot B \cdot C$$

A final example in fig. K4 shows the map for the function

$$F_4 = A \cdot B \cdot C + \bar{B} \cdot \bar{C} \cdot \bar{D} + A \cdot B \cdot \bar{D} + B \cdot C \cdot \bar{D}$$
$$+ A \cdot \bar{B} \cdot C + \bar{A} \cdot \bar{B} \cdot \bar{C} + A \cdot \bar{B} \cdot D$$
$$+ A \cdot \bar{C} \cdot D$$

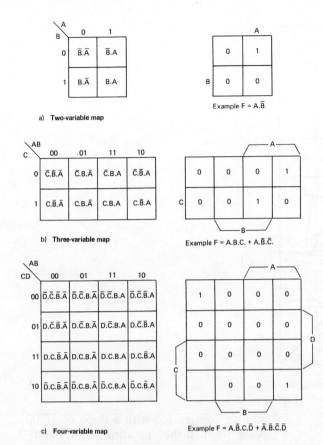

a) Two-variable map

b) Three-variable map

c) Four-variable map

Example F = A.$\bar{B}$

Example F = A.B.C. + A.$\bar{B}$.$\bar{C}$.

Example F = A.$\bar{B}$.C.$\bar{D}$ + $\bar{A}$.$\bar{B}$.$\bar{C}$.D

**Figure K1** Karnaugh maps

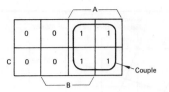

**Figure K2**  $F_1 = A \cdot B \cdot \bar{C} + A \cdot B \cdot C + A \cdot \bar{B} \cdot \bar{C}$
$+ A \cdot \bar{B} \cdot C = A$

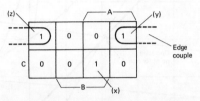

**Figure K3**  $F_2 = A \cdot \bar{C} \cdot \bar{B} + A \cdot B \cdot C + \bar{A} \cdot \bar{C} \cdot \bar{B}$
$= \bar{C} \cdot \bar{B} + A \cdot B \cdot C$

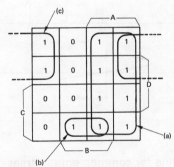

**Figure K4** Four-variable maps for

$F_4 = A \cdot B \cdot C + \bar{B} \cdot \bar{C} \cdot \bar{D} + A \cdot B \cdot \bar{D} + B \cdot C \cdot \bar{D}$
$+ A \cdot \bar{B} \cdot C + \bar{A} \cdot \bar{B} \cdot \bar{C} + A \cdot \bar{B} \cdot D + A \cdot \bar{C} \cdot D$

which simplifies to

$F_4 = A + B \cdot C \cdot \bar{D} + \bar{B} \cdot \bar{C}$

From this map three couples can be formed:
  $(a) = A$ since all cells in $A$ are coupled.
  $(b) = B \cdot C \cdot \bar{D}$ since two cells including $A$
    and $\bar{A}$ are coupled.
  $(c) = \bar{B} \cdot \bar{C}$ since four cells including $A$ and
    $\bar{A}$, and $D$ and $\bar{D}$ are coupled.
Then, $F_4$ simplifies to

$$F_4 = A + B \cdot C \cdot \bar{D} + \bar{B} \cdot \bar{C}$$

## Load Line

A more thorough analysis of the large-signal
performance of amplifiers and switching circuits
can usually be made by using a graphical, instead of a mathematical, method. The load line
is an essential feature of the graphical method,
and it is a straight line drawn, to represent the
value of the load, onto the characteristics of
the active device. The slope of the line is
determined by the value of the load and all
possible operating conditions for a particular
value are described by one load line.

Suppose a common emitter transistor stage
with collector output characteristics as shown in
fig. L1 has a load of $500 \, \Omega$ in its collector and
a $V_{CC}$ of 20 V. The load line for this $500 \, \Omega$
resistor can be constructed from two points:

Point A when $I_C$ is assumed zero. Then
$V_{CE} = V_{CC} = 20$ V.

Point B when $V_{CE}$ is assumed zero. Then
$I_C = V_{CC}/R_L = 40$ mA.

Points A and B are joined by a straight line
and all possible operating conditions for the

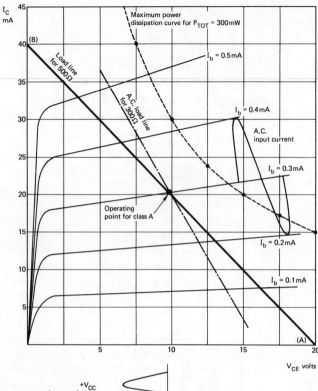

Figure L1 graph labels:

$I_C$ mA

(B)

Maximum power dissipation curve for $P_{TOT}$ = 300 mW

Load line for 500 Ω

A.C. load line for 300 Ω

$I_b$ = 0.5 mA

$I_b$ = 0.4 mA

A.C. input current

$I_b$ = 0.3 mA

Operating point for class A

$I_b$ = 0.2 mA

$I_b$ = 0.1 mA

(A)

$V_{CE}$ volts

+$V_{CC}$

500 Ω

A.C. output voltage

$v_o$

$V_{CE}$

$v_i$

*Figure L1*  **Load line for common emitter transistor stage**

stage are described by this line. For example, if $I_B = 0.1$ mA, then $V_{CE} = 16.2$ V and $I_C = 7.5$ mA, or when $I_B = 0.5$ mA, then $V_{CE} = 3.5$ V and $I_C = 33$ mA.

If the stage had to be operated in Class A then the most suitable operating point would be when $V_{CE} = 10$ V, $I_C = 20$ mA and $I_B = 0.3$ mA. This allows the maximum output voltage swing with the minimum amount of distortion. All points on the load line should be outside the maximum power dissipation curve and this is the case for a $P_{tot}$ of 300 mW.

A dynamic or a.c. load line is one that takes into account the effect of any a.c. coupled load. If, in the example, the output were capacitively coupled to an external load of 750 Ω, then the effective a.c. load is 300 Ω. The operating point is still determined by the d.c. load line but the

a.c. load line, constructed with a slope of 1/300 and passing through the operating point, determines the gain of the stage.

## Maintenance

The purpose of maintenance is to achieve a satisfactory level of system availability at reasonable cost and maximum efficiency. Availability is defined as

$$\text{Availability} = \frac{\text{MTBF}}{\text{MTBF} + \text{MTTR}}$$

where MTBF is the mean time between failures and MTTR is the mean time to repair.

To achieve high levels of availability, i.e. those approaching unity, the MTTR value must be low and this implies that the system can be maintained relatively easily. Maintainability is defined as the probability that a system that has failed will be restored to a full working condition within a given time period. The mean time to repair and the repair rate ($\mu$) are measures of maintainability:

$\mu = 1/\text{MTTR}$

Maintainability $M(t) = 1 - e^{-\mu t} = 1 - e^{-t/\text{MTTR}}$

where $t$ is the time allowed for the maintenance action.

*Example* In a system the average time to repair any fault is 2 hours. Calculate the value of maintainability for a time of 4 hours.

$$M(t) = 1 - e^{-t/\text{MTTR}} = 1 - e^{-4/2}$$
$$= 1 - 0.135 = 0.865$$

Therefore the probability $M$ of the system being returned to a working state within 4 hours is 0.865 (86.5%).

In the same way that values for system reliability can be calculated for a given operational time, the value of maintainability can also be predicted. In both cases it is the probability of success that is calculated for the given time. The time $t$ for reliability is the system operational period, while $t$ for maintainability is the allowed maintenance time. Prediction of maintainability involves establishing a value for the system MTTR.

MTTR is the average of the times taken to repair any fault in the system, and accurate assessment of this is understandably difficult. The designer can aim for a low value of MTTR by paying close attention to the accessibility of components, building in fault display panels, and by providing internal test facilities.

The maintenance policy adopted for a particular system will depend upon several factors, such as the type of system, its location and operating and environmental conditions, the required levels of reliability and availability, the standard of skilled maintenance staff, and the provision of spares. For certain types of system the maintenance policy may include in its programme details of recalibration and preventive maintenance actions. Recalibration, often carried out at 90 day intervals on measuring instruments such as the oscilloscope and d.v.m., is really a type of preventive maintenance, since the recalibration task is to first check the amount of drift of some parameter or characteristic from the specified figure and then to correct for any partial failure that may have caused the measuring instrument's performance to be outside its tolerance limits. But, in practice, no components are replaced. True PREVENTIVE MAINTENANCE is a policy of replacing components or parts of a system that are nearing the end of their life, and are

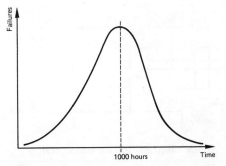

*Figure M1* Possible failure rate of filament lamps

therefore wearing out. The replacement is carried out before the component actually fails. Failures of components entering the wear-out period or subjected to continuous wear are not random and can be predicted. The reliability of a system can therefore be improved by replacing items that are wearing out before they fail. Examples of this are: components with moving parts that are continuously in use such as servo potentiometers, motors and motor brushes; or contacts on relays and switches, especially those subjected to arcing when switching inductive or capacitive loads. Filament lamps are yet another example. A typical failure graph is shown in fig. M1; in this case the failures follow a Gaussian distribution with the peak failures occurring at 1000 hours.

CORRECTIVE MAINTENANCE or "replace as failed" is the service action that is normally required for the majority of electronic systems since, during the useful life of the system, failures of the component parts will be entirely random. In this case failures cannot be predicted and cannot therefore be prevented by service checks.

There are three distinct phases in the corrective maintenance task:

1) *Fault detection* The presence of a fault must be established and all symptoms accurately noted. This means that a functional test, checking the system's actual performance against its specification, must be made. Only by doing this will a full set of fault symptoms be obtained.

2) *Fault location* The task now is to narrow down the search for the cause of the fault, first to one block (or sub-system) within the system, and finally to one component within that block.

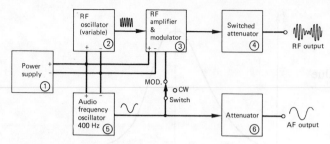

*Figure M2*  Block diagram for an r.f. signal generator

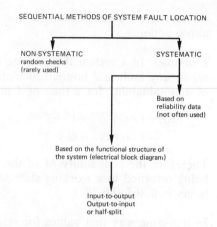

*Figure M3*

This task is simplified by the use of one or a mixture of fault location methods outlined in the next section.

   3) *Fault rectification*  The faulty component or part is repaired or replaced. A functional check must then be carried out on the whole system.

## Locating faults in systems

When a component fails in an individual circuit a certain set of fault symptoms result. These symptoms, often unique to the fault, will be changes in circuit operation and in d.c. bias levels and changes in output signals. By interpreting the symptoms it is possible to pin-point the faulty component. With a complete system, however, the task of finding a single faulty component among several thousand is made more difficult because of the size and complexity of the system. The problem can be tackled by considering the system in block diagram form.

**1**  The system is divided up into several functioning circuit blocks and, by measurement, the portion or block that has failed can be located, and then detailed measurements can be made on that block to find the actual faulty component. The block diagram is an essential aid to system fault location and in addition assists in helping the understanding of the operation of complex systems. Consider the block diagram for an r.f. signal generator shown in fig. M2. In its basic form there are six blocks to consider. Suppose, for example, that the r.f. output is correct in both the modulated and continuous wave switch positions, but that no 400 Hz a.f. is present. Then the fault must be in the attenuator or its connections. On the other hand if the generator failed giving no outputs at all, the fault would almost certainly be in the

power supply. This is because it is unlikely, although possible, that both oscillators would fail simultaneously.

   These two examples demonstrate the use of the block diagram and the sort of logical approach that is required for fault location. However, because two outputs are present and a switch can be used to modify the state of one output, the location of a faulty block is relatively simple. For more complex systems some general method must be used.

**2**  One powerful method which is being increasingly used is NON-SEQUENTIAL fault location. This uses automatic testing, based for example on the theoretical analysis of the transfer characteristics (response of output to input) of the system. Such a method is better suited to computer-aided fault analysis rather than the individual service engineer. The faulty system, linked to computer-controlled test gear, would have its whole system checked and results would be matched with those resulting from typical fault conditions held in the computer store.

   The individual service engineer, when faced with a faulty system, usually has to select one or a mixture of SEQUENTIAL fault location methods. The possible sequential methods are shown in fig. M3.

**3**  The most popular systematic fault location methods are

   *a*) Input to output
   *b*) Output to input
   *c*) Half-split.

The first two methods are fairly straightforward. A suitable input signal, if required, is

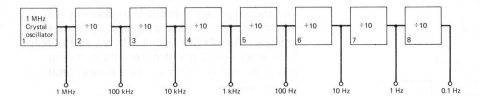

**Figure M4** Frequency-divider chain; example of half-split method of fault location

injected into the first block and then measurements are made sequentially at the output of each block in turn, either from input to the output or vice versa, until the faulty block is found. The HALF-SPLIT method on the other hand is extremely useful when the system is made up of a large number of blocks in series. As a good example, consider the frequency divider chain of a digital frequency meter shown in fig. M4. Here the frequency of a stable crystal-controlled oscillator is divided down by decade counters to give the various timing pulses. With the eight blocks shown it is possible to divide the unit into two equal halves (half-split), test to decide which half is working correctly, then split the non-working section into half again to locate the fault. Assuming that block (7) has failed; the sequence of tests would be as follows:

1) Split the whole into half by measuring the output from block (4). The output from (4) will be found to be correct at 1 kHz, showing that the fault lies somewhere in blocks (5) to (8).

2) Split blocks (5) to (8) in half by measuring the output from block (6). Again this output will be found to be correct at 10 Hz.

3) Split blocks (7) and (8) by measuring output of block (7). There will be no output proving that the fault is in block (7).

For the input to output or output to input method, the number of checks on a series system is given by the formula

$$C = \frac{1}{2n}(n-1)(n+2)$$

and for half-split by

$$C = 3.32 \log_{10} n$$

where $n$ is the number of blocks or units and $C$ is the mean number of measurements required. Note that these formulae apply to series-connected circuits only.

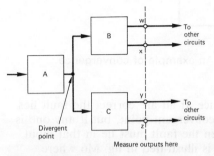

**Figure M5** Typical divergent arrangement within a system

4  Most systems do not consist of only series-connected blocks but possess parallel branches and possibly feedback loops. The connections that complicate fault location methods are

*a*) Divergence: an output from one block feeding two or more units.

*b*) Convergence: two or more input lines feeding a circuit block.

*c*) Feedback: which may be used to modify the characteristic of the system or in fact be a sustaining network.

DIVERGENCE is a commonly encountered situation. Two examples exist in fig. M2. The power supply has to supply d.c. power to blocks 2, 3 and 5, and the output from the audio frequency oscillator has to feed its 400 Hz sine wave signal to both blocks 3 and 6. The rule for any divergent arrangement is to check each output in turn and then to continue the search for a faulty block in the area that is common to the incorrect outputs. A possible divergent arrangement is shown in fig. M5. Suppose signals for w, x and z are correct but y is incorrect, then the fault must be situated in block C.

The usual arrangement for CONVERGENCE is that two or more inputs are required at a particular circuit block for the output of that block to be correct. This is similar to the AND function in digital logic circuits and is referred to as *summative*. All such inputs must be checked one by one at the point

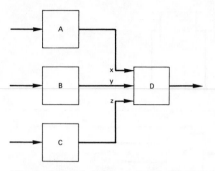

*Figure M6*  An example of convergence

of convergence. If all are correct the fault lies beyond the convergent point, but if any one is incorrect then the fault must lie in that input circuit. This is illustrated in fig. M6 where block D requires the three inputs x, y and z for correct operation. If, for example, all three inputs are correct then the fault can only be in block D. But if input y is incorrect then the fault lies in the circuits producing the y signal.

**5**  Systems with FEEDBACK LOOPS, which connect the output of some block with the input of an earlier block via some network, present one of the more difficult problems in fault location. The output signal, or some portion of the output, is fed back in some way to the input of an earlier block, thus causing a closed loop to exist round the system. This makes it more difficult to locate any faulty block within the loop since the outputs of all blocks will probably appear at fault. This is similar to a completely d.c. coupled system where an incorrect voltage at one point causes all other voltages to be incorrect.

First of all the type of feedback being used in the system and its purpose must be understood. The feedback may only be used to *modify* the characteristics of the system as is the case in automatic gain control circuits used in superhet radio receivers; or the feedback may be totally essential for an output to exist. This latter type of feedback is called *sustaining*, since a feedback signal has to be present to maintain an output of either some oscillation or fixed level. Sustaining feedback is used in many position-control systems, where the feedback signal, proportional to the position of some output device, is used to cancel the effect of an input reference level. As the output motor is driven, the feedback signal moves towards the same value as the reference input

and the error signal reduces to zero. In this way the output is fixed and held at the desired position. Any fault causing a break in the feedback loop would result in the output being driven to one or other of the extreme limits, i.e. to reach its end stop.

Having decided the type of feedback, the way in which it is connected, and its purpose, then the best course of action can be taken to locate any fault. With modifying feedback it may be possible to break the feedback loop, thus enabling each block to be tested separately without a fault signal being fed round the loop. The feedback is best disconnected at the input block end, but naturally care has to be taken with any changes like this since the feedback may be providing both d.c. bias and an a.c. modifying signal. In this case the a.c. portion of the feedback can be eliminated simply by decoupling to ground via a suitable capacitor.

If sustaining type feedback is disconnected from the input, it may be possible to inject a suitable signal in its place and then check circuit block outputs, including of course the feedback element. Since such a wide variety of feedback circuits exist, no standard rule for fault location can be used. Knowledge of the system, understanding of the operation, and a logical approach are essential. For an example consider the block diagram of a motor speed control system shown in fig. M7. The speed of the d.c. motor is set by the level from the reference supply and held constant by the feedback applied via the tachogenerator. A tachogenerator is a device that produces a d.c. output voltage proportional to its speed of rotation. When the motor has reached the desired speed, the d.c. feedback signal from the tachogenerator balances the input reference voltage. The difference signal, after amplification, is just sufficient to keep the motor running constantly at the desired speed. Imagine a break in the feedback lead. This would cause the feedback signal to the comparator to be zero and the motor would tend to run at maximum speed irrespective of the setting of the reference voltage. Faults causing the motor to run at maximum speed could be in either the comparator, controller, the tachogenerator (open circuit) or an open circuit feedback line. To locate the fault the sequence of tests would be:

*a*) Measure tachogenerator output. A relatively large d.c. output should be present since

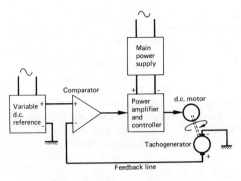

(a) SPEED CONTROL SYSTEM

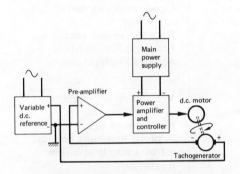

(b) ALTERNATIVE METHOD FOR CONNECTING FEEDBACK

**Figure M7** Feedback systems

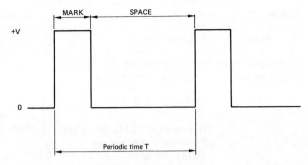

**Figure M8** Mark-to-space ratio

the motor is running at high speed. If this is correct then:

b) Measure feedback signal at the inverting input of the comparator. This should be the same as the d.c. level measured in a. If this is correct then:

c) Check comparator output, which should only be a low-value d.c. level when the variable reference is set to near minimum. If this last test is correct then the fault can only lie in the power amplifier and controller.

To illustrate the changes in fault conditions if the feedback is connected differently, study the block diagram in fig. M7b. In this case the d.c. output from the tachogenerator is connected in series with the input reference supply. When a reference d.c. is applied the motor runs, and as its speed increases the output signal from the tachogenerator also increases. This signal is subtracted from the reference d.c. input so that the pre-amplifier receives an input that is just sufficient to keep the motor running at the required speed. The operating characteristics for this method of connection will be almost identical to that of the previous circuit, but a fault

causing a break in the feedback line will result in zero input to the preamplifier and the motor will not run at all. If such a fault did occur, a simple test to check system operation would be to inject a small d.c. voltage at the preamplifier input. If the motor then runs, the fault must lie in the tachogenerator, the feedback links, or in the input reference supply.

## Mark-to-Space Ratio

Used in the specifications of pulse waveforms to indicate the ratio of pulse width (mark) to the time interval between pulses (space) (fig. M8). Thus a square wave has a mark-to-space ratio of unity.

▶ *Duty cycle*

## Measurement

Measurement techniques are an essential part of electronics since it is only by measurement that a design is verified or faults in a system located. Three standard test instruments are

Multi-range meters (analogue and digital)
Signal generators
Oscilloscopes.

Correct use of one, or a combination, of these instruments can speed the test process. This implies an understanding of the limitations of the instrument in terms of its accuracy, resolution, loading effect and bandwidth. For example the loading effect of an analogue voltmeter (multi-range meter on volts range) is appreciable, especially if voltages are being measured in a circuit which has high resistance. In fig. M9 the voltage across $R_2$ in the potential divider should be 13.33 V. If an analogue meter of 20 kΩ/volt, switched to the 10 V d.c. range, is connected across $R_2$, the meter will indicate about 10 V. This is because the resis-

**Table M1**   *Properties of general-purpose measuring instruments*

| Instruments | Typical use | Accuracy | Comments |
|---|---|---|---|
| Analogue multi-range meter | Measurement of d.c. and a.c. voltage and current<br><br>Ohms range 1 Ω to 20 MΩ<br><br>Bandwidth: 15 Hz to 15 kHz<br>Impedance d.c. ranges<br>20 kΩ per volt<br>  a.c. ranges<br>  2 kΩ per volt | ±1% f.s.d.<br><br>$\begin{cases} \pm3\% \text{ zero to midscale} \\ \pm5\% \text{ midscale to } \frac{2}{3}\text{f.s.d.} \\ \pm10\% \frac{2}{3}\text{f.s.d. to f.s.d.} \end{cases}$ | Robust and well-proven instrument. Good ranges (3 V to 3 kV). Loading effect must be taken into account in medium and high impedance circuits. |
| General-purpose digital multi-range meter ($3\frac{1}{2}$ digits) | Measurement of d.c. and a.c. voltage and current.<br>Ohms range 1 Ω to 20 MΩ.<br>Impedance: 10 MΩ in parallel with 100 pF.<br>Bandwidth: 45 Hz to 10 kHz | ±0.3% of reading or ±1 digit (whichever is the greater). | Good sensitivity and resolution. Easy to use and read. Loading effect can be neglected in most cases. |
| General-purpose cathode ray oscilloscope | Measurement of d.c. levels, a.c. voltage, frequency, wave-shape, rise and fall times. Comparison of time or phase relationship between signals.<br>Bandwidth: d.c. to 10 MHz.<br>Impedance: 1 MΩ in parallel with 20 pF. | ±3% amplitude and time. | Versatile, gives direct visual information on waveform of signal. A probe unit must be used for best results at higher frequencies. |

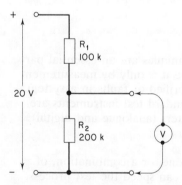

*Figure M9*   A voltmeter of 20 kΩ/V measuring the voltage output of a potential divider formed by two resistors of relatively high value. Voltmeter indicates approximately 10 V on 10 V range whereas the true output voltage is 13.3 V

tance of the meter is 200 kΩ (10 × 20 kΩ). If the next highest range of the meter is selected, say 25 V, giving a meter resistance of 500 kΩ, the meter current will be reduced and a truer indication of the voltage across $R_2$ will be given. It is usually wise to select the highest

range consistent with required accuracy when using an analogue meter to measure voltages in high-resistance circuits. Alternatively a digital meter should be used. These have an input resistance of 10 MΩ or higher so that the loading effect is much reduced. Digital meters also give greater accuracy and are particularly useful when small changes in a level have to be detected. Table M1 shows the general comparisons between the three commonly used measuring instruments.

The CATHODE RAY OSCILLOSCOPE (c.r.o.) is a versatile and extremely useful instrument. With it, it is possible to measure both d.c. values and a.c. waveforms. The sensitivity is usually high, typically 10 mV/div, and the loading effect is slight since the input impedance is usually greater than 1 MΩ. The frequency, shape, and time period of a single waveform can be determined, or waveforms can be displayed in time or phase relationship to one another. This is easily achieved either with a single-beam c.r.o or dual-beam type since the reference signal can be used to

directly trigger the c.r.o.'s time base. The accuracy of both Y (amplitude) and X (time) channels is at best ±3%. At low frequencies the voltage signal to be measured can be taken direct to the Y-input via suitable wires or a coaxial cable. For high frequencies, to avoid the possibility of signal degradation, a fully screened probe should be used. This is because a simple coaxial lead will behave like a badly matched transmission line between the test point and the c.r.o. Y-input, causing attenuation and phase distortion. The cable capacitance of coaxial lead is typically 50 pF per metre and this will be placed in parallel with the c.r.o.'s input capacitance across the test point further degrading the signal. The use of a properly compensated probe unit will reduce these effects considerably. A simple probe is basically a resistive attenuator with capacitive compensation as shown in fig. M.10.

SIGNAL or FUNCTION GENERATORS are used in maintenance when it is required to inject some suitable test signal into the system. The complexity and performance characteristics of the instrument are usually dictated by the system under test, but a very useful aid is a small hand-held signal injector. For analogue systems this is usually a simple fixed-frequency battery-powered oscillator running at 1 kHz, with its output available from a metal prod and a lead with crocodile clip provided for earth connection. A simple device like this can be easily constructed and is always available for use since it is carried around in the pocket. In the same way logic pulsers and logic state sensors can be built and used for checking digital systems. The design for a 1 kHz signal injector can be extended to create a *continuity tester*. This is a battery-powered 1 kHz oscillator with

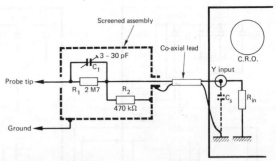

**Figure M10**  Passive probe unit for an oscilloscope; probe gives 10:1 attenuation.

Time constant $R_1 C_1$ is made equal to the input time constant of the c.r.o. $R_{in}/R_2 C_S$, where $C_S$ is co-axial lead capacitance plus input capacitance of c.r.o. When properly adjusted, the probe presents low capacitance to the circuit being measured ($C_1 \cong 10$ pF) and it operates as a simple resistive divider

an audible output provided by a small loud speaker. When the two output leads are shorted together, or connected via the low resistance of a cable wire, the oscillator's output is fed to the speaker. A small tester like this will prove very useful in checking the continuity of cables, connecting wires, and p.c.b. tracks.

A stable frequency standard unit can be effective in calibrating a c.r.o. or other instrument or as a reference for measurements. Fig. M11 shows a simple arrangement using TTL logic for generating square waves at spot frequencies of 1 MHz, 100 kHz, 10 kHz and 1 kHz. Other dividers can be added to give intermediate or lower frequencies.

Pulse generators with outputs that are logic-compatible are essential for checking digital systems. A relatively inexpensive circuit based

**Figure M11**  Frequency-standard unit

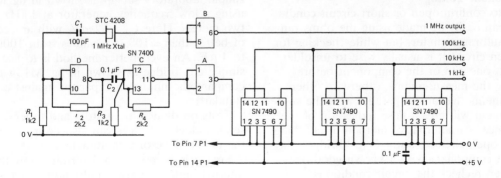

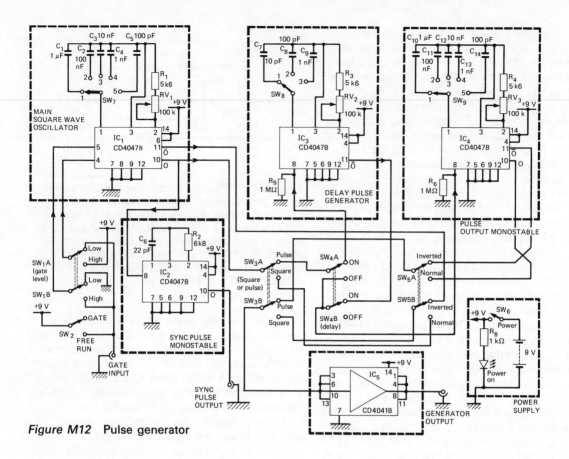

*Figure M12* Pulse generator

on the 4047B mono/astable CMOS i.c. is shown in fig. M12. It has five frequency ranges selected by $SW_7$ from 2.5 Hz up to 500 kHz. The output pulses can be delayed from the sync. pulse from 1 μs to 200 μs and the output pulse widths, selected by $SW_9$, can be set from 1 μs to 200 ms

### Component Testing

Tests to confirm open or short circuit conditions can easily be made using the ohms range of a multirange meter, but while checking for an open circuit it is usually wise to unsolder and lift one end of the component before making the measurements, otherwise other components that are in parallel with the suspect component will give a false indication of the resistance. An alternative method of checking for an open circuit resistor is to "bridge" the suspect component with a known good one, and then recheck the circuit conditions.

"Leaky" capacitors can also be tested using an ohmmeter, again by disconnecting one end

of the capacitor from the circuit. A good electrolytic should indicate a low resistance initially as the capacitor charges, but the resistance should rapidly increase to approach infinity. Open circuit capacitors are best confirmed by placing another capacitor of the same value in parallel and checking circuit operation, or by removing the capacitor and testing it on a simple laboratory set-up as shown in fig. M13$a$ using a low frequency generator at 1 kHz and two meters. Here $C_x = I/2\pi f V_0$ with an accuracy of better than ±10% for values from 1000 pF to 1 μF. An even better method is to use a simple a.c. bridge as shown in fig. M13$b$ to compare the unknown capacitor against a standard.

Tests on diodes, transistors and other semiconductor devices can also be made using the ohms range of a multimeter.

First it is necessary to determine how the internal battery in your multimeter is connected. For example, in one typical instrument the common terminal (marked black) has a

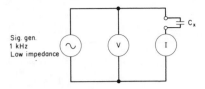

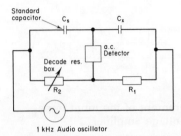

**Figure M13a** A simple laboratory set-up to measure capacitance

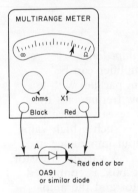

1 kHz Audio oscillator

**Figure M13b** Direct capacitance bridge. The detector may be headphones, an oscilloscope, or sensitive a.c. meter.

At balance $C_X = \dfrac{R_2}{R_1} C_S$

**Figure M14** Using a semiconductor diode to determine the polarity of a multirange meter when switched to the ohms range.

The meter measures a low resistance, indicating that the black terminal is connected to the positive plate of the internal battery.

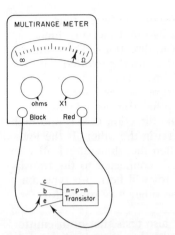

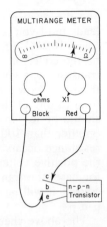

**Figure M15a** Measuring the junction resistance of an n-p-n transistor with a multirange meter. Forward bias on base emitter. A low resistance (typically less than 1 kΩ) should be indicated.

**Figure M15b** Forward bias on base collector. A low resistance (less than 1 kΩ) should be indicated.

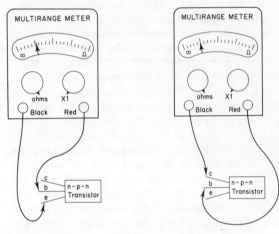

**Figure M15c** Reverse bias on emitter base. A high resistance (greater than 100 kΩ) should be indicated

**Figure M15d** Reverse bias on collector base. A high resistance (greater than 100 kΩ) should be indicated

positive voltage on the resistance range. If you do not know the connections for the particular meter you are using, the polarity can be determined by connecting the multimeter (on ohms range) to an electronic voltmeter, or by measuring the forward and reverse resistance

of a semiconductor diode of known polarity (see fig. M14.)

Having established the ohmmeter lead polarity, you can discover a great deal about a transistor. First identify the device leads if not known (see fig. M15). Measure the forward

and the reverse resistance between pairs of leads until you find two that measure high (over 100 kΩ) in both directions. These must be the collector and emitter (provided the transistor is a good one). The remaining lead is the base. Now measure the resistance from base to one of the other transistor leads; it should be low in one direction (1 kΩ) and high (greater than 100 kΩ) in the other. If the low resistance occurs when the ohmmeter lead with the positive voltage is connected to the transistor base, the transistor will be n-p-n type. Of course, it will be the other way round for p-n-p.

The above check also tests that both emitter/base and collector/base junctions in the transistor are good. If either junction shows up high resistance in both directions, it is broken down.

### Testing amplifiers: basic measurements
**1** *Measurement of gain*
The layout of the measuring circuit is shown in fig. M16. Suppose the amplifier's voltage gain at a frequency of 1 kHz is required. First the signal generator is set to give an output of say 500 mV at 1 kHz, with the attenuator switched to zero dB. This signal, at the amplifier input (point A), is connected to the Y-input of the oscilloscope and the oscilloscope controls are adjusted so that the trace displayed uses a large portion of the screen and has its peaks just on graticule lines. The oscilloscope leads are then moved to the amplifier output (point B) and, leaving the oscilloscope controls as set, the attenuation is increased until the output is exactly the same height as with the first measurement. The gain of the amplifier is now equal to the setting of the switched attenuator. The advantage of this method is that the measurement does not depend upon the accuracy of the oscilloscope. If the variable attenuator has switched ranges down to 0.1 dB, then the result will be obtained to within ±0.1 dB.
**2** *Measurement of frequency response and bandwidth*
Using the same set-up as in fig. M16, the gain of the amplifier can be found at any frequency. The gain, in dB, is then plotted against a frequency on linear/log graph paper. For an audio amplifier 4 cycles of log would be required to cover the frequency range 10 Hz to 100 kHz.

The bandwidth can be quickly determined by noting the two frequencies at which the gain falls by 3 dB from the mid-frequency gain.

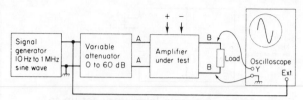

*Figure M16* Laboratory set-up to measure the voltage gain of an amplifier

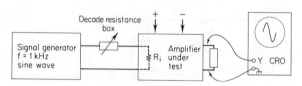

*Figure M17* Measurement of the input impedance of an audio voltage amplifier

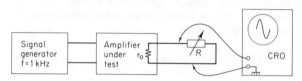

*Figure M18* Measurement of output impedance of a voltage amplifier

**3** *Measurement of input impedance*
The input circuit of an amplifier can be represented by a resistor in parallel with a low-value capacitor. At low frequencies the input impedance is mostly resistive since the reactance of the capacitor is such a high value. A circuit for measuring input impedance at 1 kHz is shown in fig. M17. A variable resistor, usually a decade resistance box, is connected between the signal generator and the amplifier input. This resistor is set to zero and the amplifier output is connected to the measuring instrument, an oscilloscope or a.c. meter.
The controls are set so that a large deflection is indicated. The resistance of the decade box is then increased until the indicated output signal falls by exactly a half. Since the resistance box and the amplifier input impedance form a potential divider when the output is halved, the setting of the decade resistance box is equal to the input resistance.
**4** *Measurement of output resistance*
The circuit shown in fig. M18 is used for this measurement. The technique is similar to that

of measuring the input impedance. A signal frequency of 1 kHz is used and initially $R_L$ is disconnected and a large deflection obtained on the oscilloscope. The external load $R_L$ is then connected and reduced in value until the output falls by exactly a half. The value of $R_L$ at which this occurs is equal to the resistance.

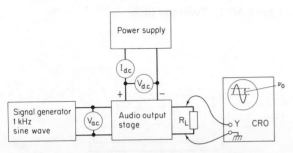

*Figure M19* Measurement of power output, efficiency and sensitivity of an audio output stage.
$R_L$ is a wire-wound resistor of the same value as the loud-speaker impedance

**5** *Measurement of power output, efficiency, and sensitivity for an audio amplifier*
For these measurements the loudspeaker should be replaced by a wire-wound load resistor of the same value as the loudspeaker impedance, and the tests should be carried out at a frequency where the loudspeaker impedance would be mostly resistive, typically about 1 kHz. The diagram for the measurement is shown in fig. M19. The wattage rating of the load resistor should be higher than that of the maximum output power. The input voltage should be adjusted until the output signal indicated by the oscilloscope is a maximum undistorted level. This is when there is no clipping of the positive and negative excursions of the output signal. Naturally if a distortion meter is available then a more accurate check on distortion levels can be made. Then the maximum output power should be recorded without exceeding the manufacturers specified value of harmonic distortion.

$$\text{Power output} = V_O^2 / R_L$$

where $V_O$ is the r.m.s. value of the output signal.

$$\text{r.m.s.} = \frac{\text{peak to peak value}}{2\sqrt{2}}$$

The efficiency of the amplifier can be checked by measuring the d.c. power taken by the amplifier from the supply.

$$\text{D.C. power} = V_{dc}I_{dc}$$

$$\text{Power efficiency} = \frac{\text{r.m.s. output power}}{\text{d.c. input power}} \times 100\%$$

The *sensitivity* of the amplifier is the input voltage required at the input to produce maximum undistorted output power.

**Transient testing of amplifiers**
All the tests previously described are made using an input signal at one frequency. By applying pulses or square waves to an amplifier it is possible to acquire information about the amplifiers frequency response, phase distortion, and any tendency to instability.

A square wave is made up of a series of pure sine wave components, which are a fundamental, having the same periodic time as the square wave, and all odd harmonics. Thus by applying a square wave or pulse to an amplifier, a large range of signals at different frequencies have to be amplified by the same ratio and without phase shift if the output is to be a perfect replica of the input.

For testing low frequency amplifiers, a square wave of 40 Hz or 1 kHz is suitable and the output signal can be observed on an oscilloscope. Departure from squareness in the output signal gives a good indication of the transient distortion that is present in the amplifier. Various conditions are shown in fig. M20.

**Distortion measurements**
Various types of distortion can affect the shape of the output signal from an amplifier.

Measurement of distortion levels is usually made using a distortion meter, an instrument which sums the power in all the harmonics and gives the result as a percentage of the output power. This gives the value of the *total harmonic distortion* resulting from amplitude and non-linear distortion, but does not include frequency, phase or intermodulation distortion. A frequency of 1 kHz is normally used for this measurement.

Total harmonic distortion can also be measured by passing the output voltage signal through a filter which attenuates the measurement frequency (1 kHz) but passes all harmonics. A good circuit for this is a

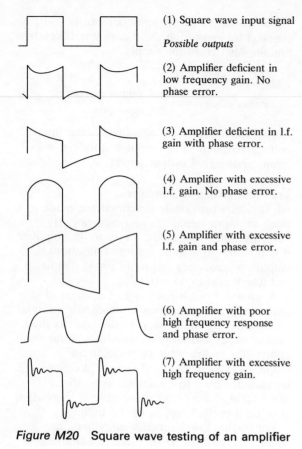

(1) Square wave input signal

*Possible outputs*

(2) Amplifier deficient in low frequency gain. No phase error.

(3) Amplifier deficient in l.f. gain with phase error.

(4) Amplifier with excessive l.f. gain. No phase error.

(5) Amplifier with excessive l.f. gain and phase error.

(6) Amplifier with poor high frequency response and phase error.

(7) Amplifier with excessive high frequency gain.

**Figure M20**   Square wave testing of an amplifier

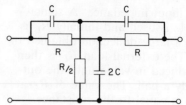

**Figure M21**   Twin-tee filter

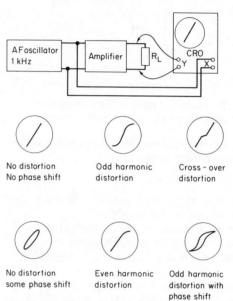

No distortion
No phase shift

Odd harmonic distortion

Cross - over distortion

No distortion
some phase shift

Even harmonic distortion

Odd harmonic distortion with phase shift

**Figure M22**   Method of displaying distortion using CRO

twin-tee filter as shown in fig. M21 since this has maximum attenuation at one frequency. The output can be measured using a sensitive r.m.s. millivoltmeter.

Intermodulation distortion can be measured by feeding two signals of 400 Hz and 1 kHz into the amplifier usually with a ratio of about 4:1. Then using a filter at 1 kHz the result of any intermodulation will be indicated using the method detailed previously.

A method that can be used to display amplitude distortion, phase shift distortion and harmonic distortion for an audio amplifier is shown in fig. M22. The signal generator set to 1 kHz is fed to the amplifier input at a suitably low level and to the X-input of the oscilloscope. The output from the amplifier is fed to the Y-input of the oscilloscope. The oscilloscope trace will be a straight line at an angle of 45° if the amplifier output is undistorted. Naturally a high-quality oscilloscope must be

used for this test, since any non-linearity in the X and Y oscilloscope amplifiers will also be displayed. Various outputs for different types of distortion are shown in fig. M22 also.

### Testing power supply circuits
The main parameters are the following:
*a*) D.C. output voltage
*b*) Available d.c. output current
*c*) Output ripple voltage at full load
*d*) Stabilization against mains supply changes
*e*) Regulation from zero to full load
These can all be measured using a standard test set-up as shown in fig. M23*a*.

The d.c. output voltage should be measured, and if necessary adjusted, when the unit is fully

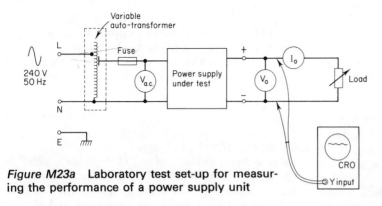

Figure M23a Laboratory test set-up for measuring the performance of a power supply unit

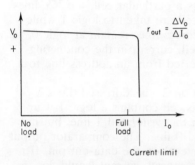

Figure M23b Typical load regulation plot for a power unit with a current limit. Between no load and full load current, the change of output voltage should be very small

loaded. However it is sometimes advisable to measure the output on a low load and then gradually increase the load current to maximum. There should, of course, be little change in the output voltage.

The peak-to-peak ripple amplitude can be checked best by measuring at the output with an oscilloscope. A sensitive a.c. range must be selected because the ripple should be quite low, typically less than 20 mV.

Measurement of stabilization and regulation requires that any small change in d.c. output be carefully noted, and therefore a digital voltmeter is often necessary. For stabilisation measurement, the unit should be fully loaded and the change in d.c. output voltage noted for say a ±10% change in the a.c. input. The mains input can be varied using an adjustable auto-transformer as shown. Then if, for example, the d.c. output changed by 50 mV from

10 V, i.e. an output change of 0.5%, then the line stabilisation would be 40:1.

Load regulation is measured, keeping the a.c. input constant, by noting the change in output when the load is varied from zero to full load.

$$\text{Load regulation} = \frac{\text{Change in d.c. output}}{\text{D.C. output on no load}} \times 100\%$$

For example suppose the output changed by 20 mV from 10 V. The load regulation is

$$\frac{20 \times 10^{-3}}{10} \times 100\% = 0.2\%$$

To obtain fuller information on a power supply's performance it is often necessary to plot the load regulation curve. This is a plot of output voltage against load current. A typical result for a unit with current limiting is shown in fig. M23b.

## Memory

An electronic memory is simply any device that accepts and stores information, but the subject becomes more complex when all the different types of memory in common use are considered. Here we are only concerned with memories designed to hold digital information, i.e. data that is stored as words consisting of sets of 1s and 0s rather than in analogue form. The term "memory" has now come to mean an arrangement of bit-storage devices which have a capacity large enough to hold programs or data. Therefore a memory has two essential properties:

1) The location where every bit of information is stored must have a unique address.

2) It has to be possible to read out the state of every location in the memory.

An electronic digital memory is characterized by

*a*) Capacity — the total number of store locations.

*b*) Organisation — the way in which the store locations are arranged. For example 256 locations can be organised as

$256 \times 1$   $128 \times 2$   $64 \times 4$   $32 \times 8$   $16 \times 16$

*c*) Method of addressing — this can be random, serial or cyclic.

*d*) Access time — the speed with which a location can be read.

*e*) General use — either to hold data, program, fixed program as in a ROM, or as a backing store for bulk storage.

The types of store are

RAM *random-access memory* This usually has very fast access and is one in which a location can be accessed in the same time as any other location. Thus all locations are equally accessible and the time required for selecting and locating a piece of data is constant. With this type, data can be read out or new data written in. Used for current program and data.

ROM *read-only memory* A storage device in which the data is fixed and not erasable. Used for fixed program, constants, character generators.

PROM *programmable read-only memory* This is a ROM that can be programmed by the user. The i.c. is made with each memory cell set to give a logic 1 output. Fusible links connecting cell outputs can be "burned" open to give a required store pattern. Once this programming has been carried out, the data is fixed.

EPROM *erasable programmable read-only memory* A PROM in which the data pattern can be erased, usually by exposing the internal cells to ultra-violet light, and a new pattern then set up.

BACKING STORE Large-capacity memories used to maintain records of programs or data in large volume. Typical examples are magnetic tape and discs.

A memory is referred to as VOLATILE if the stored data is lost when the power supply to the memory is switched off. Most semiconductor RAMs are therefore volatile whereas a ROM, which has a fixed pattern, is non-volatile.

A further distinction has to be made between STATIC and DYNAMIC memories. The former will hold data as long as d.c. power

is applied, but the dynamic type must be provided with clock pulses in order to keep the stored pattern refreshed. The advantage of a dynamic RAM over a static type is that it consumes far less power for the same memory capacity. Dynamic RAMS are sometimes referred to as DRAMS.

**1** *Bipolar RAM (TTL type)*
Each location consists of a TTL bistable which can store a 1 or 0 depending on which transistor is conducting. The cells, 16 shown (fig. M24) in a 4 by 4 array, are connected to XY address lines and to two common write/read lines. To address a particular cell, the XY lines unique to that cell are taken to logic 1 while all other XY address lines remain at logic 0. In the addressed cell, current in the conducting transistor is diverted from an address line to a read line.

Suppose location 2,1 is addressed (lines $Y_2$, $X_1$) and that this cell contains a logic 1. Current will flow in the write/read 1 line, but not in the write/read 0 line. The comparator output will give a high level on the data-out pin. If a logic 0 had been stored, current would flow in the write/read 0 line to give a low output.

Data can be written into a location by taking the cell address lines to logic 1 and then applying a WRITE command. One of the write lines will go low depending on the data in being a 1 or 0 and this causes current to flow in one of the transistors in the bistable while cutting off current in the other.

*a*) Readout is non-destructive.

*b*) The memory is volatile.

*c*) Access time is fast — typically 50 ns for Schottky TTL.

Other bipolar types are ECL and $I^2L$.

**2** *MOS RAM* (static type)
These can be either n or p type. Fig. M25 shows a p-type using all enhancement mode MOSFET. One cell consists of a bistable with XY address lines and two digit sense lines. To address a particular cell, the appropriate XY address lines are taken to ground. This causes $Tr_5$, $Tr_6$, $Tr_7$ and $Tr_8$ to conduct. For read-out, depending on the state of the bistable, current will flow either in digit line $D_0$ or in $\bar{D}_0$. This differential digit line current is sensed by a comparator as shown.

**Table M2** *Memory devices* (*typical figures*)

| Type | Capacity in bits | Type of access | Access time | Use main memory in computers |
|---|---|---|---|---|
| Core | up to $10^6$ | random | $1\ \mu s$ | Backing store (non-volatile) |
| Drum | $10^7$ | cyclic | 10 ms to 30 ms | |
| Disc | $10^9$ | cyclic | 10 ms to 200 ms | |
| Magnetic tape | $10^8$ | serial | several minutes | |
| Magnetic bubble | 250K | random | $10\ \mu s$ to $100\ \mu s$ | Program and data, non-volatile |
| Semiconductor bipolar | 1K RAM* 8K PROM | random | 50 ns | RAM, ROM, PROM, static. |
| Semiconductor MOS | 64K* | random | 200 ns | RAM, ROM, PROM, EPROM static and dynamic. |

* One i.c. package.

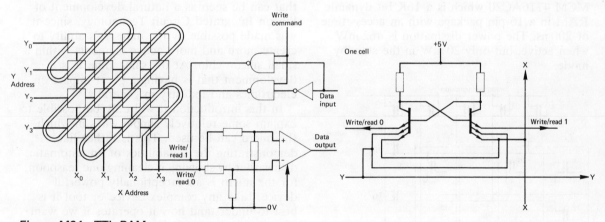

Figure M24   TTL RAM

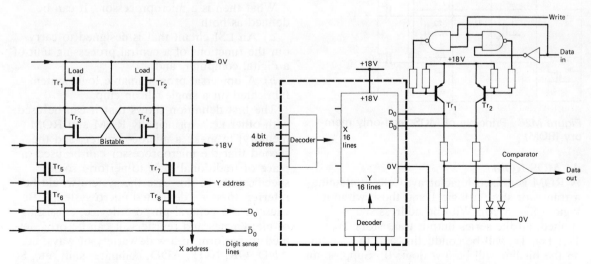

*Figure M25a*   MOS RAM: basic cell structure

*Figure M25b*   Organisation of 256-bit MOS RAM

To write in, the cell is addressed and the Write command taken to logic 1. Either $Tr_1$ or $Tr_2$ (the external transistors) conducts depending on the level of the data input, so that one digit line switches to +18 V while the other remains at ground. This forces the bistable to assume the new state.

a) Readout is non-destructive.
b) The memory is volatile.
c) Access time is typically 200 ns.

### 3 MOS dynamic RAM

This must be supplied with continuous clock signals to refresh the data. The bits of information are held on the gate capacitance of a MOSFET between clock pulses. The advantages are that only three transistors are required for a storage cell and overall power consumption is very low. A typical refresh rate is every 2 ms. A good example is the Motorola MCM 4116AC20 which is a 16K bit dynamic RAM in a 16-pin package with an access time of 200 ns. The power dissipation is 462 mW when active, but only 20 mW in the standby mode.

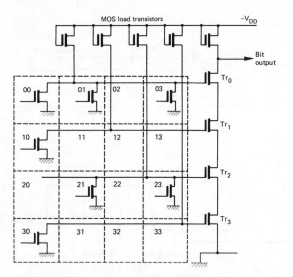

*Figure M26* Principle of MOS read-only memory (ROM)

### 4 MOS ROM

A ROM is a fixed logic array. Cells containing a transistor store a logic 1 and those without a logic 0 (fig. M26). With no address signal applied, all the series output transistors $Tr_0$, $Tr_1$, $Tr_2$, $Tr_3$ will be conducting and the output on the bit line will be low (logic 0). Suppose an address is then applied to 2.1. The transistor at

cell 2.1 will conduct taking the gate of $Tr_2$ low. $Tr_2$ turns off and the bit line output goes high to logic 1, showing that a logic 1 is stored at 2.1. If a new address is applied, say at 2.2, where there is no transistor, $Tr_2$ remains conducting and the output remains low. In the diagram the address lines have been omitted for clarity.

▶ *Microprocessors and microcomputers*

## Microprocessors and Microcomputers

Although there are many books and excellent articles on microprocessors and microcomputers an introduction to the subject must be included here because the "micro" can now be considered to be at the centre of all modern electronics.

The microprocessor is one of the innovations that can be seen as a natural development of Silicon Integrated Circuit Technology, since it was made possible by the industry's ability to create more and more active elements within a small silicon chip. At the same time it is a development that is having a great impact on electronics and industry in general.

In this introduction it will only be possible to explain some relatively simple ways in which the micro can be used. This is rather like demonstrating the capabilities of an automatic dishwasher by showing it cleaning one teaspoon; for the micro is an exceptionally powerful device. Like any complex device or tool it is best to understand how it operates if we want to make the most efficient use of it.

What then is a microprocessor? It can be defined as both

a) An LSI circuit that is designed to carry out the functions of a central processing unit of a digital computer, and

b) A universal programmable logic system fabricated on a single silicon chip.

The first definition shows that it can be used with other i.c.'s (principally RAM and ROM memory) to make a microcomputer; and the second that the microprocessor can be used in place of traditional logic to perform some specific task. In this case the system would be referred to as a "dedicated micro system". Because it is a general-purpose device, consisting of many gates and registers, it can be programmed to perform in a wide variety of ways, i.e. AND, OR, NOT, ADD, compare, shift, etc. So instead of designing a hardwired logic system

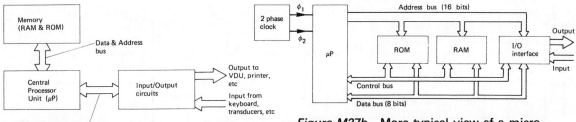

Figure M27a  Simple view of system using a microprocessor

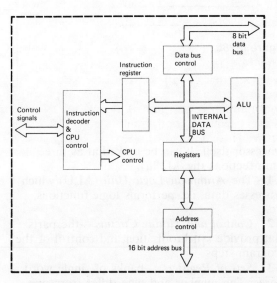

*Figure M27b*  More typical view of a microprocessor system

*Figure M28a*  Internal structure of a typical 8-bit microprocessor

as a solution to a particular problem, a designer can instead programme a microprocessor to do the job. This versatility is the reason why the microprocessor's area of application is continually expanding. In general a microprocessor will be used in place of traditional logic where

Many different functions are required.

The system must be flexible and capable of expansion.

Many logic functions have to be stored.

Microprocessors are made as 4, 8, 12 or 16 bit devices, the 8 and 16 bit types now being the industry standard. The types of technology used are

1  Metal Oxide Silicon (MOSFET)
   i.e. NMOS, PMOS, and CMOS.
2  Bipolar transistor
   $I^2L$, ECL, Schottky.

Some type numbers of commonly used 8-bit micros are the

| | |
|---|---|
| 8080 (Intel) | 6502 (Rockwell) |
| Z80 (Zilog) | 6800 (Motorola) |

The microprocessor, consisting of an arithmetic logic unit (ALU), sets of registers and control circuits, cannot be used by itself to create a system. The essential support chips are shown in the simplified view of a system in fig. M27*a*. This can form either a dedicated controller or a microcomputer since both memory and input/output devices are included. The memory holds both data and program, and the instructions in the program are read sequentially by the microprocessor from the memory. A more typical view of a microcomputer system is as shown in fig. M27*b*. A crystal-controlled clock circuit (with the oscillator section often internal to the micro) is provided to synchronise all operations. The memory is shown separated into ROM and RAM. The

ROM, a fixed memory i.c., holds the system control program. This control program, termed the *monitor*, sets up the initial conditions for the system and ensures that the microprocessor is in a ready state for any new program. The RAM holds data and current program, i.e. the set of instructions that can be altered by the user.

The connections between the various parts of the system are made by groups of parallel connecting leads, each group called a bus. These, for an 8-bit micro, are

*a*) 16-bit wide address bus — 16 connections that allow the micro to address up to 65 536 memory locations with a binary coded word.

*b*) 8-bit wide data bus — this is bidirectional allowing data to be sent to and from the micro.

*c*) CPU and system control bus.

Before looking further at the operation of the system consider the internal structure (referred to as the architecture) of the micro-

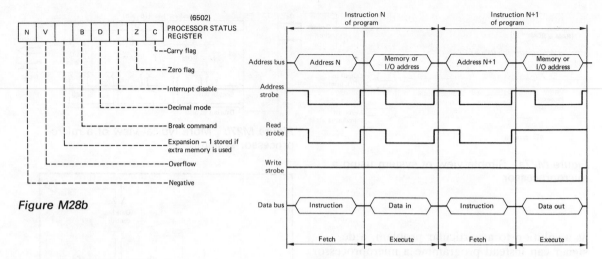

Figure M28b

*Figure M29* Timing diagram in microprocessor system

processor itself. It can be looked at as three main sections (fig. M28*a*):

**1** The *Arithmetic/Logic Unit* (ALU) which processes data, i.e. performs logic functions, etc.

**2** *Control and Timing Circuits*—the parts that provide synchronisation and control of the program steps.

**3** *Sets of Registers* — banks of single word stores. The number and type differ from one microprocessor device to another but generally the following registers are the main types:

*Address register* holds the address of the memory location being accessed.

*Instruction register* holds an instruction that defines the program operation.

*Data register* stores the data that is to be operated on.

*Accumulator(s)* hold the result of an operation while other operations continue.

*Program counter* which will hold the 16-bit binary address of the next memory location to be accessed. Having been initially set to the address at which the program commences, the program counter automatically increments after each "fetch" operation to formulate the next address.

*Status register* — sometimes called the flag register or condition code register—indicates conditions that may have occurred during the execution of the previous instruction. For example the Zero Flag would be set by the microprocessor if, during any data movement or calculation operation, the 8-bit result in the

accumulator is zero. (The other flags for a 6502 processor are shown in the fig. M28*b*.)

As mentioned previously all operations are synchronised by a clock generator. The microprocessor is organised to read an instruction from the program memory during the Fetch phase, and to process, send or receive data during the Execute phase. Each phase may take several machine cycles and, to prevent incorrect transfer of information, the control within the microprocessor generates timing or strobe signals. These are

1) Memory Read (RD). When active the microprocessor is ready to read data from a memory or input/output location. This strobe therefore gates data onto the micro's data bus.

2) Memory Write (WR). When active it shows that valid data is available from the microprocessor to be stored in a memory location or output section that has been addressed.

3) Memory Request (MREQ). When active this strobe indicates that the address bus is holding a valid address for a memory/read/write operation.

Typical timing waveforms are as shown in fig. M29 and the operation of the system is as follows.

The system starts in the Fetch phase. A valid address is available on the address bus and the data at this address, representing a coded program instruction, is strobed onto the data bus. Inside the microprocessor this data is routed to

the instruction register and then decoded to control the operation of the ALU. The program counter then moves on to the next address and the data held at this location is then strobed onto the data bus. This data is to be operated on in accordance with the previously decoded instruction, i.e. the ALU may be required to ADD this data to the contents of the accumulator. The data is therefore transferred to the data register and the instruction is executed. On the next cycle a new instruction is fetched, and another operation is performed. In this way the operations continue through the program.

## Programs

The term SOFTWARE describes the program, i.e. the series of instructions to be given to the microprocessor. Any program that has been proven and then fixed in a ROM is then referred to as FIRMWARE. The monitor program for any system is therefore Firmware.

Every microprocessor has an Instruction Set which is a list of the operations that the device will carry out, and the size of the instruction set is a measure of the versatility of the microprocessor. As seen previously, a coded instruction of the microprocessor must be a pattern of 0s and 1s—in other words a binary number, and this code is known as MACHINE LANGUAGE (or OBJECT CODE). Writing programs in Machine Language is difficult and tedious and it is therefore rarely used. Instead the MACHINE CODE (sometimes called the OPERATION CODE) is used. This is similar to Object Code but is written in HEXADECIMAL instead of binary. Hex. numbers are to the base of 16 and use letters in place of numerals for the decimal numbers 10, 11, 12, 13, 14 and 15 as shown in the next table. Hex. numbers are much more manageable than binary.

Thus $27$ decimal $= 1B_{hex}$

since $27 = 16 + 11$.

and $FA_{hex} = 15(16) + 10 = 250$ decimal

Converting from binary to hex. is even simpler.

*Example* $1110100110 = 11, 1010, 0110$

$$\downarrow \quad \downarrow \quad \downarrow$$

$$3 \quad 10 = A \quad 6$$

$$= 3A6_{hex}$$

| Decimal | Hexadecimal | Binary |
|---|---|---|
| 0 | 0 | 0000 |
| 1 | 1 | 0001 |
| 2 | 2 | 0010 |
| 3 | 3 | 0011 |
| 4 | 4 | 0100 |
| 5 | 5 | 0101 |
| 6 | 6 | 0110 |
| 7 | 7 | 0111 |
| 8 | 8 | 1000 |
| 9 | 9 | 1001 |
| 10 | A | 1010 |
| 11 | B | 1011 |
| 12 | C | 1100 |
| 13 | D | 1101 |
| 14 | E | 1110 |
| 15 | F | 1111 |

Writing the machine code program in hex. greatly simplifies the job but there are two other levels at which programs can be written:

Assembly language

and High-level language (Basic, Algol, Fortran, etc).

These are intended to make program writing even easier, but have to be converted into machine language for the microprocessor.

An Assembler converts Assembly Language into Machine Language (object code).

A Compiler is used to convert a High-level Language into Machine Language.

In most electronic control situations the microprocessor has to be programmed using the machine code. Consider a very simple example, using the 6502 micro instruction set, for adding two decimal numbers. (This is the trivial example mentioned at the beginning of this section.)

*Program for adding two decimal numbers* (p. 162) The program (op-code instructions) would be loaded into the microcomputer via a keyboard or other input device and stored in a free portion of RAM, starting in this case at the arbitrary address of 0030. With two decimal numbers placed in memory location 0050 and 0051 (say 20 and 15), and the program correctly run, the result (35) would appear in memory location 0052. There is not space here to go further into programming and the best way to learn the subject is to get practice with a microcomputer system. The above example, which could form part of a larger more detailed program, is only given to show how

| Address | Assembly language | Machine code (hex) | Comments |
|---------|-------------------|--------------------|----------|
| 0030 | CLC | 18 | Clear carry. |
| 0031 | SED | F8 | Set decimal mode. |
| 0032 | LDA | AD | Load Accumulator with contents of 0050. |
| 0033 | | 50 ⎱ | Contents of 0050. |
| 0034 | | 00 ⎰ | |
| 0035 | ADC | 6D | Add memory content of 0051 to Accumulator. |
| 0036 | | 51 ⎱ | Contents of 0051. |
| 0037 | | 00 ⎰ | |
| 0038 | STA | 8D | Store content of Accumulator at 0052. |
| 0039 | | 52 ⎱ | Location 0052 |
| 003A | | 00 ⎰ | (the result). |
| 003B | JMP | 4C | Jump to monitor |
| 003C | | X ⎱ | Location of start |
| 003D | | X ⎰ | of monitor program. |

the microprocessor would be instructed, using machine code, to carry out the various operations.

Microprocessors are used in a wide variety of applications, including robots, automatic test equipment, control systems, and multifunction test instruments.

An example of a control application is the continuous monitoring and control of the environment in a group of greenhouses. Each greenhouse is provided with sensors for temperature, humidity, the amount of direct sunlight, etc, and with output transducers such as heaters, ventilators, motor-operated shades and mist sprays. The outputs of the various sensors are sampled at time intervals decided by the microprocessor program and converted from analogue to digital. The digital output corresponding to, say, the level of humidity in one greenhouse is then compared by the microprocessor, under program control, with a reference word held in the system memory. This reference word corresponds to the desired humidity level and, if the input is low, the microprocessor then outputs a signal to operate a mist spray. In a similar way the microprocessor, via the program, checks and controls temperature (causing heaters or fans and ventilation to operate) and all other parameters within the greenhouses. Apart from the obvious advantage of providing continuous control, the microprocessor system offers flexibility since the program can be altered to suit the number of greenhouses and the different environments required in each house.

Suggested further reading:
*Microprocessor Fundamentals* by F. Halsall/P. F. Lister (Pitman).
*Microcomputer Design* by C. A. Ogdin (Prentice-Hall).

## Miller Effect

Any inverting amplifier, such as a bipolar transistor connected in common emitter mode, or a FET in common source, will have some small feedback capacitance between its output and input terminals. In a transistor this feedback capacitance is $C_{b'c}$ and in a FET it is $C_{gd}$.

This capacitance is amplified by the gain of the device and appears as a much larger value in parallel with any other input capacitance.

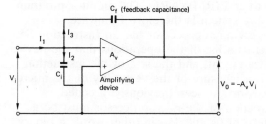

**Figure M30**

This amplification of feedback capacitance is referred to as Miller Effect. For good high-frequency response the total input capacitance must be kept low. In fig. M30

$$I_1 = I_2 + I_3$$
$$I_2 = V_i(j\omega C_i)$$
$$I_3 = (V_i - A_v V_i)j\omega C_f = V_i(1 - A_v)j\omega C_f$$
$$\therefore \quad I_1 = V_i[j\omega C_1 + j\omega C_f(1 - A_v)]$$

The portion in the square brackets represents the equivalent input capacitance $C_{eq}$, i.e.

$$C_{eq} = C_1 + C_f(1 - A_v)$$

For example, suppose $C_1 = 22 \text{ pF}$, $C_f = 5 \text{ pF}$ and $A_v = -50$, then

$$C_{eq} = 22 + 5(51) = 277 \text{ pF}$$

The effect is put to good use in the generation of ramp and triangle waveforms.
▶ *Cascode*    ▶ *Waveforms*

## Monostable

The monostable multivibrator, sometimes called a "one shot", is a circuit that is widely used for generating an output pulse of fixed width and amplitude. This output pulse is only produced when the circuit is triggered into operation by a narrow input pulse. The monostable can be made using discrete components or is available in an integrated circuit package. The most common form for producing the circuit using discrete components is shown in fig. M31, but it should be noted that there are several variations of this. These include emitter coupled and complementary types.

**1** The basic circuit can be seen to consist of a two-stage amplifier with resistive coupling from output to input. As the name suggests, the circuit has one fixed stable state. This is with no input trigger pulse, when $Tr_2$ is on and $Tr_1$ is off. $Tr_2$ conducts because it has forward bias provided by $R_T$. This resistor has a value low enough to provide sufficient base current to just drive $Tr_2$ into saturation. The collector voltage of $Tr_2$ will then be approximately 0.1 V, and this ensures that $Tr_1$ is held off in a non-conducting state. The circuit can be switched into a "quasi-stable" condition by applying a positive pulse to $Tr_1$ base. This need only be of short duration, as its purpose is merely to trigger the circuit into operation. The pulse causes $Tr_1$ to conduct and its collector voltage falls. This change in voltage is coupled via $C_T$ to $Tr_2$ base. Remember that the voltage across a capacitor cannot change instantaneously, so a change of voltage on one plate appears as an equal change on the opposite plate. The voltage on $Tr_2$ base therefore goes negative and this turns off $Tr_2$. The collector voltage rises towards $V_{CC}$ and because of the positive feedback via $R_3$, $Tr_1$ is forced to conduct more. Very rapidly the circuit switches its state, making $Tr_1$ on and $Tr_2$ off, but this is not a permanently stable condition. The base of $Tr_2$ is negative, while $Tr_1$ collector is at approximately +0.1 V, so the capacitor $C_T$ now charges via $R_T$ and $Tr_1$ causing its right-hand plate to move from $-V_{CC}$ towards $+V_{CC}$. When this voltage reaches nearly +0.6 V, $Tr_2$ begins to conduct again, and the circuit rapidly switches back to its stable state. The output from $Tr_2$ collector is thus a positive pulse of amplitude approximately $V_{CC}$ and with a defined width. The width of the pulse is determined by the time constant $C_T R_T$ and

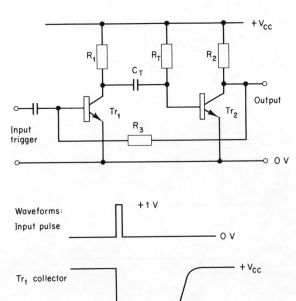

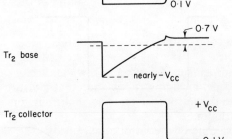

*Figure M31*   Basic transistor monostable circuit

approximately

$$t_d \simeq 0.7 C_T R_T$$

**2** Op-amps can also be wired to give the monostable function as in fig. M32. Positive feedback is provided by $R_1$ and $R_2$ which, in the absence of a trigger input, forces the output to be at $+V_{o(sat)}$. This is a stable condition, since the voltage across the timing capacitor $C_T$ is clamped at +0.6 V by the diode $D_2$. When a negative trigger input is applied to the non-inverting input, the output switches rapidly to $-V_{o(sat)}$. A portion of this negative level appears across $R_1$ and is applied to the non-inverting input. Capacitor $C_T$ now discharges via $R_T$ towards $-V_{o(sat)}$, and, when the voltage across $C_T$ is just more negative than the voltage across $R_1$, the output switches back to its stable state of $+V_{o(sat)}$. This circuit will have a relatively poor recovery time since the timing

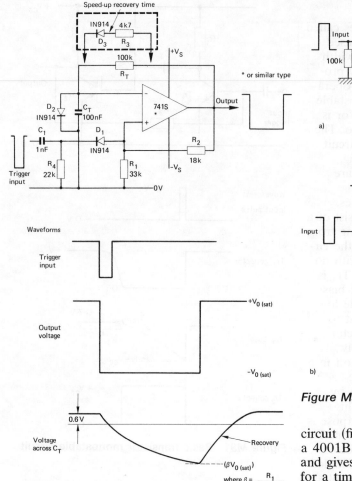

*Figure M32* Monostable using an op-amp

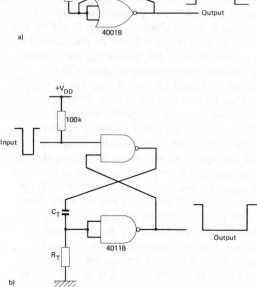

*Figure M33* CMOS monostable circuits

capacitor must be recharged to +0.6 V. This can be improved by using a speed-up circuit as shown in the inset. Assuming that $\pm V_{o(sat)} \gg 0.6$ V then the pulse width is given by

$$T \simeq C_T R_T \log_e 1/(1-\beta)$$

where $\beta = R_1/(R_1 + R_2)$. If $\beta = 0.63$, then $T \simeq C_T R_T$.

With the values shown in the circuit, $T \simeq$ 10 ms. Monostables based on op-amps will not have fast rise and fall times because of slew rate limitations and it is best to use them for generating relatively wide pulses, i.e. in the range 0.1 ms to 1 sec.

**3** CMOS gates with their very high input resistance are ideal for creating low-cost reasonable-performance monostables. The

circuit (fig. M33*a*) using two NOR gates from a 4001B i.c. is triggered from a positive input and gives an output pulse that switches high for a time period set by the *CR* network. If $R_T$ has a value of 1.5 MΩ then the output pulse width is approximately 1 sec per μF of $C_T$. In the NAND circuit, using two of the gates from a 4011B i.c. (fig. M33*b*) the trigger is a low input which causes the output to switch low for the defined time period. The stability of pulse width with power supply and with changes in i.c. is not high, but the circuits do possess the advantage of simplicity. Another CMOS monostable using the 4013D bistable is shown in the section on ▶ *Touch control switches*.

For output pulse widths in the range from 10 μs up to minutes, the 555 range of timers give excellent performance (▶ *Timers*). For pulse widths less than 10 μs the specialised monostable i.c.s (TTL CMOS, and ECL types) are usually preferred since they are totally compatible with the type of logic used; possess multiple inputs; have very fast rise and fall times; and have good stability. The most popular types include

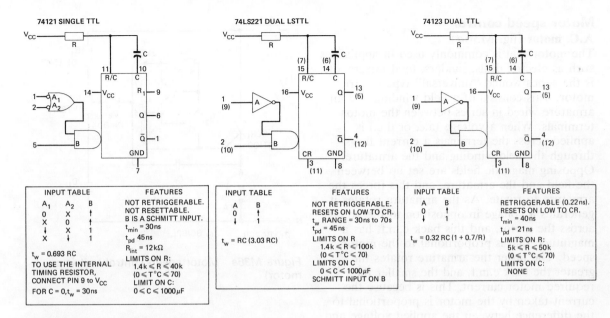

**74121 SINGLE TTL**

**74LS221 DUAL LSTTL**

**74123 DUAL TTL**

| INPUT TABLE | | | FEATURES |
|---|---|---|---|
| $A_1$ | $A_2$ | B | NOT RETRIGGERABLE. |
| 0 | X | ↑ | NOT RESETTABLE. |
| X | 0 | ↑ | B IS A SCHMITT INPUT. |
| ↓ | X | 1 | $t_{min}$ = 30ns |
| X | ↓ | 1 | $t_{pd}$ 45ns |
| | | | $R_{int}$ = 12kΩ |

$t_w$ = 0.693 RC
TO USE THE INTERNAL TIMING RESISTOR, CONNECT PIN 9 TO $V_{CC}$
FOR C = 0, $t_w$ = 30ns

LIMITS ON R:
1.4k ≤ R ≤ 40k
(0 ≤ T°C ≤ 70)
LIMIT ON C:
0 ≤ C ≤ 1000μF

| INPUT TABLE | | FEATURES |
|---|---|---|
| A | B | NOT RETRIGGERABLE. |
| 0 | ↑ | RESETS ON LOW TO CR. |
| ↓ | 1 | $t_w$ RANGE = 30ns to 70s |
| | | $t_{pd}$ = 45ns |

$t_w$ = RC (3.03 RC)

LIMITS ON R
1.4k ≤ R ≤ 100k
(0 ≤ T°C ≤ 70)
LIMITS ON C
0 ≤ C ≤ 1000μF
SCHMITT INPUT ON B

| INPUT TABLE | | FEATURES |
|---|---|---|
| A | B | RETRIGGERABLE (0.22ns). |
| 0 | ↑ | RESETS ON LOW TO CR. |
| ↓ | 1 | $t_{min}$ = 40ns |
| | | $t_{pd}$ = 21ns |

$t_w$ = 0.32 RC (1 + 0.7/R)

LIMITS ON R:
5k ≤ R ≤ 50k
(0 ≤ T°C ≤ 70)
LIMITS ON C:
NONE

*Figure M34* **TTL and CMOS monostable circuits for low pulse widths**

CMOS  4047B single
          4528B dual
TTL  74121 single
    74123 dual
    74LS221 dual low-power Schottky TTL
These are shown in pin-out form in fig. M34.

I.C. monostables can be triggered using a high-to-low or a low-to-high transition and usually at least two inputs are provided. For example the TTL 74121 mono has two A-type inputs and one B-type input. An A input has to be a TTL-conditioned signal and will trigger the mono with a high-to-low transition when the B input is high. The B input, on the other hand, has Schmitt circuitry and will trigger the mono with a low-to-high transition as slow as 1 volt per sec while either A input is held low.

The mono's output pulse width set is set by an external resistor $R_T$ and capacitor $C_T$. The value of resistor that can be fitted has a lower and upper limit and in practice it is best to use medium values. Long pulses would require high-value capacitors which may mean using an electrolytic type. A high-quality tantalum is then the best choice.

Retriggering is a useful feature provided to some i.c.s and means that, should a second trigger pulse arrive while the output is still high, then the mono will still respond to this

second pulse and remain high. In this way a train of input trigger pulses will hold the output high giving an output pulse with a very long duration. Retriggering can be useful in such applications as missing pulse detectors.

Monostables are useful in instruments such as pulse generators. [▶ *Measurements* for an example using the 4047B.] Fig. M35 shows the use of a TTL 74123 dual mono as a test oscillator in which $RV_1$ controls the period and $RV_2$ the output pulse width. This simple effective circuit can be useful in testing logic systems.

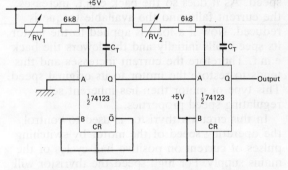

*Figure M35* **Simple pulse generator with independently adjustable period and pulse width using a 74123 TTL dual monostable**

## Motor speed control

### A.C. motor (fig. M36a)

The motor that is commonly used in appliances such as electric drills, sanders, food mixers, etc. is the series wound "universal" type electric motor. This consists of a field winding and an armature wired in series between the motor terminals. When a voltage (a.c. or d.c.) is applied across the terminals a current flows through the field winding and the armature. Opposing magnetic fields are set up between the field and the armature, and this forces the armature to rotate. As the armature rotates it generates a voltage in opposition to the voltage across the motor, and this back e.m.f. has a magnitude that is proportional to the motor speed. The faster the armature rotates the greater the back e.m.f. and the smaller the required motor current. This is because the current taken by the motor is proportional to the difference between the applied voltage and the back e.m.f. When the motor is first started a large current is taken since the back e.m.f. is zero. This means that a high torque is developed and the motor rapidly increases its speed. As it does so the back e.m.f. increases, the current falls, and the available torque is reduced. Now if a load is applied to the motor its speed falls initially and this lowers the back e.m.f. Therefore the current increases and this tends to restore the motor to its original speed. This type of motor then has inherent self-regulating speed properties.

In this circuit a thyristor is used to control the operating speed of the motor by switching pulses of current on positive half-cycles of the mains supply. For high speed the thyristor will be triggered on very early in each positive half-cycle, and by increasing the phase shift of the gate trigger signal the thyristor will be triggered later and this reduces the motor's speed.

Each positive half-cycle of the supply causes a current to flow through the potential divider network $R_1$, $RV_1$, and $D_1$. An attenuated positive half-cycle of the supply therefore appears on the wiper of $RV_1$ and this positive voltage charges $C_1$ via $D_2$. In fact $C_1$ stores a charge that is proportional to the difference between $RV_1$ wiper and the voltage on the cathode of the thyristor. The latter voltage is of course the speed-dependent back e.m.f. of the motor. When the voltage across $C_1$ exceeds about 3 V, $Tr_1$ and $Tr_2$, which are wired as a regenerative switch, are triggered into conduction and

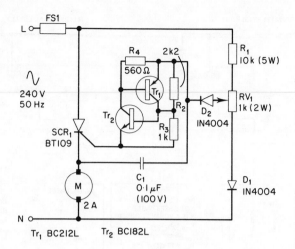

*Figure M36a*  Motor speed control unit (a.c. motor)

supply a pulse of current to the thyristor gate. The thyristor conducts to supply power to the motor. On the negative half-cycle of the mains the thyristor naturally turns off. One of the features of the universal-type motor is that by operating from a half-wave rectified supply only about 20% of the power from an equivalent full-wave system is lost. This fact enables quite efficient and relatively cheap half-wave control systems to be used.

The circuit achieves low speed operation when the wiper of $RV_1$ is moved towards the anode of $D_1$. This means that a smaller fraction of the positive half-cycle is applied across $C_1$, and since the other end of $C_1$ is connected to the motor back e.m.f. the transistor trigger switches at a later stage in the cycle. This reduces the power switched to the motor by the thyristor and therefore reduces the motor's speed.

At very low speed, "skip-cycling" takes place, when the thyristor delivers power in a portion of one half-cycle out of, say, five. This occurs because the gate signal cannot, in this circuit, be later than 90° in each positive half-cycle. So when the thyristor fires, it supplies at least a quarter-cycle of power to the motor. This causes the motor to accelerate and to increase the back e.m.f. Consequently the voltage across $C_1$ on the next half-cycle will be too small to trigger $Tr_1$ and $Tr_2$, and the thyristor will remain off. The thyristor will only conduct again when the speed and back

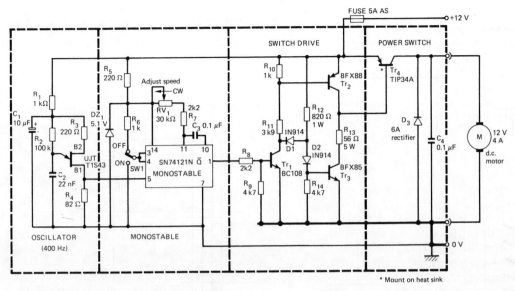

FUSE 5A AS

+12 V

SWITCH DRIVE

POWER SWITCH

OSCILLATOR
(400 Hz)

MONOSTABLE

* Mount on heat sink

*Figure M36b* D.C. motor speed control

e.m.f. have fallen to the original value and this may take several cycles. At first sight this would seem to be a disadvantage, but when the motor's speed is averaged out over a period of say a quarter of a second one can see that a fairly constant low speed is maintained.

Regulation at high or low speeds is achieved since any sudden change of load must produce a drop in motor speed with a consequent fall in the back e.m.f. This means that on the next positive half-cycle the thyristor receives a gate pulse earlier in the cycle and this switches on the thyristor which delivers more power to the motor than previously.

### D.C. motor (fig. M36b)
The speed of permanent magnet d.c. motors can be controlled by varying the voltage applied to the motor. With a fixed supply this can be achieved by placing a variable resistor of high power rating in series with the motor, but this is inefficient since power is wasted in the form of heat loss in the resistor. A better method is to vary the amount of power applied by using a pulse width modulated switching circuit — the switch in series with the motor being on for a shorter period than it is off to give low speed, and vice versa for high speed. The switching rate, or frequency, is kept constant but the duty cycle is varied.

In this system a unijunction oscillator running at a frequency of about 400 Hz applies

narrow positive pulses to the Schmitt input of a 74121N TTL monostable i.c. The width of the negative going (from logic 1 to logic 0) pulses from the $\bar{Q}$ output of the monostable i.c. is controlled by the potentiometer $RV_1$. The pulse width can be varied from about 0.1 msec up to about 2 msec by rotating $RV_1$ clockwise. The monostable can be inhibited from operation by taking pins 3 and 4 to logic 1 by $SW_1$ which will cause the $\bar{Q}$ output to be held in a high state. $SW_1$ is therefore used as an on/off control for the motor. The pulses from the monostable are then fed to a power switch ($Tr_4$) via a switch drive circuit. The purpose of the switch drive is to ensure that $Tr_4$ is switched rapidly between the two possible states of either hard on and saturated, or fully off. This is essential so that power dissipation in the switch is kept to a low level.

While the output $\bar{Q}$ is high, $Tr_1$ conducts and its collector will be low at about 0.2 V. Under these conditions $Tr_3$ is held off since its base current supply is diverted through $D_1$ into $Tr_1$ collector. At the same time $Tr_2$ is conducting, ensuring that the power switch $Tr_4$ is held off with its base clamped to its emitter via $Tr_2$. When the $\bar{Q}$ output of the monostable goes low, $Tr_1$ turns off thus turning off $Tr_2$. $Tr_3$ now switches on with base current supplied by $R_{12}$ and this switches $Tr_4$ hard on with its base current supplied by $R_{13}$. The motor receives nearly 12 V and current passes for the pulse

**167**

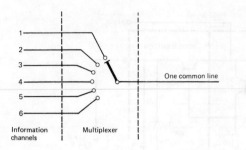

One common line

Information channels | Multiplexer

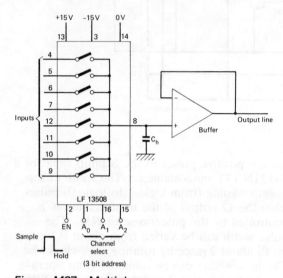

Figure M37 Multiplexer

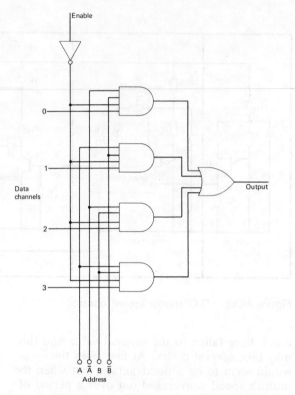

Figure M38 A 4-to-1 line digital multiplexer

period. On the trailing edge of the pulse $Tr_4$ turns off but $D_3$ conducts to limit any transients across the motor.

▶ *Ultrasonics*

## Multiplexer

A multiplexer (time division type) is an electronic circuit that performs the function of a fast rotary switch. It connects several information channels, *one at a time*, to a common line. The use of a multiplexer in a data acquisition system enables several sources of information to be sent along the common line and therefore reduces the number of connections required in any particular application. Analogue multiplexers can be built up using CMOS or JFET analogue switches or special-purpose i.c.s such as the LF 13508. A simple example of a multiplexer followed by a sample and hold is shown in fig. M37. Digital multiplexers, sometimes referred to as DATA SELECTORS, are

also available in i.c. form. In fig. M38 a 4 to 1 line digital system is shown. The select channel address can be driven from a counter.

Data that is transmitted in multiplexed form has to be decoded at the receiver by a demultiplexer, in other words separated out into the correct sequence of channel information.

## Negative Feedback

An amplifier is said to have negative feedback when a portion of its output signal is fed back to oppose the input signal. In fig. N1 an amplifier, with gain $A_o$ before negative feedback is applied, has a portion of its output signal fed back in series with the input so that the effective input to the amplifier's input terminals is reduced. At medium frequencies, where any additional phase shifts are negligible,

$$V_f = \beta V_o$$

where $\beta$ is the fractional gain of the feedback network.

At the input $\quad V_i = V_s + V_f$

$$\therefore \quad V_i = V_s + \beta V_o$$

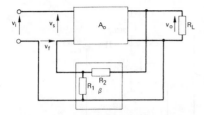

*Figure N1*  **Principle of negative feedback**

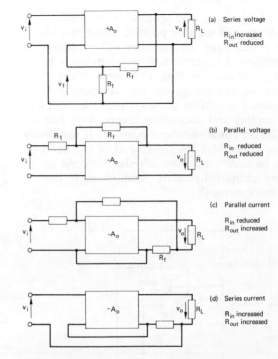

*Figure N2*  **Methods of applying negative feedback**

(a) Series voltage
$R_{in}$ increased
$R_{out}$ reduced

(b) Parallel voltage
$R_{in}$ reduced
$R_{out}$ reduced

(c) Parallel current
$R_{in}$ reduced
$R_{out}$ increased

(d) Series current
$R_{in}$ increased
$R_{out}$ increased

But $\quad A_0 = V_o/V_s \quad$ giving $\quad V_s = V_o/A_o$
Therefore

$$V_i = \frac{V_o}{A_o} + \beta V_o = V_o\left(\frac{1}{A_o} + \beta\right)$$

$$\therefore \quad \frac{V_0}{V_i} = \frac{A_0}{1 + A_0\beta}$$

The overall gain with negative feedback, called the CLOSED LOOP GAIN $A_c$ is given by

$$A_c = V_o/V_i = A_o/(1 + A_o\beta)$$

where $A_o$ is called the OPEN LOOP GAIN, i.e. the gain before the negative feedback
 loop is closed.
and $A_o\beta$ is called the LOOP GAIN.
 If the loop gain is much greater than unity, then

$$A_c \simeq 1/\beta$$

This result is important, for its means that the gain with negative feedback applied is now solely dependent on the components in the feedback loop. In the example

$$\beta = R_1/(R_1 + R_2)$$

Thus if $R_1 = 1\,\text{k}\Omega$ and $R_2 = 10\,\text{k}\Omega$, $\beta = 1/11$.
 Provided that $A_o\beta \gg 1$, then $A_c = 11$.
 This condition will result if $A_o$ has a value of 1000 or more.
 The advantage of negative feedback in stabilising the gain of an amplifier can be seen from the following where $\beta$ is assumed to be 1/11.

| $A_o$ | $A_c$ |
|-------|-------|
| 1000  | 10.88 |
| 500   | 10.76 |

A 50% change in open loop gain only results in a 1.1% change in closed loop gain.
 Negative feedback is widely used for the following reasons:

*a*) It stabilises the gain of the circuit, making the gain independent of changes due to ageing of components, component replacement, temperature and power supply lines.
 *b*) The frequency response is improved and the bandwidth increased.
 *c*) The way in which the feedback signal is derived from the output and applied to the input can be used to modify the input and output impedance of the circuit.
 *d*) Internally generated noise and non-linear distortion will be reduced.
 A manufacturer making, say, wideband linear amplifiers, can by designing with negative feedback ensure that all amplifiers produced have nearly the same characteristics.
 There are four types of negative feedback which depend upon the way in which the feedback signal is derived and how it is applied to the input. These are shown in fig. N2.
 Desirable as negative feedback is, there is always a practical limit to the amount of loop gain $A_o\beta$ which may be applied in any particular circuit. This is due to the inevitable phase shifts within the loop which at some frequency, either at the low-frequency or high-frequency

end of the band, add up to an extra 180°. Feedback will then be positive and if the loop gain is still unity the amplifier will oscillate. Instability of negative feedback amplifiers is studied by using ▶ *Bode Plots* or ▶ *Nyquist diagrams*. The design methods to avoid low-frequency instability are to reduce the number of coupling and decoupling components and to use direct coupling where possible. If instability exists at the high-frequency end of the band it may be necessary to modify the response by fitting additional reactive components in the feedback network.

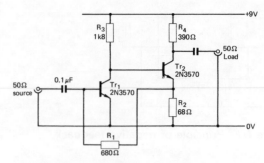

**Figure N3b**  Two-stage amplifier : parallel current feedback

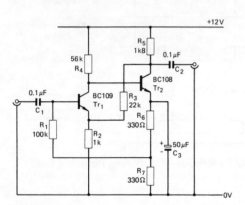

**Figure N3a**  Two-stage amplifier : series voltage feedback

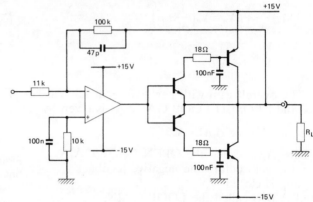

**Figure N3c**  Power amplifier : parallel voltage feedback

**1**  *Two-stage amplifier with series-voltage feedback* (fig. N3a). The output voltage at $Tr_2$ collector appears across the potential divider $R_3$, $R_2$ so that a portion is applied to the emitter of $Tr_1$. This opposes the input voltage giving negative feedback.

$$\beta = R_2/(R_2 + R_3)$$

and provided that $A_0\beta$ is $\gg 1$

$$A_c \simeq 1/\beta = (R_2 + R_3)/R_2 = 23 \quad (27\,\text{dB})$$

Direct coupling is used with a d.c. feedback path via $R_1$ which assists in stabilising the operating point. $R_1$ determines the input resistance of the stage.

**2**  *Two-stage amplifier using parallel current feedback* (fig. N3b). This very simple arrangement gives a voltage gain of 11 and an upper cut-off frequency of 100 MHz. A portion of the output current is fed back and applied in parallel with the input. The two resistors that determine the current gain are $R_1$ and $R_2$. Since 1/11th of the output current is fed back,

the overall current gain is very nearly 11. With the source and output resistive loads identical, the voltage gain will also be 11.

**3**  *Power amplifier with parallel voltage feedback* (fig. N3c). This class B 20 W amplifier has a sensitivity of 1 V r.m.s. and uses an op-amp followed by a class B power booster stage. The cross-over distortion is almost eliminated by the negative feedback which is via the 100 kΩ.

  ▶ *Amplifier*      ▶ *Op-amp*

## Noise

Electrical noise is defined as any unwanted signal which is present at the output of a system or at any part within the system. It is particularly important in communications receivers that unwanted signals are kept to a minimum, otherwise the required output information may be lost within the noise. Noise is a source of error in both analogue and digital systems but the latter is much more tolerant of an electri-

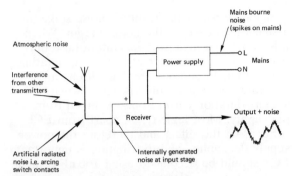

*Figure N4*  Sources of noise

cally noisy environment because in a digital system a wanted signal is either a logic 1 or a logic 0. The difference between these two logic states gives a barrier to noise and is referred to as the noise margin. This is discussed later, but first consider the sources and effects of noise on a purely analogue system such as a communications receiver (fig. N4).

The external noise affecting the receiver can have several origins. Artificial or man-made sources of noise are, for example, arcing contacts on switches or relays controlling heavy loads such as motors. The spark will give off an electromagnetic radiated signal which is picked up by the aerial. Alternatively the interference may be carried along the mains lead since the heavy loads being switched on and off produce large spikes on the mains which can then be transmitted through the systems' power supply. Another source of noise is interference from other radio transmitters. This is called second channel and image channel interference. The effects of artificial sources can be minimised either by suppression at source (i.e. preventing arcing at switch contacts) or by filters and special shields at the receiver. Second and image channel interference can be reduced by improved selectivity in the first stage in the receiver.

There are also natural sources of noise, referred to as atmospheric noise, such as static noise from space and electrical discharges during storms.

The signal arriving at the input of the system will therefore have a small noise superimposed. The receiver itself now adds more noise in the process of selecting and amplifying the wanted information. Internal noise is mostly the result of that produced in the first stage and is caused by noise from resistors and semiconductor devices.

## Internal noise

**1** *Thermal agitation or resistance noise*
Produced by the random motion of free electrons in a conductor.

R.M.S. noise voltage in a conductor is given by

$$V_n = \sqrt{[4kTBR]}$$

where  $k$ = Boltzmann's constant $1.38 \times 10^{-23}$ J/°K
   $T$ = Temperature of conductor in degrees Kelvin
   $B$ = Bandwidth in Hz over which the noise is measured
   $R$ = Resistance of conductor in circuit.

*Example*  The noise voltage produced by a 100 kΩ resistor at a temperature of 20°C and over a bandwidth of 100 kHz is

$$V_n = \sqrt{[4 \times 1.38 \times 10^{-23} \times 293 \times 100 \times 10^3 \times 100 \times 10^3]}$$
$$= \sqrt{[162 \times 10^{-12}]} = 12.73 \ \mu V$$

The available noise power from any resistor is $P_n = kTB$.

**2** *Noise in bipolar transistors*
This has several components:

*a*) *Thermal agitation noise*, developed mostly in the base spreading resistance $r_{bb'}$ of the device, given by

$$V_n = \sqrt{[4kTBr_{bb'}]}$$

*b*) *Partition noise* resulting from the random variations of the emitter current division between base and collector.

*c*) *Shot noise* caused by the random arrival and departure of charge carriers by diffusion across the p-n junction.

*d*) *Flicker noise* (1/*f* noise) resulting from changes in the conductivity of the semiconductor material and changes in its surface conduction. This noise is inversely proportional to frequency and is usually negligible above 1 kHz.

To achieve low noise figures from a bipolar transistor it is operated at low values of collector current (a few microamps) and at low voltage.

**3** *Noise in FETs*
Since a FET is a unipolar device it is inherently less noisy than a bipolar transistor. Only one type of charge carrier is used and only one current flows. The three sources of noise are

*a*) Shot noise, resulting from the changes in the small leakage currents in the gate-to-source junction.

**171**

*b*) Thermal agitation noise developed in the channel resistance of the device.

*c*) Flicker noise.

## Signal-to-noise ratio

For a quoted input signal power, over a defined bandwidth, the signal-to-noise ratio in an amplifier or receiver is given by

$$\text{S/N ratio} = \frac{\text{Average wanted signal power}}{\text{Average noise power present}} = \frac{P_s}{P_n}$$

This is usually expressed in dB as

$$\text{S/N ratio} = 10 \log_{10}(P_s/P_n)$$

*Example* At a frequency of 10 kHz the average wanted signal power at the input is 800 $\mu$W and the average noise power present is 6 $\mu$W. What is the input signal-to-noise ratio?

$$\begin{aligned}\text{Input S/N ratio} &= 10 \log_{10}(800/6) \\ &= 21.25 \text{ dB at } 10 \text{ kHz}\end{aligned}$$

In electronics, voltage ratios are also often used:

$$\text{S/N ratio} = 20 \log_{10}(V_s/V_n) \text{ dB}$$

## Noise factor (B.S. 3860)

Used to specify the noisiness of an amplifier or device, noise factor is

$$F = \frac{\text{Total noise power out}}{\text{Power gain} \times \text{Noise power}}$$
$$\text{due to source resistor}$$

$$= P_N/GP_n$$

But since $G = P_{s(o)}/P_{s(i)}$ where $P_s$ is signal power,

$$F = \frac{P_{n(o)}}{\dfrac{P_{s(o)}}{P_{s(i)}} \cdot P_n} = \frac{P_{s(i)}/P_{n(i)}}{P_{s(o)}/P_{n(o)}}$$

$$= \frac{\text{Signal/noise ratio at input}}{\text{Signal/noise ratio at output}}$$

Noise figure $= 10 \log_{10} F$ dB.

Thus if the noise figure for a device at a particular frequency is say 3 dB and the input signal-to-noise ratio is 100:1 (20 dB), then the resulting signal-to-noise ratio at the output will be 3 dB less at 17 dB (a ratio of 50:1).

## Noise in digital systems

The same sources of external noise such as arcing contacts and mains bourne spikes can affect a digital system if the resulting noise spike on a signal lead exceeds the noise margin. When this occurs the logic will generate a false output which is usually worse than not generating any output. In addition the logic itself can generate power line noise as gates switch and a short-duration current pulse is taken from the supply. Most logic types, apart from ECL, suffer from this effect and therefore the power supply decoupling and distribution is important. I.C.s should be decoupled using 100 nF ceramic capacitors wired directly across the i.c. supply pins. If possible a ground plane should be used to give a low-inductance earth return. Other sources of internal noise are cross-talk when the signal on one track is coupled to an adjacent track, and reflections from mismatched lines. For cross-talk

$$V_{in} = V_s \frac{1}{\left(1.5 + \dfrac{Z_m}{Z_0}\right)\left(1 + \dfrac{Z_1}{Z_0}\right)}$$

where $V_{in}$ is the induced voltage between the two parallel tracks

$V_s$ is the voltage swing of the logic

$Z_1$ is the output impedance of gate 1

$Z_0$ is the line impedance.

$Z_m$ is the mutual coupling impedance.

Careful design can eliminate the effects of internal logic generated noise, and external sources can be effectively stopped from affecting the logic by the use of mains filters, screening and special filters on the input lines. The higher the noise margin the better the immunity of the logic to noise. Manufacturers usually quote d.c. values of noise margin giving typical and worst-case values. Taking TTL as an example, the typical noise margin will be the difference between the voltage level from the output of a gate and the threshold of the gate input it is driving (fig. N5). Using this criterion the best logic 1 or high-state noise margin is 1.9 V, whereas the logic 0 or low-state noise margin is 1.2 V. However the typical noise margin is 1 V in both cases. The worst-case d.c. noise immunity has to take into account the minimum and maximum values of output levels and input threshold. The maximum value of logic 0 output is 400 mV and the minimum value of threshold ($V_{10}$) is 800 mV giving a worst-case noise margin of 400 mV.

▶ *High-threshold logic*

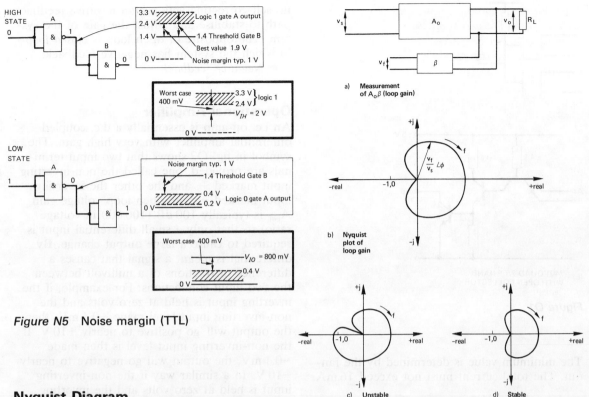

*Figure N5* Noise margin (TTL)

a) **Measurement of $A_o\beta$ (loop gain)**

b) **Nyquist plot of loop gain**

c) **Unstable**          d) **Stable**

*Figure N6* Nyquist plot for a.c. coupled amplifier

## Nyquist Diagram

To obtain stability in a negative feedback amplifier (or control system) the phase shifts occurring in the amplifier must not approach 180° while the loop gain $A_o\beta$ is greater than unity. If this does occur then the feedback becomes positive and the amplifier will oscillate. A Nyquist diagram is one method of investigating the stability of negative feedback systems by plotting the magnitude and phase of the loop gain on a graph with polar co-ordinates. This can be achieved by breaking the feedback loop, injecting a signal at the input, and then by measuring the amplitude and phase angle of the feedback signal $V_f$ over the whole frequency range.

$$\text{Loop gain } A_o\beta = \frac{v_f}{v_s}\underline{/\phi}$$

A typical plot for an a.c. coupled amplifier is shown in fig. N6.

Nyquist's stability criterion states that a negative feedback system will be conditional stable if the plot of the loop gain does not enclose the point $(-1, 0)$ while the frequency is increasing.

▶ *Bode plot*     ▶ *Negative feedback*

## Open Collector Gate

The standard TTL gate circuit has a totem pole output stage that gives a low output impedance in both the high and low output states. This provides high speed and the ability to drive capacitive loads, but does not permit the gate's output to be directly wired to the outputs of other gates. If wired logic is required then special gates in the TTL range such as the 7401, 7403, 7406, must be used. These are called open collector because the upper transistor, the diode and the resistor are omitted (fig. O1). The outputs of several of these gates can be wired together with one common pull-up resistor to give the wired-AND function.

The value of the load resistor depends upon the number of open collector outputs ($m$) connected together and the required fan-out ($n$).

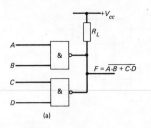

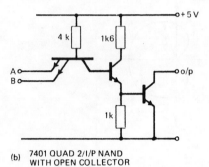

(b)  7401 QUAD 2/I/P NAND
     WITH OPEN COLLECTOR

*Figure O1*

The minimum value is determined by the fan-out. The total current must not exceed 16 mA.

$$R_{min} = \frac{V_{CC} - V_{OL}}{16 - 1.6n} \text{ k}\Omega$$

where $V_{CC}$ is +5 V and $V_{OL}$ is 0.4 V.

Thus for a fan-out of 4, $R_{min} = 480\ \Omega$.

The maximum possible value for the resistor is determined when the common output is high. The current through the resistor has to supply the leakage current to the outputs (250 $\mu$A max. each output) and the input current to each driven gate (40 $\mu$A max.).

$$R_{max} = \frac{V_{CC} - V_{OH}}{0.25m + 0.04n} \text{ k}\Omega$$

where $V_{OH} = 2.4$ V. Thus if 6 open collector gates are wired together with a fan-out of 5, the maximum value of the common resistor is

$$R_{max} = 1529\ \Omega$$

▶ *TTL*

## Open Loop

When the gain of a system is unaffected by the output it produces, for example a room heater without the controlling action of a thermostat, then the system is said to be operating under open-loop conditions. In other words there is

no self-correcting action by a negative feedback path to stabilise and adjust the gain of the system. The opposite of open loop is *closed loop* in which the system has negative feedback.

▶ *Negative feedback*

## Operational Amplifier

An i.c. op-amp is essentially a d.c. coupled differential amplifier with very high gain. The symbol in fig. O2 shows that two input terminals are provided, one called the non-inverting input marked +, and the other the inverting input marked −. The open loop voltage gain $A_{vol}$ is typically 100 dB (100 000 in voltage ratio) so that only a small differential input is required to cause a large output change. By differential is meant a signal that causes a difference of fractions of a millivolt between the two input connections. For example, if the inverting input is held at zero volts and the non-inverting input level made +0.1 mV, then the output will go positive to nearly +10 V. If the non-inverting input level is then made −0.1 mV, the output will go negative to nearly −10 V. In a similar way if the non-inverting input is held at zero volts and the inverting input level made +0.1 mV, the output will go to −10 V. The amplifier responds to the difference in voltage between the two input leads, and when this is zero the output should be nearly zero. Thus the op-amp must be provided with both a positive and negative supply rail so that the output can swing about zero.

A typical transfer characteristic is shown in fig. O2b. This shows that, when $(V_1 - V_2)$ is positive, the output goes positive. The output will saturate if $(V_1 - V_2)$ exceeds about +0.1 mV. Similarly when $(V_1 - V_2)$ is negative, the output goes negative. The characteristic has been drawn passing through zero at the point when $V_1 = V_2$. In practice some "offset" always occurs and a potentiometer has to be added to trim out or "null' any such offset voltage. This is discussed later.

The typical op-amp has a differential input stage which is supplied from a constant-current source, a second stage of amplification and d.c. level shifting, and finally a complementary class B type output state. The actual circuit diagrams shown in fig. O3 do appear rather complicated because of the d.c. coupling, use of transistors as resistors, and necessary protection circuits, but the operation as described above is fairly

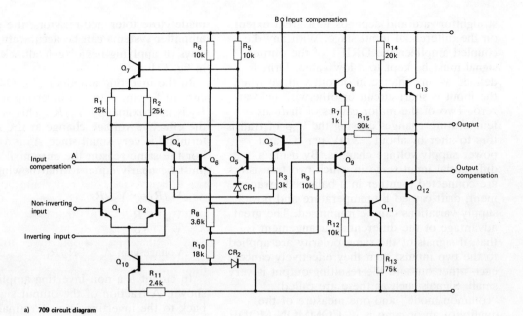

B ○ Input compensation

R₆ 10k  R₅ 10k  R₁₄ 20k

Q₇  Q₈  Q₁₃

R₁ 25k  R₂ 25k  R₇ 1k  R₁₅ 30k

Input compensation  A

Q₄  Q₆  Q₅  Q₃  Q₉

○ Output

Non-inverting input ○

Q₁  Q₂  R₃ 3k  R₉ 10k  ○ Output compensation

Inverting input ○  CR₁  Q₁₂

R₈ 3.6k  Q₁₁

R₁₂ 10k

Q₁₀  R₁₀ 18k  CR₂  R₁₃ 75k

R₁₁ 2.4k

a)  709 circuit diagram

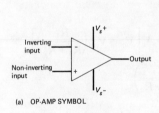

Inverting input ─|−
                    |＞── Output
Non-inverting input ─|+

$V_s^+$
$V_s^-$

(a)  OP-AMP SYMBOL

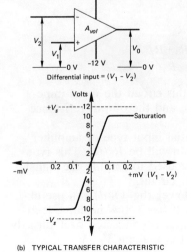

+12 V

$A_{vol}$

$V_2$  $V_1$  $V_0$

−12 V

0 V  0 V

Differential input = $(V_1 - V_2)$

Volts

$+V_s$ ----- 12
10
8
6
4
2
Saturation

−mV  0.2  0.1  0.1  0.2  +mV  $(V_1 - V_2)$

2
4
6
8
10
$-V_s$ ----- 12

(b)  TYPICAL TRANSFER CHARACTERISTIC

*Figure O2*  **Op-amp symbol and transfer characteristic**

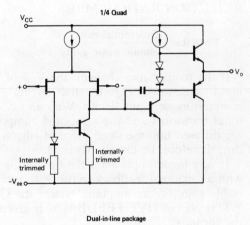

1/4 Quad

$V_{CC}$

+ ○  − 

○ $V_o$

Internally trimmed  Internally trimmed

$-V_{ee}$

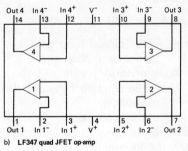

Dual-in-line package

Out 4  In 4⁻  In 4⁺  V⁻  In 3⁺  In 3⁻  Out 3
14  13  12  11  10  9  8

4  3

1  2

1  2  3  4  5  6  7
Out 1  In 1⁻  In 1⁺  V⁺  In 2⁺  In 2⁻  Out 2

b)  LF347 quad JFET op-amp

*Figure O3*  **Op-amp circuits**

straightforward and depends to a great extent on the differential input stage. With any d.c. coupled amplifier the DRIFT of the output signal must be kept to a low value. Drift is defined as any change in output voltage when the input is short-circuit or otherwise held at zero. Two of the major causes of drift are temperature changes causing the $V_{BE}$ of transistors to alter by about $-2\,\text{mV}$ per $°C$, and power supply voltage changes. By using a differential input stage in which two transistors are connected together in a balanced arrangement, drift caused by temperature and power supply variations can be minimised. The great advantage of the differential arrangement is that, if signals of the same polarity are applied to the two inputs, then they effectively cancel each other out and the resulting output is very small. Signals such as these are called "common mode" and one measure of the quality of an op-amp is its COMMON MODE REJECTION RATIO (CMRR):

$$\text{CMRR} = \frac{\text{Differential gain}}{\text{Common mode gain}}$$

If the temperature changes, the $V_{BE}$ of both input transistors change together, giving common mode input signals. With an i.c., both input transistors and the associated components are diffused into the same piece of silicon and can, therefore, be closely matched.

In any linear application the op-amp is wired with an external feedback network to give a stable gain. For an amplifier system the GAIN WITH NEGATIVE FEEDBACK is given by the formula:

$$A_c = A_o/(1 + A_o\beta)$$

where $A_c$ is the closed loop gain, i.e. gain with feedback applied

$A_o$ is open loop gain

$\beta$ is the fractional gain of the feedback network

$A_o\beta$ is the loop gain.

Since with an op-amp $A_o$ is typically 100 000, the loop gain $A_o\beta$ is usually much greater than unity. In this case the formula reduces to

$$A_c \simeq A_o/A_o\beta \simeq 1/\beta$$

This shows that the closed loop gain is then solely dependent on the component values of the feedback loop and, since these can be made close tolerance resistors, the gain of amplifier systems can be accurately set. The ways of applying negative feedback are shown in fig. O4.

In the inverting amplifier (fig. O4a), the current flowing into the inverting terminal via $R_1$ is approximately $V_{in}/R_1$. This is because the effective voltage change at the inverting terminal is very small since $A_o$ is very large. For the same reason the current flowing in $R_2$ must be nearly equal to that flowing in $R_1$.

$$V_{in}/R_1 = -V_{out}/R_2$$

$$\frac{V_{out}}{V_{in}} = -\frac{R_2}{R_1}$$

Voltage gain $A_c = -R_2/R_1$

In circuit $b$ a non-inverting amplifier is shown. A fraction of the output signal is fed back to the inverting input terminal. This opposes the input signal on the non-inverting input.

$$V_f = V_{out}\left(\frac{R_1}{R_1 + R_2}\right)$$

$$V_{out} = A_o(V_{in} - V_f) = A_o\left[V_{in} - V_{out}\left(\frac{R_1}{R_1 + R_2}\right)\right]$$

$$\frac{V_{out}}{A_o} = V_{in} - V_{out}\left(\frac{R_1}{R_1 + R_2}\right)$$

Since $A_o$ is very large $V_{out}/A_o \simeq 0$ and can be neglected. Therefore

$$V_{in} \simeq V_{out}\left(\frac{R_1}{R_1 + R_2}\right)$$

$$\text{Voltage gain} = \frac{R_1 + R_2}{R_1}$$

Note that with this circuit the input impedance is very large and the output resistance very low.

For the differential input type of amplifier the closed loop gain will be $R_2/R_1$. This type of circuit would be used to amplify the signals from, say, a bridge circuit.

The voltage follower (fig. O4d) is a useful impedance buffer. It has 100% negative feedback since the output is connected back directly to the inverting input making $V_f = V_{out}$.

$$V_{out} = A_o(V_{in} - V_f)$$

$$V_{out}/A_o = V_{in} - V_f = V_{in} - V_{out}$$

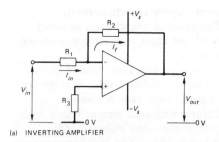

(a) INVERTING AMPLIFIER

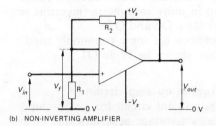

(b) NON-INVERTING AMPLIFIER

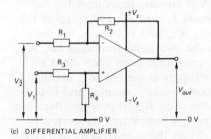

(c) DIFFERENTIAL AMPLIFIER

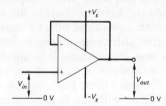

(d) VOLTAGE FOLLOWER

*Figure O4* Methods of applying negative feedback to an op-amp

(a) Voltage gain $\cong -R_2/R_1$
Input impedance $= R_1$

(b) Voltage gain $\cong \dfrac{R_1 + R_2}{R_1}$

Input impedance $= R_{in} \dfrac{A_O}{A_C}$

(c) Voltage gain $= \dfrac{R_2}{R_1}$

Usually, $R_1 = R_3$, $R_2 = R_4$

(d) Voltage gain is unity. Very high input impedance. Very low output impedance.

However, since $A_o$ is typically 100 000 $V_{out}/A_o \simeq 0$, then

$$V_{out} = V_{in}$$

In other words the output follows the input. The main advantage in this circuit is that the input impedance is very high ($>100\,M\Omega$) while the output impedance is very low, approaching less than an ohm.

## Op-Amp performance characteristics

*Supply voltage range* The maximum and minimum voltages that can be applied to the power supply pins ($+V_S$ and $-V_S$).

*Maximum differential input voltage* The maximum value of voltage that can be applied across the two input terminals.

*Open-loop gain $A_{vol}$* The low-frequency differential gain without any feedback applied. Some manufacturers specify "large-signal voltage gain" which is the open-loop voltage gain measured while the amplifier is driven so that a large undistorted output results.

*Input resistance* The resistance seen looking into the input terminals under open-loop conditions. Typical values for bipolar op-amps are in the range $1\,M\Omega$ to $10\,M\Omega$ while FET and BI-FET types have input resistances of $10^{12}\,\Omega$ or greater.

*Input bias current* The average of the currents into the two input terminals with the output at zero volts.

*Input offset current $I_{io}$* The difference between the currents flowing in the two input terminals with the output at zero volts.

*Temperature coefficient of input offset current $dI_{io}/dT$* The drift of input offset current with temperature. Measured in nA/°C or pA/°C.

*Input offset voltage $V_{io}$* Ideally when both inputs are grounded the output of the op-amps should also be at zero volts. In practice some offset is present. $V_{io}$ is the d.c. voltage that must be applied between the input terminals to force the quiescent d.c. output voltage to zero. Typical values are ±1 mV for bipolar op-amps and up to ±15 mV for FET input types.

*Temperature coefficient of $V_{io}$ $dV_{io}/dT$* The rate of change of input offset voltage with temperature. Usually specified in $\mu$V/°C.

*CMRR (common mode rejection ratio)* The ratio of differential to common mode gain, i.e. the ability of the amplifier to reject common mode signals.

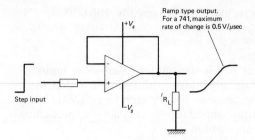

Ramp type output.
For a 741, maximum
rate of change is 0.5 V/μsec

Step input

Figure O5  Op-amp slew-rate limiting; the response to a sudden change at the input cannot be immediate

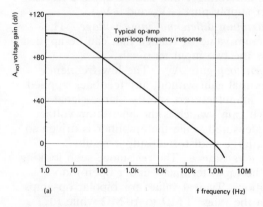

(a)

Figure O6  Op-amp frequency range: open-loop voltage gain of i.c. op-amps. Gain may be as high as 100 000 or more at dc, but can begin to drop off sharply above 100 Hz.

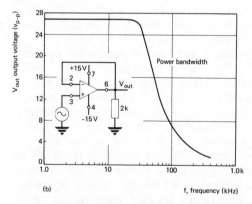

(b)

Figure O7  Power bandwidth of op-amps. For amplifiers using 15 volt supplies, the power bandwidths has come to be specified as the frequency range in which a ±10-volt swing can be accommodated at 5 percent total harmonic distortion

Slew rate is related to power bandwidth by

$$S = 2\pi f_p v_o$$

*PSRR* (power supply rejection ratio) $dV_{io}/dV_S$  The ratio of the change in input offset voltage to the change in supply voltages. Measured in μV/V.

*Slew rate*  The average time rate of change of the closed loop amplifier output voltage for a step input signal. Slew rate is usually measured between specified output levels with feedback adjusted for unity gain (fig. O5).

*Full-power bandwidth*  The frequency range of the op-amp in unity gain mode (inverting or non-inverting) (figs. O6 and O7).

The specifications for some commonly used i.c. op-amps are shown in Table O1.

## Drifts and errors in op-amp circuits

A simplified equivalent circuit for an op-amp (fig. O8) includes a voltage generator due to the offset voltage $V_{io}$; current generators for the input bias currents $I_b^+$ and $I_b^-$; and resistors (or impedances) $R_{in}$, $R_{cm}$ and $R_o$. $R_{in}$ is the resistance between the two input terminals and $R_{cm}$ is the effective resistance from each input to earth or power supply common. $R_{cm}$ is called the common mode input resistance and is usually much greater than $R_{in}$ so that its effect can in most cases be ignored. When connected in an amplifying circuit these generators and resistances cause errors and drifts with temperature. The bias currents required to operate the input transistors cause an equivalent offset voltage due to the unequal volt drops across resistors in the feedback and input leads. It is important to equalise the resistances at each of the inputs to compensate for this effect. For example for an inverting amplifier with the non-inverting input terminal connected directly to ground, the change in input error with temperature is

$$\Delta T\left[\frac{dI_b^-}{dT}R_1 + \frac{dV_{io}}{dT}\left(\frac{R_2+R_1}{R_2}\right)\right]$$

But if a resistor $R_a = R_1R_2/(R_1+R_2)$ is connected from the non-inverting input to ground the drift is reduced to

$$\Delta T\left[\frac{dI_{io}}{dT}R_1 + \frac{dV_{io}}{dT}\left(\frac{R_2+R_1}{R_2}\right)\right]$$

Most op-amps are provided with methods for zeroing input offset, called OFFSET NULL, so the minimum detectable d.c. level is then primarily determined by the temperature coefficients of $V_{io}$ and $I_{io}$. As an example, consider the case of a d.c. inverting amplifier with gain

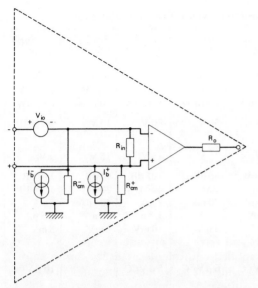

*Figure O8*  Simplified equivalent circuit for op-amp

occurs in the formulae for the output voltages for the two amplifier configurations as shown in figs. O9 and O10.

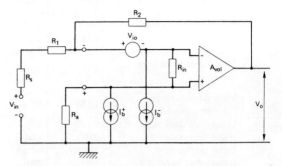

*Figure O9*  Errors for inverting amplifier

$R_a = R_1 // R_2$

$I_{io} = I_b^- - I_b^+$

$R_S \ll R_1$

$$V_0 = -\frac{R_2}{R_1}\left[V_{in} - \left(V_{io}\frac{R_2 + R_1}{R_2} + I_{io}R_1\right)\right]\frac{1}{1 + 1/\beta' A_{vol}}$$

↑ Closed loop gain    ↑ Signal    ↑ Input error    ↑ Gain factor error

of $-10$ and an input of 10 mV. If the accuracy required is 1% then the allowable input error is 100 $\mu$V. With a 741 and input resistance of say 10 k$\Omega$ the drift for a temperature change of 20°C is

$$\Delta T\left[\frac{dI_{io}}{dT}R_1 + \frac{dV_{io}}{dT}\left(\frac{R_2 + R_1}{R_2}\right)\right]$$

Drift at input $= 20\,(0.5 \times 10^{-9} \times 10^4 + 5 \times 10^{-6} \times 1.1)$
$= 210\ \mu$V

Obviously an op-amp with a better performance in terms of $dI_{io}/dT$ and $dV_{io}/dt$ is required.

Note that the above calculation does not take into account drift caused by PSRR nor any gain error.

Normally the open loop gain $A_{vol}$ is so large that the closed loop gain of an op-amp with feedback can be taken as $1/\beta$, where $\beta$ is the feedback fraction. However the input resistance will modify the value of $\beta$ since $R_{in}$ appears in parallel with $R_1$ in both inverting and non-inverting configurations. The new value of $\beta$ is therefore

$$\beta' = \frac{R_1 // R_{in}}{R_1 // R_{in} + R_2}$$

assuming $R_{in} \gg R_a$ and $R_S$.

Thus an error factor of

$$\frac{1}{1 + (1/\beta' A_{vol})}$$

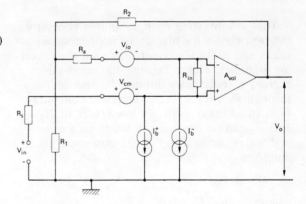

*Figure O10*  Errors for non-inverting amplifier

$R_S \ll R_{cm}$

$R_a = R_s - (R_1 // R_2)$

Common mode error $V_{cm} = V^+/\text{CMRR}$

$$V_0 = \left(\frac{R_1 + R_2}{R_1}\right)\left[V_{in} + (V_{cm} + V_{io} + I_{io}R_S)\right]\frac{1}{1 + 1/\beta' A_{vol}}$$

↑ Closed loop gain    ↑ Signal    ↑ Input error    ↑ Gain factor error

**Table O1**  *Specifications for some commonly used op-amps* (typical figures at 25°C)

| | Bipolar 709 | Bipolar 741 | Bipolar 741S | 101/301 | 108/308 | Bipolar 531 | MOSFET 3130 |
|---|---|---|---|---|---|---|---|
| Supply voltage range | ±9 V ±18 V | ±3 V ±18 V | ±5 V ±18 V | ±5 V ±18 V | ±5 V ±18 V | ±5 V ±22 V | ±5 V ±8 V |
| Max. power dissipation | 120 mW | 500 mW | 625 mW | 500 mW | 500 mW | 300 mW | 630 mW |
| Max. differential input voltage | 5 V | 30 V | 30 V | 30 V | 30 V | 15 V | ±8 V |
| Max. input voltage, either input to 0 V | 10 V | 15 V | 15 V | 15 V | 15 V | 15 V | $\pm V_S$ |
| Open-loop gain $A_{vol}$ | 93 dB | 106 dB | 100 dB | 88 dB | 102 dB | 96 dB | 110 dB |
| Input resistance | 250 k | 2 M | 1 M | 2 M | 40 M | 20 M | $1.5 \times 10^{12}$ |
| Input bias current | 300 nA | 80 nA | 200 nA | 70 nA | 7 nA | 400 nA | 5 pA |
| Input offset current $I_{io}$ | 100 nA | 20 nA | 30 nA | 3 nA | 1.5 nA | 50 nA | 0.5 pA |
| Input offset voltage $V_{io}$ | 2 mV | 1 mV | 2 mV | 2 mV | 10 mV | 2 mV | 8 mV |
| Temp. coeff. of input offset voltage $dV_{io}/dT$ | 3.3 µV/°C | 5 µV/°C | 3 µV/°C | 6 µV/°C | 5 µV/°C | — | 10 µV/°C |
| Temp. coeff. of input offset current $dI_{io}/dT$ | 0.1 nA/°C | 0.5 nA/°C | 0.5 nA/°C | 20 pA/°C | 2 pA/°C | 0.6 nA/°C | Doubles |
| CMRR | 90 dB | 90 dB | 90 dB | 90 dB | 100 dB | 100 dB | 80 dB |
| PSRR | 25 µV/V | 30 µV/V | 10 µV/V | 16 µV/V | 16 µV/V | 10 µV/V | 300 µV/V |
| Output voltage swing | ±14 V | ±13 V | ±13 V | ±14 V | ±13 V | ±15 V | ±6.5 V |
| Frequency compensation | Ext. | Int. | Int. | Ext. | Ext. | Ext. | Ext. |
| Slew rate | 12 V/µs | 0.5 V/µs | 20 V/µs | 0.4 V/µs | 0.1 V/µs | 35 V/µs | 10 V/µs |
| Full-power bandwidth | — | 10 kHz | 200 kHz | 10 kHz | 10 kHz | 500 kHz | 120 kHz |

The non-inverting configuration is generally the best choice for high-input impedance amplifiers and it has less input drift than the inverting amplifier. This is because the input offset current $I_{io}$ only flows through $R_s$ and not through $R_S + R_1$ as in the inverting configuration. In addition relatively low values of $R_1$ and $R_2$ can be used, giving lower gain error and wider bandwidth. For the non-inverting amplifier

$$R_i = R_{in}(1 + A_o\beta)$$

where $\beta = R_1/(R_1 + R_2)$.

The only drawback is that an additional input error arises due to the common mode signal, this is not present in the inverting configuration.

### Frequency response and compensation

The gain/frequency characteristic of an op-amp is another important factor in any design. The open loop voltage gain $A_{vol}$ is quoted for d.c. or very low-frequency operation. As the signal frequency is increased, so the open-loop gain falls in value. The frequency response for a 741 op-amp is shown in fig. O6, where it can be seen that the gain starts to fall off just above 10 Hz and drops at 20 dB/decade. The 741 has an internal 30 pF capacitor that causes the gain to fall off in this way. An internally compensated op-amp such as the 741 will not oscillate in any negative feedback configuration. Many other op-amps are not internally compensated and it is up to the user to provide the necessary external components. A lag or lead network ($C$ and $R$) has to be connected between particular pins or from one pin to ground. This gives the amplifier stability and prevents oscillations for the value of chosen closed-loop gain. The advantage is that a much wider bandwidth can be obtained by tailoring the open-loop frequency response to suit the particular application.

The open-loop frequency for a typical uncompensated op-amp is shown in fig. O11. In this example the first break point occurs at 10 kHz and the gain roll-off is 20 dB/decade. Then at 1 MHz a second break occurs and the roll-off rate increases to 40 dB/decade. Finally at about 10 MHz there is a third break and the

| MOSFET 3140 | JFET 355 | JFET 351 | CMOS 7611 |
|---|---|---|---|
| ±2 V | ±4 V | ±5 V | ±0.5 V |
| ±18 V | ±18 V | ±18 V | ±9 V |
| 630 mW | 500 mW | 500 mW | 250 mW |
| ±8 V | ±30 V | ±30 V | { ±0.3 V Lower than $V_s$ |
| ±$V_s$ | ±$V_s$ | +$V_s$ | { No more than ±0.3 V greater than $V_s$ |
| 100 dB | 106 dB | 110 dB | 102 dB |
| $10^{12}$ | $10^{12}$ | $10^{12}$ | $10^{12}$ |
| 10 pA | 30 pA | 50 pA | 1 pA |
| 0.5 pA | 10 pA | 25 pA | 0.5 pA |
| 5 mV | 3 mV | 5 mV | 15 mV |
| 8 μV°C | 5 μV/°C | 10 μV/°C | 25 μV/°C |

for every +20°C (approx.)

| MOSFET 3140 | JFET 355 | JFET 351 | CMOS 7611 |
|---|---|---|---|
| 90 dB | 100 dB | 100 dB | 91 dB |
| 100 μV/V | 10 μV/V | 30 μV/V | |
| ±6.5 V | ±13 V | ±13.5 V | ±4.5 V |
| Int. | Int. | Int. | Int. |
| 9 V/μs | 5 V/μs | 13 V/μs | 0.16 V/μs |
| 110 kHz | 60 kHz | 150 kHz | — |

roll-off rate goes to 60 dB/decade. Suppose this op-amp is used in a circuit with no frequency compensation, then the minimum closed-loop gain that can be used without instability is about 60 dB. If more negative feedback is applied the phase shifts will be approaching 180° and the resulting circuit will be unstable [▶ *Negative feedback*].

To get lower stable closed-loop gains a compensation network has to be fitted to modify the open loop response. One simple method is to connect a capacitor at a suitable point in the amplifier to ground to introduce a lag. This capacitor must have a value that, with a roll-off rate of 20 dB/decade, gives unity gain at 10 kHz. The response is shown as graph (x). This "brute-force" method wastes much of the useful higher-frequency performance of the op-amp. By using a composite *RC* network the amplifiers open loop response can be modified to give unity gain at 1 MHz. This is shown as graph (y) where:

$$f_{o1} = 10 \text{ kHz} = \frac{1}{2\pi R_c C_c}$$

This response is far superior to that obtained with the single capacitor and the amplifier will be stable even under 100% negative feedback conditions.

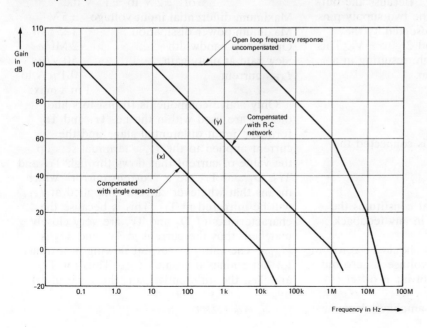

*Figure O11* Open-loop frequency for uncompensated op-amp

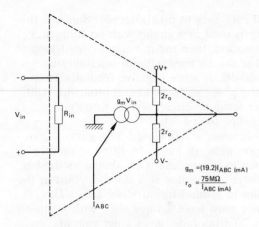

$$g_m = (19.2)I_{ABC} \text{ (mA)}$$
$$r_o = \frac{75M\Omega}{I_{ABC} \text{ (mA)}}$$

*Figure O12* Equivalent circuit of OTA

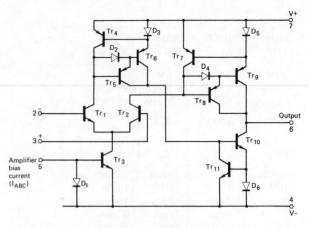

*Figure O13* Circuit of 3080 OTA

## Operational transconductance amplifier

This is a unique form of op-amp that differs from the standard device in two important features:

1) The OTA delivers an output current that is proportional to the differential input voltage.

2) An additional terminal (labelled $I_{ABC}$) is provided that controls the amplifiers transconductance $g_m$.

So the equivalent circuit for an OTA (fig. O12), while having the same input arrangement as a standard op-amp, has an output that consists of a current generator $g_m V_{in}$ in parallel with the output resistance $r_o$. Because the output current flows between the two supply pins, the output resistance is represented by the two resistors $2r_o$ to $+V_S$ and $2r_o$ to $-V_S$. This output resistance is very high, resulting in a high value of open-loop gain.

With no load $\quad V_o = g_m V_{in} r_o$

$\therefore \quad A_{vo} = V_o / V_{in} = g_m r_o$

When an external load $R_L$ is connected the gain is reduced to

$$A_v = g_m R_L \quad \text{(assuming } R_L \ll r_o)$$

It is this value of gain that constitutes the open-loop gain of an OTA in any feedback arrangement.

Ideally the output current should be zero when the differential input voltage is zero but in practice a small amount of input offset occurs.

As stated previously, the amplifiers bias current pin $I_{ABC}$ is used to set the value of $g_m$ which therefore allows the OTA's gain to be programmed externally. The $I_{ABC}$ input current can be used either to set up the OTA's characteristics or to process input signals.

Typical applications of the OTA are voltage controlled amplifiers, multiplexers, modulators, fast comparators, sample and hold circuits, active filters, and waveform generators.

The 3080 OTA i.c. will be taken as typical (see fig. O13). This has the following specification

Supply voltage range +4 V to +30 V d.c.
or ±2 V to ±15 V d.c.

| | |
|---|---|
| Maximum differential input voltage | ±5 V |
| Maximum power dissipation | 125 mW |
| Open loop bandwidth | 2 MHz |
| Slew rate at unity gain | 50 V/$\mu$s |
| $I_{ABC}$ current | 0.1 $\mu$A to 1 mA max. |

Only active components (transistors and diodes) are used within the i.c. $Tr_1$ and $Tr_2$ form the input differential stage and the current applied to the $I_{ABC}$ terminal sets up the value of current that flows through $Tr_1$ and $Tr_2$. $D_1$ and $Tr_3$ are a "current mirror" which means that whatever current is applied at $I_{ABC}$ will be mirrored in $Tr_3$. This is because the characteristics of $D_1$ and $Tr_3$ are very closely matched, thus the current in $Tr_1$ and $Tr_2$ is $\frac{1}{2}I_{ABC}$. The value of current flowing in a transistor determines its value of $g_m$. Thus for $Tr_1$ and $Tr_2$, the differential pair, the $g_m$ is given by

$$g_m = q\alpha I_c / 2KT$$

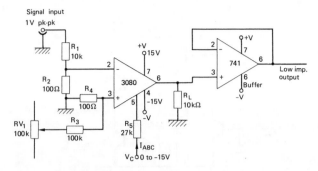

*Figure O14*  Voltage-controlled amplifier

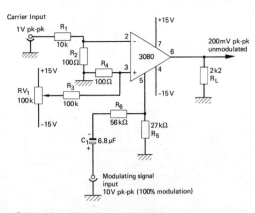

*Figure O15*  Amplitude modulator

where $\alpha I_c = I_E = I_{ABC}/2$

$$\therefore \quad g_m \simeq 19.2 I_{ABC}$$

where $g_m$ is in milliSiemens and $I_{ABC}$ is in milliamps.

With $I_{ABC} = 0.1 \, \mu A$ $\qquad g_m = 1.92 \, \text{mS}$

$I_{ABC}$ can be varied from $0.1 \, \mu A$ to $1 \, \text{mA}$ and this allows linear control of the transconductance.

The collector load of $Tr_1$ is another current mirror formed by $Tr_4$, $Tr_5$, $Tr_6$ and $D_3$, and in the same way $Tr_7$, $Tr_8$, $Tr_9$, $D_5$ form a current mirror that acts as the collector load for $Tr_2$. Darlington connected transistors are used in these two mirrors to reduce their voltage sensitivity. Finally, to give correct phase at the output, the current flowing from $Tr_6$ is fed to the mirror formed by $Tr_{11}$, $Tr_{10}$ and $D_6$. In this way push-pull output signals can be obtained from the collectors of $Tr_9$ and $Tr_{10}$. The whole circuit is biased into class A and the output can either source current via $Tr_9$ or sink current via $Tr_{10}$. Under quiescent conditions, with both inputs held at zero, the currents flowing through the output transistor are ideally equal at a value of $\frac{1}{2}I_{ABC}$ and the current taken from the $+V$ supply pin is approximately $2I_{ABC}$. A small differential input voltage signal will cause a change in the currents flowing in $Tr_1$ and $Tr_2$: $Tr_1$ current increasing and $Tr_2$ decreasing if the inverting input is made a few millivolts positive with respect to the non-inverting input and vice versa. These changes in current in the differential pair are mirrored in the output transistors $Tr_9$ and $Tr_{10}$.

**1** *Voltage-controlled amplifier* (fig. O14)

By varying the $I_{ABC}$ current the voltage gain of

an OTA can also be varied. In this example the control voltage $V_c$ sets up a value of $I_{ABC}$ through $R_5$. $I_{ABC}$ will be a maximum value of $0.5 \, \text{mA}$ when $V_c = 0 \, \text{V}$ and will be zero when $V_c = -15 \, \text{V}$. Therefore the maximum voltage gain of the OTA is

$$A_{v(max)} = g_m R_L = 10 \times 10^{-3} \times 10 \times 10^3 = 100$$

The signal input voltage is reduced by the potential divider $R_1 R_2$ to give $\pm 10 \, \text{mV}$ signals at the OTA input. By keeping the voltage level at the OTA input terminals low, the distortion (due to inherent non-linearities in the differential stage) will be less than 1%. With $R_L$ at $10 \, \text{k}\Omega$ the overall gain is therefore unity, but by using a unity gain buffer, $R_L$ can be increased in value to give higher overall gain.

$RV_1$ and the potential divider $R_3 R_4$ provide the necessary offset null facility and should be adjusted to give zero output voltage when $V_{in}$ is also zero.

**2** *Amplitude modulator* (fig. O15)

This is an extension of the previous VCA circuit to allow a low frequency (audio) modulating signal to vary the quiescent level of $I_{ABC}$. With zero modulating signal, $I_{ABC}$ is set by $R_5$ to $0.5 \, \text{mA}$ and the carrier wave input (up to $1 \, \text{MHz}$) will be unmodulated and appear at the output as a $200 \, \text{mV pk-pk}$ signal. With a modulating signal applied the $I_{ABC}$ current varies with the amplitude of the modulating signal and changes the voltage gain of the stage. The result is an output that is an amplitude-modulated wave.

**3** *Multiplexor* (fig. O16)

The $I_{ABC}$ terminal on an OTA allows the amplifier to be gated on and off. This can be useful for multiplex applications and a simple

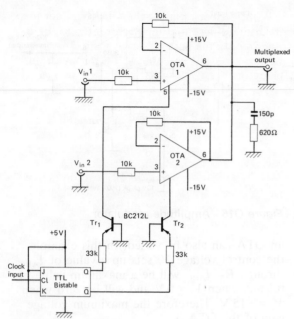

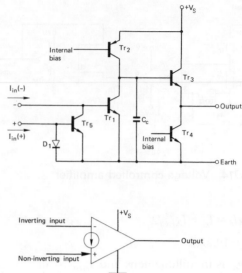

**Figure O16**  OTA multiplexer

**Figure O17**  Current-differencing amplifier (CDA)

2-channel system is shown in fig. O16. The outputs of the two OTAs are connected together and a common frequency compensation network can be used. The 10 kΩ resistors prevent excessive differential input current when an OTA is in the off state.

When the TTL bistable gives $Q = 1$, $\bar{Q} = 0$, then $Tr_1$ conducts, supplying 100 $\mu$A to the $I_{ABC}$ pin of OTA$_1$ but $Tr_2$ is switched off giving zero control current to OTA$_2$. On the next clock pulse the situation reverses so that OTA$_1$ switches off and OTA$_2$ switches on. The system can easily be extended to give a multi-channel system.

## Current-differencing amplifier
The CDA type op-amp is a low-cost amplifier that can operate from a single-ended power supply. Its important features are

a) Operation at low power from low-voltage single-ended supplies.

b) Simplicity and low cost.

c) An output voltage that is proportional to differential input current.

Typical devices are the 3401/3301 and the 3900/2900. A CDA differs considerably from the standard type of op-amp but can be used in a wide variety of similar applications. The basic circuit arrangement is shown in fig. O17.

The common emitter connected $Tr_1$ provides the high voltage gain. Its collector load is $Tr_2$ which supplies an almost constant current to $Tr_1$. This arrangement gives high gain and also minimises changes of gain with supply voltage. An internal compensation capacitor $C_c$ ensures frequency stability. The collector of $Tr_1$ is buffered to the output via the emitter follower $Tr_3$, and $Tr_4$ is an active pull-down that puts the output stage into class A and therefore gives optimum linearity. The non-inverting input is connected to the current mirror formed by $D_1$ and $Tr_5$ so that it diverts current from $Tr_1$ base and causes the output voltage to move positive. In this way the circuit is provided with differential inputs.

*Typical performance figures* ($T_{amb} = 25°C$, $V_s = +15$ V)

|  | 3401/3301 | 3900/2900 |
|---|---|---|
| Supply voltage | +5 V to 18 V | +4 V to +36 V |
| Open-loop gain | 2000 | 2800 |
| $f_T$ | 5 MHz | 2.5 MHz |
| $I_{ib}$ | 50 nA | 30 nA |
| $r_{in}$ | 1 MΩ | 1 MΩ |
| $r_o$ | 8 kΩ | 8 kΩ |
| Slew rate | 0.6 V/$\mu$s | 0.5 V/$\mu$s |

The temperature range for the 3401 and 3900 is 0° to +75°C, whereas for the 3301 and the 2900 it is −40°C to 85°C.

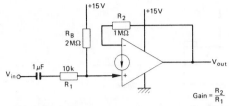

a) Non-inverting amplifier

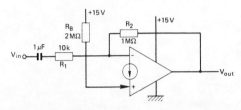

b) Inverting amplifier

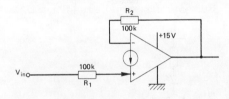

c) D.C. voltage follower

*Figure O18* CDA amplifier circuits

Basic CDA amplifier circuits are shown in fig. O18. The bias arrangement for both amplifiers is via the current mirror input. The current flowing into the non-inverting input is set by $R_B$ and is given by

$$I_b = [(V+) - V_f]/R_b$$

where $V_f$ is the voltage drop of the mirror diode $D_1$ and is typically 0.5 V.

*Figure O19* State variable filter using CDAs

Centre frequency $= 1/2\pi C_1 R_4$

$C_1 = C_2 \qquad R_3 = R_6 \qquad R_4 = R_5$

$Q = (R_2 + R_1)/2R_1$

$R_4 = R_5 = 1/2\pi f_o C_1$

$R_1 = 2R_4 \qquad R_8 = 2R_5$

Low pass output $= 2R_1/(R_2 + R_1)$

High pass output $= 2R_2/(R_2 + R_1)$

With values shown

$f_o = 250$ Hz $\qquad Q = 50$

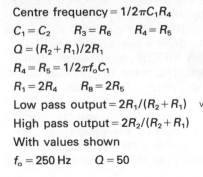

With feedback via $R_2$ the CDA will act to make the differential input current zero, or in other words the current flowing into the inverting terminal will be equal to the bias current. Therefore with $R_b = 2R_2$,

$$V_{o(dc)} = V^+/2$$

So the quiescent output voltage is held at half the supply voltage. Both the circuits are shown with a voltage gain of 100 but by changing $R_1$ lower values of gains can be achieved.

The CDA is also used extensively in active filter circuits and an example is shown in fig. O19 of a "state variable filter", a circuit that provides simultaneously low pass, high pass, and bandpass outputs.

## Opto-electronics

This phrase covers the wide range of devices and components that link light (optics) with electronics. These include:

*Displays* Light-emitting diodes (LEDs); liquid crystal displays (LCDs); filament; and gas discharge.

*Detectors* Photo-conductive cells (light-dependent resistors); silicon solar cells; photo-diodes; photo-transistors; light-activated switches; and light-activated thyristors/triacs.

*Light emitters and couplers:* infra-red sources and links; opto-isolators.

**Displays** (fig. O20)
An LED is a semiconductor diode that emits light when it is forward biased. In the same way that energy is needed to generate a hole-electron pair in a p-n junction, so energy is released when an electron recombines with a hole. When this happens in a silicon junction, the energy is in the form of heat but in some materials, particularly gallium arsenide, the

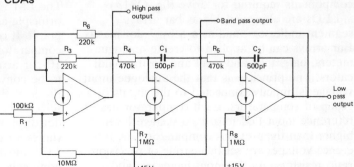

energy is released as infra-red radiation. By mixing other substances with gallium, LEDs can be made that emit visible light.

| Material | Colour | Approximate wavelength |
|---|---|---|
| Gallium arsenide phosphide (GaAsP) | red | 0.65 μm |
| Gallium phosphide (GaP) | green (red) | 0.56 μm |
| Gallium indium phosphide (GaInP) | yellow | |

All that is required to get an LED to emit light is to forward bias it with a current of a few milliamps. An LED must be driven from a constant current source or from a voltage via a series limiting resistor.

$$R_S = (V_S - V_F)/I_F$$

where $I_F$ the forward current is typically, 10 mA to 20 mA; and $V_F$ is the forward volt drop across the LED.

| LED type | $V_F$ at $I_F$ | | $V_{R(max)}$ | $I_{F(max)}$ |
|---|---|---|---|---|
| Red | 2.0 V | 10 mA | 5 V | 40 mA |
| Green | 2.2 V | 10 mA | 5 V | 40 mA |
| Yellow | 2.4 V | 10 mA | 5 V | 40 mA |

A single LED makes a useful on/off or fault indicator. The calculation for the series resistor required when driving an LED from TTL gates is shown in fig. O20. Some CMOS i.c.s such as the 4049 and 4050 can drive an LED via a resistor, or a transistor buffer, as indicated, will interface between the LED and high impedance outputs such as the 4011 gate.

Groups of LEDs can be arranged to form an array, of rows (Y) and columns (X), useful for display purposes and in reducing the number of components required for drive (fig. O21a)

LEDs are also available as bar arrays, 7-segment indicators, and as a 5 by 7 dot matrix. Bar arrays can be applied to create analogue meters, bargraphs and level and position indicators. The principle is that the analogue input voltage is directly connected to the inputs of a group of comparators. Each comparator has its reference input tied to a fixed voltage level higher than the previous comparator. The reference voltages are set by a string of precision ratio resistors. As the input increases, and

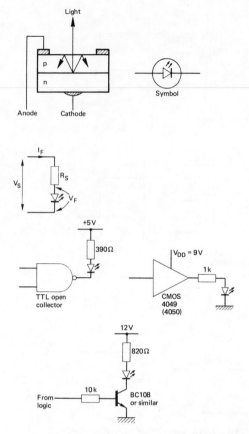

*Figure O20*   **LED construction and operation**

crosses the threshold level of each comparator, the outputs switch low to turn on the LEDs. To get a position indicator display the LEDs are wired in series, with resistors, so that only one LED lights for any particular voltage input (figs. O21b,c).

Special purpose-built LED arrays (usually 10 but some arrays go up to 30) fitted close together in line in the same package and bar driver i.c.s such as the LM 3914 and LM 3915 are available for analogue meter applications. The LM 3914 bar driver i.c. works on the same principle as described in the previous paragraph. It converts an analogue input voltage via comparators to drive a common anode LED 10-bar array. A MODE pin allows the display to be converted from a bar to a moving dot. Pin 9 of the i.c. controls the internal dot/bar selector logic and is connected to pin 11 for dot display or to pin 3 for bar display. The d.c. supply to the i.c. can be any value from 3 to 25 V and the internal reference voltage is nominally 1.28 V, but this can be programmed

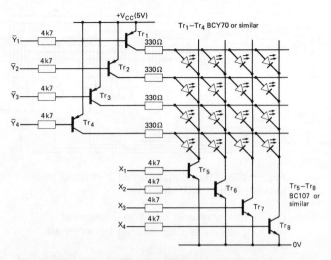

Figure O21a  X-Y display system using 4×4 LED array

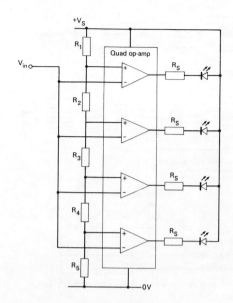

Figure O21b  Bar graph display

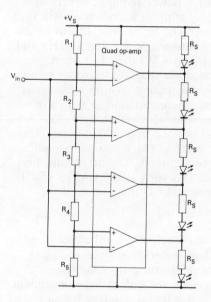

Figure O21c  Position-indicator display

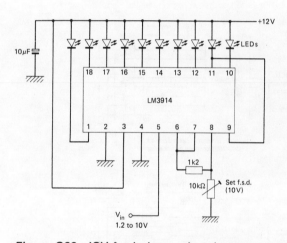

Figure O22  IOV f.s.d. dot mode voltmeter

externally to give effective reference values up to 12 V. A circuit of a 10 V f.s.d. dot mode voltmeter is shown in fig. O22. If this is converted to bar mode, by connecting pin 9 to pin 3, instead of pin 11, the power rating of the i.c. may be exceeded when all 10 LEDs are on. Each LED should then be fitted with a current limiting resistor or the LEDs can be driven from a separate low-voltage supply. The LM 3915 is similar in operation except for the fact that it has a logarithmic relationship, each LED representing a 3 dB increase at the input.

The displays commonly used for indication of the output of counters, frequency meters and microprocessors are the 7-segment types. These can be either LED, LCD, gas discharge, or filament. The format is shown in fig. O23 together with methods of direct d.c. driving for LED displays from BCD inputs.

Where several 7-segment displays are used in line to make a multi-digit indicator the most efficient drive method is to use a multiplexed (strobed) system. This allows the LEDs to be driven by a high peak current but at the same time reduces the average power dissipation because a low-duty factor is used. A multiplexed

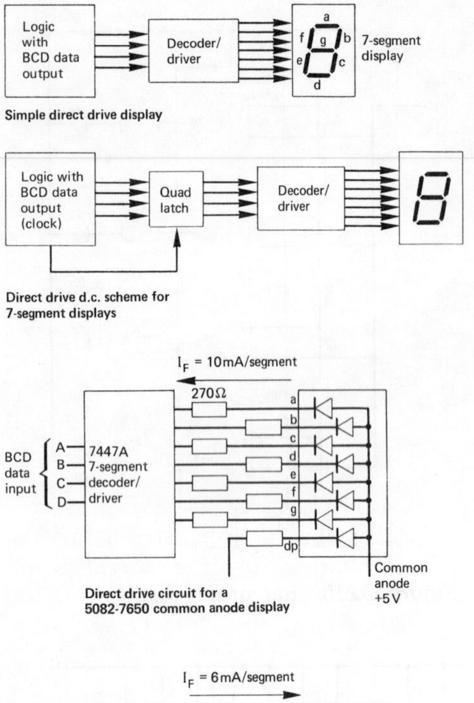

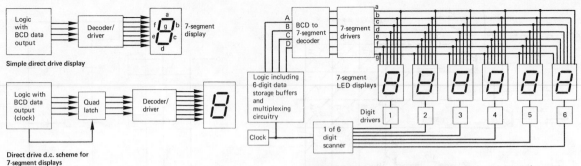

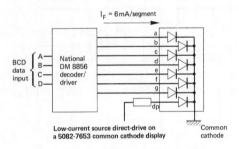

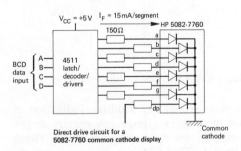

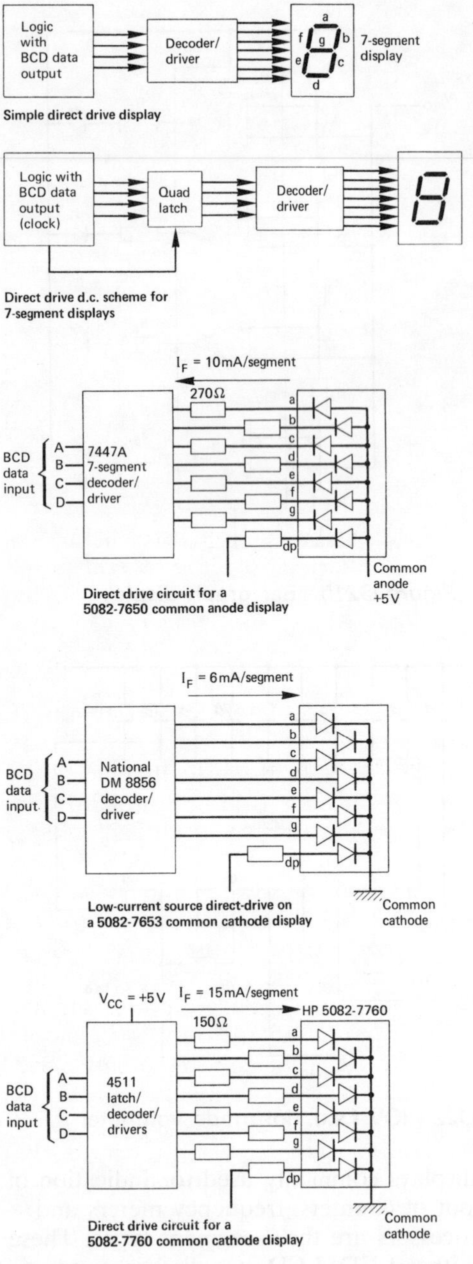

**Figure O23** Direct d.c. drive for 7-segment displays

*(Left column diagrams, top to bottom)*

Simple direct drive display

Logic with BCD data output → Decoder/driver → 7-segment display

Direct drive d.c. scheme for 7-segment displays

Logic with BCD data output (clock) → Quad latch → Decoder/driver → [7-segment display]

$I_F = 10\,mA/segment$
$270\,\Omega$
BCD data input: A B C D → 7447A 7-segment decoder/driver → a b c d e f g dp
Common anode +5 V

Direct drive circuit for a 5082-7650 common anode display

$I_F = 6\,mA/segment$
BCD data input: A B C D → National DM 8856 decoder/driver → a b c d e f g dp
Common cathode

Low-current source direct-drive on a 5082-7653 common cathode display

$V_{CC} = +5\,V$   $I_F = 15\,mA/segment$   HP 5082-7760
$150\,\Omega$
BCD data input: A B C D → 4511 latch/decoder/drivers → a b c d e f g dp
Common cathode

Direct drive circuit for a 5082-7760 common cathode display

**Figure O24** Block diagram of a strobed (multiplexed) 6-digit LED display

*(Right column top diagram)*
A B C D → BCD to 7-segment decoder → 7-segment drivers → 7-segment LED displays
Logic including 6-digit data storage buffers and multiplexing circuitry
Clock → 1 of 6 digit scanner → Digit drivers 1 2 3 4 5 6

six-digit display is shown in block form in fig. O24. The display clock, running at 1 kHz or higher, causes each 7-segment display to be selected in turn and to be connected to its appropriate input data. All like segments of each display are tied to a common bus line connected to the output of the decoder/driver i.c. In this way one BCD to 7-segment decoder/driver is time-shared between all digits.

Liquid crystal displays, which are field effect devices, are ideal for battery power equipment because the drive power required is very low. A 4-digit display with all segments on will take less than 20 $\mu$A. The input drive has to be a.c. at a frequency between 30 Hz and 100 Hz and at a voltage of 3 to 9 V r.m.s. The input voltage is applied via shaped electrodes to an organic compound retained within glass plates. The field causes the compound to switch from a liquid phase, when it is transparent, to become almost opaque in localised areas. Thus the display appears as black opaque segments on a silvered background. The drive must be a.c. and special i.c.s such as the CMOS 4543B BCD to 7-segment LCD must be used.

## Detectors

Opto-electronic detectors are either based on photo-conductive materials such as cadmium sulphide (CdS) or silicon p-n junctions.

The *photo-conductive cell*, or light-dependent resistor (l.d.r.), is a useful detector for slow and medium speed applications with a spectral response that almost matches the eye. The sensitive area of a photoconductive cell consists of a layer of cadmium sulphide deposited onto an insulating substrate. The assembly, with connecting leads, is plotted in clear epoxy resin. In absolute darkness the resistance of the cell may be as high as 10 M$\Omega$ but when light falls on the CdS track its resistance falls to a much lower value. In direct sunlight the cell may have a resistance of less than 20 $\Omega$. The light energy falling on the CdS track releases electrons within the cadmium sulphide causing this fall in resistance. Photo-conductive cells are

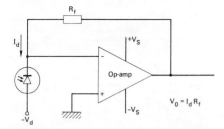

Figure O25 Photo-diode and amplifier

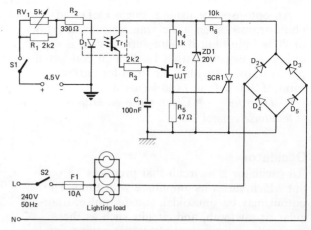

Figure O26 Lighting control circuit

widely used in light meters, slow-speed object direction, and in light-sensitive trips and alarms [▶ *Alarm circuits*].

The speed of response is not high and it may take several tens of millisecs for the cell to reach a new value of resistance when the light intensity is changed abruptly. A typical rise time is 75 ms.

For high speed applications the *photo-diode* and the *photo-transistor* are used. When light falls on a p-n junction, hole-electron pairs will be generated giving rise to an increased current. This current varies linearly with the light flux. A photo-diode is a silicon p-n junction diode fitted with a transparent window so that it is sensitive to light. It is normally operated with reverse bias since a high field results in fast response. A typical response time is 250 ns. A suitable amplifying circuit is shown in fig. O25. The diode current, caused by the light input, flows through the feedback resistor $R_f$ so that the output voltage is given by

$$V_o = I_d R_f$$

The *photo-transistor* gives a higher conversion efficiency than a photo-diode but has slower response. A small transparent window allows light to fall on the transistor's reverse biased collector/base junction. The hole-electron pairs that are generated form a small current that is injected into the base region. This is then amplified by the transistor to give a larger collector current.

A *photo-voltaic cell* is a special application of larger silicon photo-diodes to convert light energy to electrical energy. It has a high surface area to maximise light collection and gives a typical open circuit voltage of 0.55 V. A cell of about 5 cm² will provide a current of approximately 75 mA when illuminated with light energy of 1 kW/m². Cells like these can be used in series (to increase voltage) or parallel (increased current) to charge batteries.

**Opto-couplers**

An important use of light-emitting diodes and photo-sensitive detectors is in opto-isolators. In its simplest form this consists of an infra-red LED optically linked to, but electrically isolated from, a photo-transistor. The electrical isolation may be as high as 5 kV between the input and output, making the device ideal for interface applications where the control circuit has to be isolated from the main power supply. Another important advantage of opto-isolators is that they give excellent noise immunity since common mode signals at the input will not cause the LED to operate. In fig. O26 is an example of an opto-isolator used to control the trigger point of a thyristor. With switch $S_1$ made, a d.c. current is passed through the diode and the intensity is set by $RV_1$. This light intensity controls the output current from the photo-transistor. At low intensity, capacitor $C_1$ is charged relatively slowly compared with the time period of one half-cycle of the mains so that the unijunction gives an output pulse late in each half-cycle. Only a small amount of power will be dissipated in the load. Increasing the diode current by adjusting $RV_1$ will cause $C_1$ to be charged more rapidly and the thyristor will be triggered on early in each half-cycle to give higher power dissipation in the load.

An opto-isolator using a single LED and photo-transistor has a d.c. transfer current ratio of 20% (min.) but if more current gain is required an isolator with a Darlington connected photo-transistor with a d.c. transfer current ratio of 300% (min.) can be used.

▶ *Thyristor*    ▶ *Clock pulse generators*
▶ *Phase control*

## Oscillator

An oscillator is a circuit that produces an output which varies its amplitude with time. The output may be sinusoidal, square, pulse, triangular or sawtooth, and circuits such as these are used in all types of electronic equipment.

Oscillators can be constructed using components that exhibit a negative-resistance characteristic, typical of these being the tunnel diode and the unijunction transistor. However the large majority of oscillator circuits are based on an amplifier with a positive feedback loop. If a portion of an amplifier's output is fed back in phase with its input, the effective input is increased and so is the overall gain. For positive feedback

$$A_c = A_o/(1 - A_o\beta)$$

and as the loop gain $A_o\beta$ approaches unity, $A_c \rightarrow A_o/0 \rightarrow \infty$.

The high gains that result from using positive feedback can be used to maintain the amplitude of any oscillation by replacing the losses that occur in a frequency-determining network.

The requirements for a circuit to produce continuous oscillations are

a) Amplification to maintain oscillations.
b) Positive feedback.
c) A frequency-determining network.
d) A source of d.c. power.

This is shown in block form in fig. O27 and a typical example of a sine wave oscillator using a twin T filter as the frequency-determining network is shown in fig. O28. In this circuit $Tr_2$ provides the amplification to make up for the losses in the frequency-determining network and positive feedback is arranged from the output of the filter via the emitter follower $Tr_1$ back to the base of $Tr_2$. The emitter follower prevents the $CR$ network from being loaded by the relatively low input impedance of $Tr_2$.

Many other circuits can be used to generate sine waves, the most commonly used being the

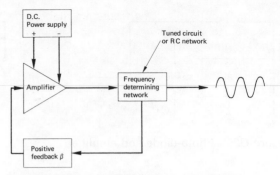

**Figure O27**  Block diagram of sine wave oscillator

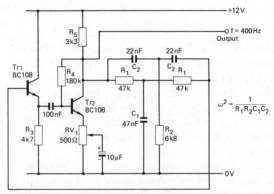

**Figure O28**  Typical oscillator showing control of amplifier gain

WIEN BRIDGE, COLPITTS and CLAPP. Each type has its own particular advantages and areas of application. The Wien bridge is a circuit that is often used in standard audio (1 Hz to 1 MHz) signal generators mainly because of its good frequency stability and the relative ease with which the frequency can be made continuously variable over quite a wide range.

The main requirements for a sine wave oscillator are

a) Frequency and frequency range.
b) Output amplitude and stability.
c) Frequency stability.
d) Purity of output, i.e. the amount of harmonic distortion present in the output waveform.

Correct amplitude and low distortion can be obtained by controlling the overall gain of the amplifier so that it just makes up for the losses in the frequency-determining network. In fig. O28 the gain is adjusted via the variable resistor in $Tr_2$ emitter which gives a small amount of negative feedback.

In some applications the frequency stability is required to be as high as possible. Changes in the frequency of an oscillator are caused by several factors. In the long term the drifts of component values and parameters as they age will cause a corresponding change in frequency, but the short term stability (24 hours) will be affected by

*a*) Variations in the load presented to the oscillator output. This can be reduced by using a buffer amplifier.

*b*) Changes in d.c. power supply voltage, as these will alter the parameters of the amplifier and change values of components. This can be virtually eliminated by using a very well stabilised and regulated power supply.

*c*) Changes in component values with temperature. This particularly affects the components that are used to determine the frequency. All passive components change their value with temperature; a metal oxide resistor having a temperature coefficient of about $\pm 250\,\text{ppm}/^\circ\text{C}$ and a polyester capacitor of about $-170\,\text{ppm}^\circ\text{C}$.

Frequency changes with temperature can be minimised by choosing components with the lowest temperature coefficients or by matching the temperature coefficients so that cancellation is achieved. It is obviously important to locate sensitive components away from heat-generating areas in a system and if necessary to mount these critical components in a temperature-controlled enclosure. By using such techniques it is possible to achieve frequency stabilities approaching $\pm 0.1\%/^\circ\text{C}$, but if better stability is required a crystal oscillator must be used.

▶ *Crystal oscillator*

## Parameter

A parameter is one of the constants, for a particular set of operating conditions, that is used to define the performance of a device. The current gain $h_{fe}$ of a bipolar transistor is an example of a parameter, and a value for $h_{fe}$ at stated operating conditions of voltage $V_{CE}$, current $I_C$, temperature $T^\circ\text{C}$, and frequency $1\,\text{kHz}$ is given by a manufacturer on the transistor's data sheet. It enables comparisons to be made between types and gives essential information for effective design.

There are many other parameters that are necessary to define a device and in general an active device like a bipolar transistor or FET is

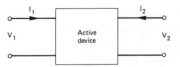

*Figure P1* "Black box" treatment of any active device for analysis

described using a parameter set.

Any four-terminal network can be treated as a "black box" and then the particular parameter set can be seen to be related to the input and output currents and voltages (fig. P1). Note that a three-terminal device like a transistor can be analysed in this way because one lead is common.

Three of the common ways of defining the small-signal operation of devices are

## 1 The OPEN-CIRCUIT IMPEDANCE PARAMETERS ($Z$-parameters)

$$v_1 = i_1 Z_{11} + i_2 Z_{12}$$
$$v_2 = i_1 Z_{21} + i_2 Z_{22}$$

or $\quad v_1 = i_1 Z_i + i_2 Z_r \qquad (1)$

and $\quad v_2 = i_1 Z_f + i_2 Z_o \qquad (2)$

Here the subscripts $i = $ input, $r = $ reverse
$\qquad\qquad\qquad f = $ forward, $o = $ output

By definition the impedance parameters are measured by open circuiting either the output or the input.

Thus, with the output open circuit, $i_2 = 0$.
Therefore, from equation (1)

$\quad Z_i = v_1/i_1 = $ input impedance (ohms)

and from equation (2)

$\quad Z_f = v_2/i_1 = $ forward transfer impedance (ohms)

With the input open circuit, $i_1 = 0$.
Therefore, from equation (1)

$\quad Z_r = v_1/i_2 = $ reverse transfer impedance (ohms)

and from equation (2)

$\quad Z_o = v_2/i_2 = $ output impedance (ohms)

$Z$-parameters are most useful in analyzing low-impedance devices.

## 2 The SHORT CIRCUIT ADMITTANCE PARAMETERS ($y$-parameters)

$$i_1 = v_1 y_{11} + v_2 y_{12}$$
$$i_2 = v_1 y_{21} + v_2 y_{22}$$

or $\quad i_1 = v_1 y_i + v_2 y_r \quad$ (1)

and $\quad i_2 = v_1 y_f + v_2 y_o \quad$ (2)

The y-parameters are measured by short circuiting either the output or the input.

Thus with the output short circuit $v_2 = 0$. Therefore, from equation (1)

$$y_i = i_1/v_1 = \text{input admittance (Siemens)}$$

and from equation (2)

$$y_o = i_2/v_2 = \text{output admittance (Siemens)}$$

These parameters prove most useful for describing high-impedance devices since it is easier to short circuit a high impedance than a low impedance. Thus FETs are described using y-parameters.

In common source mode the four parameters are given the additional subscript $s$ and are

$$y_{is} = \text{input admittance} = \left.\frac{di_g}{dv_{gs}}\right|_{v_{ds}=0}$$

$$y_{fs} = \text{forward transfer admittance} = \left.\frac{di_d}{dv_{gs}}\right|_{v_{ds}=0}$$

$$y_{rs} = \text{reverse transfer admittance} = \left.\frac{di_d}{dv_{ds}}\right|_{v_{gs}=0}$$

$$y_{os} = \text{output admittance} = \left.\frac{di_d}{dv_{ds}}\right|_{v_{gs}=0}$$

Note that stating $v_{gs} = 0$ or $v_{ds} = 0$ means an a.c. short circuit. In other words bias voltages are required and either $v_{gs}$ or $v_{ds}$ are held constant for the measurements.

**3  HYBRID PARAMETERS (h-parameters)**
The bipolar transistor is a low input and high output impedance device and is therefore best described by a mixture, or hybrid, set of parameters.

$$v_1 = i_1 h_{11} + v_2 h_{12}$$

$$i_2 = i_1 h_{21} + v_2 h_{22}$$

or $\quad v_1 = i_1 h_i + v_2 h_r \quad$ (1)

and $\quad i_2 = i_1 h_f + v_2 h_o \quad$ (2)

Thus when the output is short circuit $v_2 = 0$.
From equation (1)

$$h_i = v_1/i_1 = \text{input impedance (ohms)}$$

and from equation (2)

$$h_f = i_2/i_1 = \text{forward current transfer ratio}$$

With the input open circuit, $i_1 = 0$.
Therefore, from equation (1)

$$h_r = v_1/v_2 = \text{reverse voltage transfer ratio}$$

and from equation (2)

$$h_o = i_2/v_2 = \text{output admittance (Siemens)}$$

In common emitter mode the four h-parameters are

$$h_{ie} = \text{input impedance} = \left.\frac{dv_{be}}{di_b}\right|_{v_{ce}=0} \text{(ohms)}$$

$$h_{fe} = \text{forward current transfer ratio} = \left.\frac{di_c}{di_b}\right|_{v_{ce}=0}$$

$$h_{re} = \text{reverse voltage transfer ratio} = \left.\frac{dv_{be}}{dv_{ce}}\right|_{i_b=0}$$

$$h_{oe} = \text{output admittance} = \left.\frac{di_c}{dv_{ce}}\right|_{i_b=0} \text{(Siemens)}$$

These four are quoted on the manufacturer's data sheet.

The parameters can then be used in the analysis of a circuit by replacing the active device by its equivalent circuit.

▶ *Amplifiers*    ▶ *Equivalent circuit*
▶ *Field effect transistor*    ▶ *Transistor*

## Parity

Parity is a simple and effective method of error detection and indication, and consists of an extra single bit of information that is added to a system word. This extra bit, called the parity bit, is chosen so that the transmitted word contains either an even number of 1s (even parity) or an odd number of 1s (odd parity). Before transmission each word is examined by a parity generator which calculates whether a 0 or 1 is to be placed in the parity bit. Following transmission the parity word can be checked to determine if a failure or error has occurred. The parity detection circuit examines the transmitted word to see if the correct parity still exists. If there has been an error this can be indicated and if necessary the information can be retransmitted. This type of scheme can only detect the presence of a single error in a word. If two errors are present, parity will be maintained and no error will be indicated. Simple parity schemes can therefore only detect odd numbers of errors. More complicated schemes such as the Hamming parity single error detec-

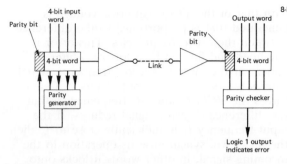

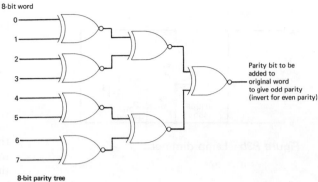

Parity bit to be added to original word to give odd parity (invert for even parity)

8-bit parity tree

**Figure P2** Principle of error detection using parity

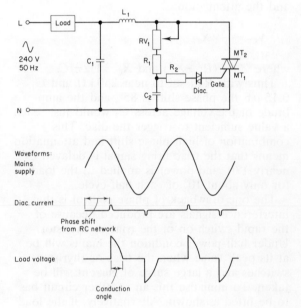

**Figure P3a** Triac full-wave a.c. power controller

tion and correction method are used but these require additional parity bits; a 4-bit word has to be provided with 3 Hamming parity bits.

Parity generators and checkers are built up using Exclusive-OR gates and Exclusive-NOR gates to form parity trees as shown in fig. P2. Here seven Exclusive-NOR gates are connected to form an 8-bit parity tree. An Exclusive-NOR gate will give a 0 output if, and only if, one of its two inputs is at logic 1. Thus the output of the parity tree will be in the 0 state if there is an odd number of 1s in the input word. This can be inverted and placed in the parity bit of the word to be transmitted to give even parity.

## Phase Control

One of the important uses of thyristors and triacs is in the smooth control of a.c. power, such as lamp dimmers and motor speed controllers. Fig. P3 shows a full-wave a.c. power control circuit in which the trigger pulse for the triac is derived from a $CR$ phase shifting network. The average power dissipated by the load can be varied by adjusting the time position of the gate trigger pulse relative to the a.c. supply waveform. On each half-cycle of the a.c. supply, $C_1$ is charged by the combination of $R_1$ and $RV_1$. When the potential across $C_1$ exceeds the breakover voltage of the diac, typically 30 V, the diac conducts briefly to supply a gate pulse to the triac. The triac rapidly switches from an off (non-conducting) to the on (conducting) state and the supply voltage is applied across the load.

The $CR$ network provides both a variable potential divider and phase shifting action, and by adjusting $RV_1$ the time position of the gate

trigger pulse can be varied. For example, with $RV_1$ set to minimum, very little phase shift and attenuation takes place and the voltage across $C_2$ almost follows the a.c. input. The diac conducts very early in each half-cycle so that almost full power is applied to the load. On the other hand, with $RV_1$ near maximum, the voltage across $C_1$ is a much-attenuated version of the a.c. and is also phase shifted by almost 90°.

The phase angle between the voltage across $C_1$ and the supply is given by

$$\tan \theta = \omega C_2 R \quad (\omega = 2\pi f)$$

**193**

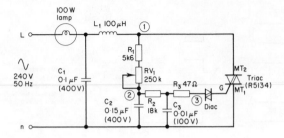

*Figure P3b* Lamp dimmer

and the attenuation

$$V_{C2} = \frac{V_s}{Z} \times X_{C2}$$

where $Z = \sqrt{(R^2 + X_C^2)}$ and $X_C = 1/2\pi fC$.

Thus with $RV_1$ set to near 250 k$\Omega$ and $C_1$ 0.15 $\mu$F the phase shift is 85°, and the amplitude of the voltage across $C_2$ would just reach a value sufficient to trigger the diac. This combination of high phase shift and attenuation means that the triac firing signal is delayed by nearly 170° and power is applied to the load for only about 10° of each half-cycle.

The one drawback of phase control is that interference signals are produced because of the rapid switch-on of the triac or thyristor. Under half-power condition the mains will be at its peak value when the triac or thyristor switches and a large surge of current will be taken. To minimise this an r.f. filter circuit has to be fitted as shown. Alternatively, if the load has a relatively slow response time, burst firing can be used.

▶ *Thyristor*  ▶ *Burst firing*

## Phase Locked Loop

This is basically (fig. P4) a feedback control system made up of a phase detector, a low pass filter, and a voltage controlled oscillator. The VCO is an oscillator whose frequency will vary from its free running value when a d.c. voltage is applied. The analysis of a PLL is beyond the scope of this book but the operation is fairly straightforward. With no input signal applied, the output voltage will be zero and the VCO will free run at a frequency determined by the external components $R_1C_1$.

When an input signal of frequency $f_1$ is applied, the phase comparator circuit compares the phase and frequency of the incoming signal

with that of the VCO. An error voltage is generated that is proportional to the difference between these two frequencies. This error is amplified and filtered by a low pass filter to appear at the output as a low frequency signal. This is fed back to the input of the VCO and forces the VCO to alter its frequency so that the difference or error signal reduces. If the input frequency $f_1$ is sufficiently close to $f_o$ then the VCO will synchronise its operation to the incoming signal, in other words it locks onto the input frequency. Once in lock the VCO frequency is almost identical to the input frequency except for a small phase difference. This small phase difference is essential so that a d.c. output is produced that keeps the VCO frequency equal to the input. If the input frequency or phase changes slightly the d.c. output will follow or track this change.

A PLL can, therefore, be used as an FM demodulator, or for FM telemetry, and for FSK receivers.

## Planar Process

The planar process resulted from the early research work carried out on the diffusion of impurities into silicon and it solved many of the problems concerned with the manufacture of transistors and diodes in volume, at lower cost, and with reliable characteristics. Since the process is precise and accurate, more refined transistor designs became possible and transistors with cut-off frequencies ($f_T$) of 1 GHz could be manufactured. The basis of the process depends on the fact that a thin coating of silicon dioxide ($S_1O_2$), grown in the surface of a silicon slice, forms a very effective seal against moisture and impurities. If selected areas of this oxide are removed, in other words "windows" created, then impurities can be forced into exposed silicon by diffusion to create individual p or n regions. Planar processing is a continuous repetition of three basic sequences:

*a*) Oxidation—the formation of silicon dioxide over the whole surface of the silicon slice.

*b*) "Window" opening by photo-resist techniques via an accurate mask.

*c*) Solid-state diffusion to make selected areas p or n type.

Consider the manufacturing steps required to make discrete transistors (n-p-n) (fig. P5):

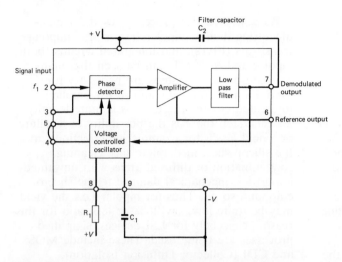

*Figure P4* Basic phase locked loop

1) Slices of heavily doped n-type silicon are cut, lapped and chemically polished. The slice diameter can be from 25 mm up to 75 mm and is typically 0.3 mm thick. On a large slice several thousand transistors can be made at the same time.

2) An epitaxial layer of lightly doped n-type silicon is deposited onto the surface of the slice. It is in this epitaxial layer that the active parts of the transistor will be created. This layer, typically 7 $\mu$m to 12 $\mu$m thick, is deposited by exposing the slice to an atmosphere of silicon-tetrachloride at a temperature of about 1200°C.

3) The slice is heated to about 1100°C in steam for 1 to 2 hours and a skin of silicon dioxide forms over the surface of the slice. The thickness of the oxide is typically 0.5 $\mu$m to 1 $\mu$m.

4) Photo-resist stage. Slices are spun-coated with a film of resist material and this is then dried by baking. The first mask is brought into contact with the slice and ultra-violet is exposed to those areas of the photo-resist not covered by the mask. The unexposed photo-resist is removed with a solvent and the silicon dioxide thus revealed is chemically dissolved away, leaving a "window" to the epitaxial layer. Finally the rest of the photo resist is stripped off. All in all five process steps.

5) Base diffusion—an impurity, in this case boron since the base is to be p-type—is first deposited on the surface of the slice and is then diffused into the silicon via the exposed

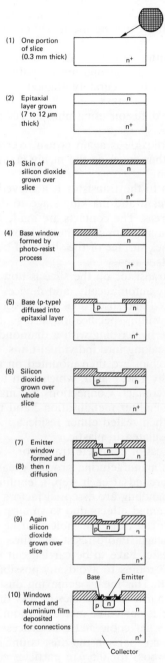

(1) One portion of slice (0.3 mm thick)

(2) Epitaxial layer grown (7 to 12 $\mu$m thick)

(3) Skin of silicon dioxide grown over slice

(4) Base window formed by photo-resist process

(5) Base (p-type) diffused into epitaxial layer

(6) Silicon dioxide grown over whole slice

(7) Emitter window formed and
(8) then n diffusion

(9) Again silicon dioxide grown over slice

(10) Windows formed and aluminium film deposited for connections

*Figure P5* Discrete planar transistor

window by heating the slice in a furnace at 1100°C for between 1 and 2 hours. Since the process of solid state diffusion is slow it is readily controllable and therefore precise. When the required base is achieved the slices are removed from the furnace and the excess boron removed chemically from the surface.

6) Silicon dioxide is again grown over the whole slice thus covering the newly created base region.

7) The emitter window is defined by a photo-resist process and a second mask. Note that this mask must be accurately aligned with respect to the first mask.

8) Emitter diffusion using phosphorus as the impurity.

9) Silicon dioxide is again grown over the whole slice, thus covering the newly created emitter region.

10) Access to the transistor is achieved by a third photo-resist and masking stage to define the contact areas. The contacts are made by evaporating aluminium onto these exposed areas. The back surface of the slice will be the collector contact area.

11) Each transistor on the slice is now tested, usually automatically, and devices which fail are marked and discarded. Transistors that pass, i.e. are within specification, are separated from the slice by scribing with a diamond stylus and breaking into individual units.

12) These chips are then mounted on a header, usually gold plated, which then forms the collector contact. Connections are made to the base and emitter metallisations, and the transistor is then sealed either inside a metal can or encapsulated in plastic.

13) Final test of complete transistor.

During the manufacturing process great care has to be exercised at each step. Cleanliness and careful handling are essential factors in avoiding unwanted effects due to small particles of dust which can cause pin-holes during the photo-resist stage. Photo-resist and slice preparation processes have to be carried out in "super-clean" rooms to avoid any possibility of contamination. This means achieving dust counts for 0.5 $\mu$m size particles of less than 35 000 particles per cubic metre in rooms and 3500 particles per cubic metre in processing cabinets. For comparison the dust count in an uncontrolled room for 0.5 $\mu$m particles could be approximately $2 \times 10^7$ particles per cubic metre. All liquids used in the various stages such as water, solvents, photo-resist solutions and acids must also be filtered to remove any contaminating particles.

The quality of the masks is also of the utmost importance. These must be very accurate in detail and dimensions and have as few defects as possible.

As described this process is used for manufacturing diodes, bipolar transistors, unipolar transistors (FETs), and integrated circuits (both bipolar and MOS). It can be seen that many separate processing steps are necessary for even a single transistor and so for bipolar integrated circuits the process is quite lengthy. At each step there is the possibility of a failure or "yield loss", for example imperfections in the silicon slice, incomplete slice cleaning, contamination of diffused areas with unwanted impurities, mechanical damage to the slice or chip, and so on. Thus for bipolar i.c.s the yield may be quite low, say 20% or less, and for this reason, especially for LSI devices, simplified processes are being used. These include MOS and CDI (Collector Diffusion Isolation).

Returning to the planar transistor, apart from the improvement in manufacture the device itself possesses several advantages over the alloy-junction type, notably

a) Low leakage current

b) Accurate control of base width giving high $f_T$.

c) Low collector resistance and low saturation voltage $V_{CE(sat)}$.

These transistors can, then, be used for fast switching applications and digital logic circuits.

▶ *Integrated circuit*　　▶ *Transistor*

## Potentiometer

These useful components consist basically of a track of some kind of resistive material to which a moveable wiper makes contact. The most simple construction is shown in fig. P6. Naturally the methods of manufacture vary considerably but potentiometers can be grouped under three main headings depending on the resistive material used:

a) *Carbon*　Either moulded carbon composition giving a solid track, or a coating of carbon plus insulating filler onto a substrate.

b) *Wire-wound*　Nichrome or other resistance-wire wound onto a suitable insulating former.

c) *Cermet*　A thick film resistance coating on a ceramic substrate.

Several different types of resistor are made, from single turn either open or enclosed slide type, up to multi-turn. The component may be intended only as a resistance that requires to be preset and therefore adjusted only a few times during its operational life, or as a control that is required to be continually varied over

**Table P1** *Applications of variable resistors*

| Type | Example of application | Selection tolerance | Linearity | Stability | Expected number of rotations | Turns |
|---|---|---|---|---|---|---|
| Preset or trimmer | Adjustment of fixed pulse width from a monostable | ±20% | not important | high ±2% | less than 50 | Single or multi-turn |
| General purpose control (panel mounting) | Brilliance control on c.r.o. | ±20% | ±10% | medium ±10% | 10 000 | Single |
| Precision control (panel mounting) | Calibrated output voltage control from a laboratory power supply | ±3% or better | ±0.5% | high ±0.5% | 50 000 | Single or multi-turn |

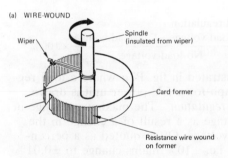

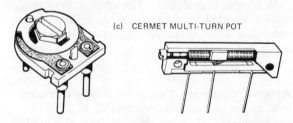

*Figure P6*  **Basic potentiometer constructions**

the whole of its track. The latter must be robust, stable and capable of many thousands of rotations before failing. Generally the requirement for a potentiometer falls into one of the following three categories:

Preset or trimmer
General purpose control
Precision control.

Typical examples together with required performances are given in Table P1.

A typical construction for a cermet is shown in fig. P6. The metal/ceramic resistance track is housed in a dust-proof plastic case and the end leads and wiper connections brought out for mounting on a plastic-coated base. The wiper, which may be multifingered, is usually a precious metal of say gold-plated bronze, and travels along the resistance track via a stainless steel leadscrew. A slipping clutch action is included so that the wiper assembly idles at both ends of the track to prevent damage from over-adjustment.

Table P2 gives information on the comparative performance of the three types of resistive track.

## Power Supply

A power unit of some type is essential for the operation of electronic instruments and systems. The power to drive a system or instrument may of course be supplied from batteries, but more usually it is derived from the single phase a.c. mains. The purpose of the power unit in this case is to accept the local mains supply (240 V rms at 50 Hz in the UK and 110 V rms at 60 Hz in the USA) and convert it into a form that is suitable for the internal circuits of the system or instrument. In the majority of cases this means converting the a.c. mains into a fixed stable d.c. voltage. The d.c. output has to remain substantially constant

**Table P2**  *Variable resistors – comparison table*

| Type | Construction | Law | Value | Tolerance | Power rating | Temperature coefficient | Stability | Life expectancy typical |
|------|-------------|-----|-------|-----------|--------------|------------------------|-----------|------------------------|
| Carbon composition (moulded or coated track) | Single turn or preset. Slide type. (Can be ganged.) | Linear or log | 100 to 10 M | ±20% | 0.5 W to 2 W | ±700 ppm below 100 K. ±1000 ppm above 100 K. | ±20% | 20 000 rotations |
| Wire-wound General-purpose Precision | Single or multi-turn control or preset. (Ganged.) Multi-turn (Helipot). | Linear sine cosine | 10 to 100 k | ±5% ±3% | 3 W | 100 ppm/°C 50 ppm/°C | ±5% ±2% | 20 000 to 100 000 rotations |
| Cermet | Single or multi-turn preset. | Linear | 10 to 500 k | ±10% | 1 W | ±200 ppm/°C | ±5% | 500 cycles |

against changes in load current, mains input and temperature. In addition to this are the requirements of isolation and possibly automatic overload and overvoltage protection. The power unit must effectively isolate the internal circuits from the raw mains, and usually has to provide an automatic current limit or trip if an overload or short occurs. If, in the event of a power supply fault, the d.c. output voltage rises above a maximum safe value for the internal circuits, then the power must be automatically disconnected.

Two main methods are used to provide regulated and stabilised d.c. voltages. The commonly used type has been the linear series regulator and this still predominates for modest power requirements. Increasingly, for higher power requirements, switched mode power units (SMPU) are being introduced. A switched system is more efficient, wastes less heat, and therefore takes up less space than a conventional linear regulator.

**Important parameters**

1) *Range*  The maximum and minimum limits of the output voltage and output current of a power supply.

2) *Load regulation*  The maximum change in output voltage due to a change in load current from no load to full load. The percentage regulation of a power supply is given by the formula:

% load regulation

$$= \frac{\text{No-load voltage} - \text{Full-load voltage}}{\text{No-load voltage}} \times 100\%$$

This is illustrated in fig. P7a where a load regulation graph for a 5 V power unit is drawn.

3) *Line regulation*  The maximum change in output voltage as a result of a change in the a.c. input voltage. Often quoted as a percentage ratio, i.e. ±10% mains change to ±0.01% in output voltage.

4) *Output impedance*  The change in output voltage divided by a small change in load current at some specified frequency (100 kHz is typical).

$$Z_{out} = \delta V_o / \delta I_L$$

At low frequencies, that is for slowly changing load currents, the resistive part of $Z_{out}$ predominates. $R_{out}$ can be read from the load regulation graph (see fig. P7a) and for a reasonable power unit should be at most a few hundred milliohms.

5) *Ripple and noise*  The peak-to-peak or rms value of any alternating or random signal superimposed on the d.c. output voltage with all external operating and environmental parameters held constant. Ripple may be quoted at full load or alternatively at some specified value of load current.

6) *Transient response*  The time taken for the d.c. output voltage to recover to within

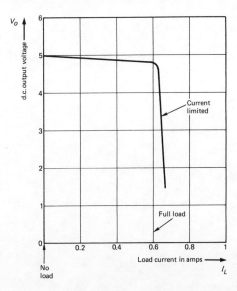

Figure P7a Example of a load regulation curve for a stabilised power supply

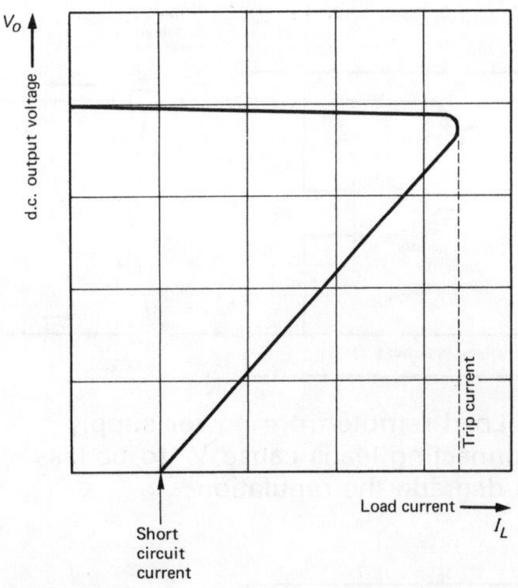

Figure P7b Foldback current limiting characteristic

$$\text{Load regulation} = \frac{\text{No-load voltage} - \text{full-load voltage}}{\text{No-load voltage}} \times 100\%$$

$$= \frac{5 - 4.8}{5} \times 100\%$$

Regulation $= 4\%$

$$r_{\text{out}} = \frac{\delta V_o}{\delta I_o} = 0.33 \text{ ohms}$$

10 mV of its steady state value following the sudden application of full load.

7) *Temperature coefficient*   The percentage change in d.c. output voltage with temperature at fixed values of a.c. mains input and load current.

8) *Stability*   The change in output voltage with time, assuming the unit has reached thermal equilibrium and that the a.c. input voltage, the load current, and the ambient temperature are all held constant.

9) *Efficiency*   The ratio of output power to input power expressed as a percentage. For example suppose a power supply of 24 V when loaded to 1.2 A requires an input current of 200 mA from the 240 V mains. Then

$$\text{Efficiency} = \left[\frac{V_o I_L}{V_{ac} I_{ac}}\right] \times 100\%$$

$$= \frac{24 \times 1.2}{240 \times 0.2} \times 100 = 60\%$$

10) *Current limiting*   A method used to protect power supply components and the circuits supplied by the power unit from damage caused by an overload current. The maximum steady state output current is limited to some safe value (see fig. P7a).

11) *Foldback current limiting*   An improvement over simple current limiting. If a preset trip value of load current is exceeded the power supply switches to limit the current to a much lower value (fig. P7b).

In some situations a power unit may be required to supply its load via fairly long lengths of connecting lead as in fig. P8a. As the load current flows along the supply and return wires, a volt drop will be set up causing the voltage across the load to be less than the voltage at the power supply terminals and consequently to have degraded regulation. One technique used to improve this is called REMOTE SENSING in which two extra leads are used to compensate for the effects of supply lead resistance (fig. P8b). In effect the technique causes the supply lead resistance to be included within the feedback loop of the regulator. This gives optimum regulation at the load rather than at the power supply output terminals. The current carried by the two sense wires is very small so light gauge wire is used.

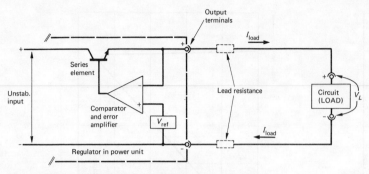

*Figure P8a* Load remote from power supply terminals; connecting leads cause $V_L$ to be less than $V_O$ and degrade the regulation

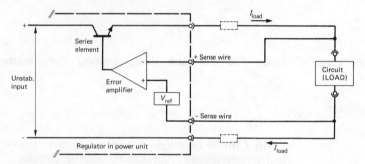

*Figure P8b* Remote sensing; includes connecting leads within feedback of regulator and therefore compensates for lead resistance

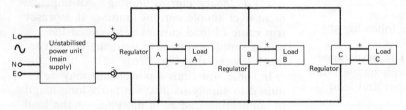

*Figure P9* Use of "point of load" regulators

However, because the two sense wires form the input of the comparator circuit, they have to be shielded to prevent pick up of interference. In practice a shielded pair is used and the shield or screen is connected to earth at the power supply end only. Note that the technique of remote sensing can only be used to give optimum regulation across one load. If the power supply is used to feed a large number of loads in parallel, then some other technique has to be used. Now that i.c. regulators are readily available, and comparatively cheap, the use of "point of load" regulators or remote regulators is increasing. A simple example is shown in fig. P9 where each load is provided with its own regulator circuit. The main power unit supplying the three separate regulators is often unstabilised.

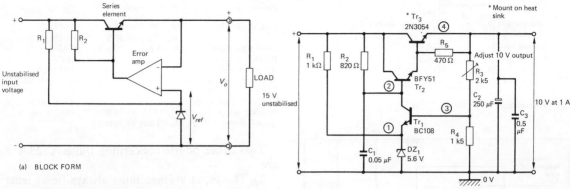

*Figure P10* **Basic diagram of series regulator**

(a) BLOCK FORM

(b) TYPICAL CIRCUIT USING DISCRETE COMPONENTS

## Linear regulators

In order to provide improved performance the unstabilised power supply must be followed by some form of regulator. The linear series regulator is a circuit commonly used for medium power requirements and even quite simple circuits are capable of excellent performance. Basically it is a high gain control circuit that continuously monitors the d.c. output voltage and automatically corrects the output to hold it constant irrespective of changes in load current and unstabilised input voltage. As shown in fig. P10a the output is compared with a stable reference voltage and any difference, or error, between the output and reference is amplified and fed to the base of the series control element. The series element is a power transistor connected as an emitter follower providing a low output impedance to drive the load. The performance of the circuit depends upon the stability of the voltage reference source and the gain of the error amplifier.

A typical example of a series regulator using discrete components is shown in fig. P10b. This unit should provide 10 V at 1 A from a 15 V unstabilised supply. Here the control element is formed by the Darlington connection of $Tr_2$ and $Tr_3$. The full load current of 1 A flows through $Tr_3$ and, since its value of current gain $h_{FE}$ may be relatively low, the base current required by $Tr_3$ may be as high as 40 mA. This current is supplied by $Tr_2$, which then itself only requires a base current of between 1 and 2 mA. The error amplifier is $Tr_1$, the inverting input being the base and the non-inverting input the emitter. The latter is held constant by

the 5.6 V zener. Under normal conditions the base voltage of $Tr_1$ will be about 0.6 V higher than its emitter at 6.2 V. Therefore, if the voltage across $R_4$ is 6.2 V then, if $R_3$ is adjusted to a value of 1 kΩ, the total volt drop across $R_3$ and $R_4$ should be 10 V. If the output voltage drops in value, a portion of this fall appears on $Tr_1$ base. Since $Tr_1$ emitter is held constant by the zener reference voltage, the base/emitter voltage of $Tr_1$ will decrease in value. $Tr_1$ collector voltage rises increasing the forward bias to $Tr_2$ and $Tr_3$ which thus tends to correct the output voltage. This process is, of course, automatic.

Many modern power supplies now use MONOLITHIC I.C. REGULATORS. There are naturally several different i.c.s available, but it would not be useful to detail them all here. Instead we shall look at one of the most popular, relatively inexpensive and versatile i.c. regulators, the μA 723. This is available in a 14-pin DIL (μA 723A) encapsulation or as a metal can version (μA 723L) with ten leads. The pin configuration for the 14 pin DIL together with the equivalent circuit are shown in fig. P11. The internal circuitry contains a reference supply, error amplifier, series pass transistor, and a current limiting transistor. The connections to the various sections are brought out to the i.c. pins allowing the user flexibility in designing a regulator to suit his requirements. The stable, temperature-compensated, voltage reference source gives a voltage at pin 6 of 7.15 V ±0.2 V, and this can be used either directly connected to the non-inverting input or via a potential divider. Two basic circuits

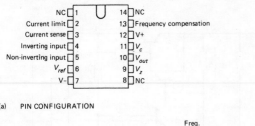

(a)   PIN CONFIGURATION

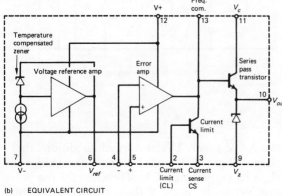

(b)   EQUIVALENT CIRCUIT

*Figure P11*   µA 723A i.c. regulator

| Parameter | Test conditions | Typical value |
|---|---|---|
| Line regulation | $V_{in}$ =12 V to 15 V | 0.01% |
|  | $V_{in}$ =12 V to 40 V | 0.02% |
| Load regulation | $I_L$ =1 mA to 50 mA | 0.03% |
| Ripple rejection |  | 86 dB |
| Reference voltage |  | $7.15 \pm 0.2$ V |
| Long-term stability |  | 0.1% per 1000 hrs |
| Input voltage range |  | 9.5 V to 40 V |
| Output voltage range |  | 2 V to 37 V |
| Average temp. | $V_{in}$ =12 V to 15 V | 0.002% per °C |
| coefficient | $I_L$ = 1 mA to 50 mA |  |

Two other points concerning the µA723A are

1) The input voltage must always be at least 3 V greater than the output voltage. This, of course, means that some power is dissipated as heat in the series pass transistor.

2) A low-value capacitor has to be connected from the frequency compensation pin to the inverting input. This ensures that the circuit does not oscillate at high frequencies.

### Switched power supplies

Switched power systems and switched mode regulators are used for their high efficiency. Intensive development has taken place over the last few years to produce power supplies of maximum efficiency and small size and weight. Many of these circuits are developments from the basic invertor shown in fig. P13. An invertor is a device that converts d.c. to a.c. In the circuit this is achieved by the switches $S_1$ and $S_2$ which alternately reverse the d.c. connection to the transformer primary. The transformer has to be centre-tapped. On one half cycle current flows through the top half of the primary winding and on the other, when the switches change, the current flows in the opposite direction through the lower half of the primary. The result is that a.c. will be produced at the secondary. The switches are usually special-purpose transistors or thyristors driven by some form of square wave or pulse oscillator. Another method is to have feedback windings on the primary so that the invertor transistors form a self-oscillating circuit. The frequency of the switching signal, especially if the invertor is used as part of a regulator, is typically in the range 5 kHz to 25 kHz. A high frequency is used because the transformer and any subsequent filter components will be relatively small. Manufacturers and users also prefer a frequency that is just above the audio range (15 kHz) for obvious reasons. The upper limit of operating frequency is set by the core losses

are shown in fig. P12. The first gives output voltages from 2 to 7 V and the second output voltages from 7 V to 37 V. In the first circuit a current limiting resistor $R_{sc}$, of 10 Ω, is shown. This will limit the maximum output current to 65 mA.

The maximum current that can be taken via the series pass transistor is 150 mA, but the safe maximum current in a particular application depends upon the value of the unregulated input. At an ambient temperature of 25°C the maximum power dissipation of the IC is 660 mW. Therefore the safe current limit when the output is short circuited is given by

$$\text{Maximum current limit} = P_{max}/V_S$$

Thus if $V_S$, the unregulated input, is 20 V, the maximum safe current under short circuit conditions will be 33 mA. $R_{sc}$ should be made a 22 Ω and then the current limit will be about 30 mA.

The output current of the regulator can be increased by using an external power transistor. The internal pass transistor then supplies the base current to the external transistor. An example in fig. P12 shows this.

Briefly the specification for the µA 723A is as follows:

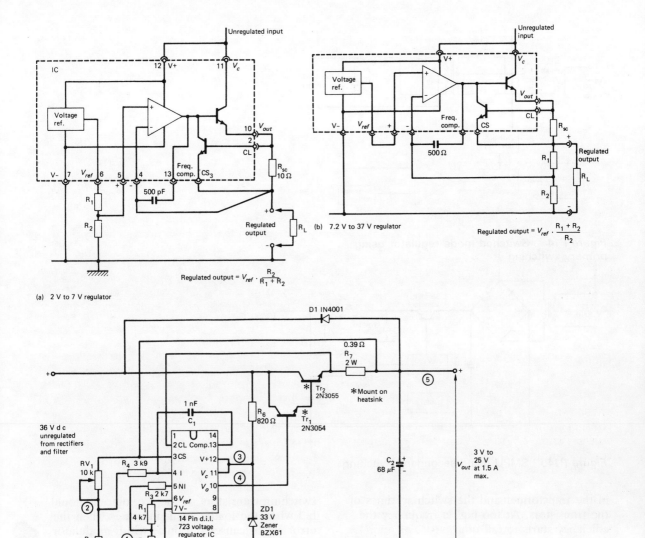

Unregulated input

(a) 2 V to 7 V regulator

Regulated output = $V_{ref} \cdot \dfrac{R_2}{R_1 + R_2}$

(b) 7.2 V to 37 V regulator

Regulated output = $V_{ref} \cdot \dfrac{R_1 + R_2}{R_2}$

**Figure P12 Basic uses of μA 723A**

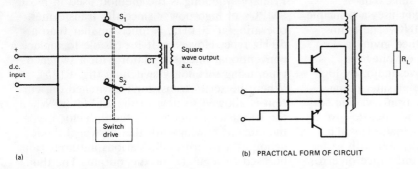

**Figure P13 Basic invertor circuit**

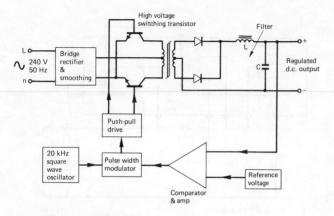

Figure P14a  Switched mode regulator using primary switching

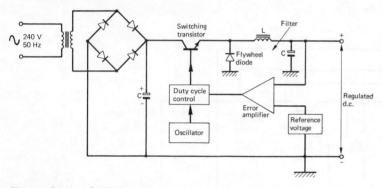

Figure P14b  SMPU using secondary switching

in the transformer and the switching times of the transistors. At too high a frequency the efficiency starts to fall off.

By following an invertor with a rectifier circuit and filter, a convertor, that is d.c. to d.c., is created. If a feedback loop is then added that senses the d.c. output, compares it with a reference level, and feeds a signal that can modify the switching time of the transistors, a type of switched regulator results. This is shown in fig. P14a. This circuit uses a principle called PRIMARY SWITCHING. The mains supply is rectified and smoothed giving a d.c. level of about 340 V. This d.c. voltage is switched, at a frequency above audio, by high voltage transistors to provide an alternating waveform to the transformer primary. The secondary a.c. is rectified and smoothed to give a d.c. output voltage across the load. This d.c. output is regulated by comparing it with a zener reference supply. The difference or error signal is used to alter the duty cycle of the

switching transistors. If the d.c. output should fall when the load current increases, then the error signal causes the pulse width modulator to switch the transistors on for a longer time than they are off during each cycle of the 20 kHz oscillator. More power is provided via the transformer to the load and the output voltage rises to very nearly its previous value. The opposite occurs if the load current is reduced. Primary switching is the method used in most SMPUs of high power since the transformer, operating at 20 kHz, is much smaller than a 50 Hz type. However, it is possible to replace a conventional linear regulator with a switched type, using secondary switching (fig. P14b). When the series transistor is switched on, current is allowed to flow to the LC filter. When the transistor switches off, the inductor keeps the current flowing with the "flywheel diode" acting as a return path. Various methods can be used to regulate the d.c. output. The duty cycle of the switching waveform, or the

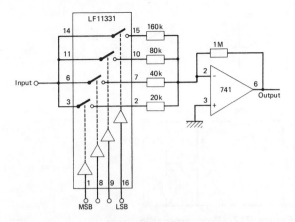

Figure P15  Programmable gain op-amp

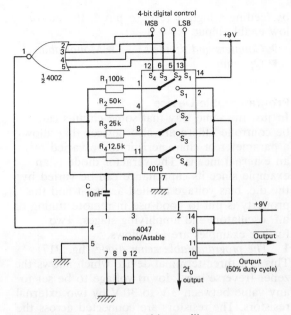

Figure P16  Programmable oscillator

frequency of the oscillator, can be varied, or a mixture of both methods. Since the transistor is being operated as a switch, it is either off or on, and in both these cases the power dissipated by the transistor will be low.

Although SMPUs are more efficient and take up less space than linear regulators, the SMPU cannot match the regulation performance of the linear circuit. SMPUs find their main use in units supplying large currents at low and medium voltages.

▶ *Rectifier circuits*

## Programmable Circuit

This is a circuit in which some parameter or characteristic can be changed by an external control signal. This ability to program in a new value of the parameter or characteristic of the circuit is a very useful feature. It allows the user to select, for example, a particular precise value of frequency from an oscillator, to set the gain of an amplifier, to alter the characteristics of a filter, or to adjust the output voltage from a power supply. Programming can be remote from the circuit and can be carried out by switches adjusted by the operator or by a digital command word from a computer. A few examples that follow show how this can be achieved.

Specialised i.c.s are available such as

Programmable timers (2240, 2250, and 8260)
Programmable op-amps (4250, 776, HA-2400/2404/2405)
Programmable counters (4018)
Programmable attenuators.

[PROMS (programmable read-only memories) are dealt with under *Memories*.]

The following two examples are of programmable circuits that can be built up using commonly available components.

**1** *Programmable gain op-amp* (fig. P15)
The gain of this inverting amplifier is programmed by a 4-bit digital word. Any bit high in this programming control causes the associated switch within the JFET i.c. to operate and to connect a resistor into circuit. The resistors are shown binary-weighted but any other weighting can be chosen. The gain can be varied in 15 steps from 6.25 (digital programming = 0001) to 93.75 (digital programming = 1111).
**2** *Programmable square wave oscillator* (fig. P16)
The CMOS 4047 mono/stable i.c. is used here in its astable mode with a fixed value of capacitor and binary weighted timing resistors. Combination of these resistors are switched into circuit by the digitally controlled 4016 CMOS i.c.

The square wave output from pin 10 (and its complement on pin 11) can, with the values shown, be varied from about 300 Hz in 15 steps up to 4.5 kHz. A useful feature is that twice the output frequency is available from pin 13. The 4-input NOR gate ensures that the 4047 is disabled when the digital input is 0000

by feeding a logic 1 level to pin 4, the active low enable input.

▶ *Counters and dividers*   ▶ *Monostable*
▶ *Op-amp*

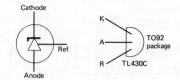

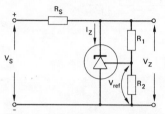

## Programmable Device

In just the same way that some circuits can be controlled there are also devices that allow a parameter or characteristic to be varied by an external means. The varactor diode is an example since its capacitance can be varied by the d.c. bias voltage applied across it and this property is put to good use in remote tuning of h.f. oscillators and amplifying stages. Two further examples are as follows.

**1** *The programmable zener diode* (fig. P17)
This is a three-terminal device which allows the zener (reverse breakdown) voltage to be set to any value between 3 V to 30 V by two external resistors. The resistors are connected across the diode and their common connection is taken to the ref. lead of the device. The internal reference voltage is typically 2.7 V and the reference input current is less than 10 $\mu$A. The zener voltage is then given by

$$V_Z = V_{ref.}(R_1 + R_2)/R_2$$

A device such as the TL 430C has the following specification:
($T_{amb}$ = 25°C)
Max zener voltage $V_Z$ = 30 V
Diode current $I_Z$ = 2 mA min., 100 mA max.
Power dissipation $P_Z$ = 775 mW (free air)
Temp. coefficient ±50 ppm/°C
Slope resistance 3 $\Omega$ max.
▶ *Zener diode*

**2** *The programmable unijunction transistor* (fig. P18)
This device is similar in operation to a unijunction but is really a thyristor with an anode gate. This gate allows the forward breakover voltage of the p-n-p-n thyristor to be controlled. While the anode voltage is less than the control voltage applied to the gate, the device remains in a high-resistance off-state. As soon as the anode voltage exceeds the gate voltage by a small amount (approximately 0.7 V), the device turns on, giving a very low resistance between anode and cathode. The negative resistance characteristic makes it useful in creating simple oscillators and timers, which can be programmed by the voltage applied to the gate. In the circuit example, using a BRY39

**Figure P17**   **Programmable zener**

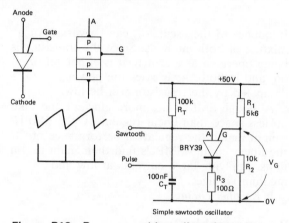

**Figure P18**   **Programmable unijunction**

PUT, the resistors $R_1$ and $R_2$ form a potential divider setting the gate voltage at 31.5 V which is about 63% of $V_S$. Capacitor $C_T$ is charged via $R_T$ until the voltage at the anode just exceeds 32 V. Then the PUT switches to its low resistance state. $C_T$ is discharged rapidly via $R_3$ to give a short-duration pulse. When $C_T$ is discharged, the PUT turns off and the cycle continues. Sawtooth type waves are obtained from the anode with a frequency of about 100 Hz.

$$f = 1/C_T R_T \quad \text{with} \quad (R_1 + R_2)/R_2 = 1.59$$

▶ *Thyristor*   ▶ *Unijunction transistor*

## Pulse Waveform

A pulse occurs when a voltage or current changes its level abruptly to some new value and then, after a time interval, returns to its original level. Much of modern electronics is concerned with pulse waveforms so there has

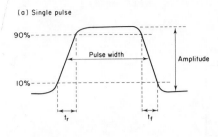

(a) Single pulse

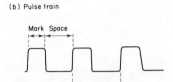

(b) Pulse train

**Figure P19**  Pulse wave form

In(*b*), pulse repetition frequency = $1/T$ pulses per sec (Hz)

to be a common method of specifying the main features of a pulse. This is shown in fig. P19.

**1** The *rise time*, measured on the leading edge, irrespective of the polarity of the pulse, is the time taken for the amplitude to change from 10% to 90% of full pulse amplitude.

**2** The *fall time*, measured on the trailing edge, is the time taken for the amplitude to change from 90% to 10%.

**3** *Pulse width* is usually measured at 50% amplitude.

A pulse may be distorted within a system or by transmission, the typical effects being to produce overshoot, ringing, or sag. One circuit that is useful in reshaping pulses that have been degraded during transmission is the Schmitt trigger.

## Rectifier Circuits and Unstabilised D.C. Supplies

In practically all power supply circuits, some form of rectifier is required to convert the alternating voltage and current into unidirectional voltage and current. To give an unstabilised d.c. output some filter circuit then has to be used to smooth out the pulsating d.c.. In the diagrams (fig. R1) the forward voltage drop across the diodes is assumed to be very small in comparison to the peak a.c. of the transformer secondary. In practice

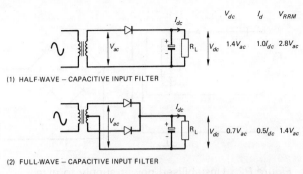

(1) HALF-WAVE – CAPACITIVE INPUT FILTER

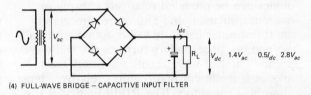

(2) FULL-WAVE – CAPACITIVE INPUT FILTER

(3) FULL-WAVE – CHOKE INPUT FILTER

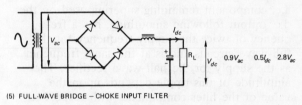

(4) FULL-WAVE BRIDGE – CAPACITIVE INPUT FILTER

(5) FULL-WAVE BRIDGE – CHOKE INPUT FILTER

**Figure R1**  Basic rectifier circuits

$I_d$ = average current through diode

$V_{ac}$ = rms secondary voltage

$V_{RRM}$ = maximum repetitive peak reverse voltage

about 0.7 V should be subtracted from the d.c. outputs of the half-wave and full-wave circuits, and about 1.4 V for the full-wave bridge circuit. The maximum repetitive peak reverse voltage $V_{RRM}$, formally called peak inverse voltage, is the voltage across one diode when it is reversed. Thus $V_{RRM}$ is the sum of the d.c. output voltage and the peak value of the a.c. secondary voltage. Note that when the secondary voltage of a transformer is quoted it is the rms value that is given. Naturally the diodes must be able to withstand the reverse voltage without breakdown.

The rectifier circuit almost universally used is the full-wave bridge since the silicon

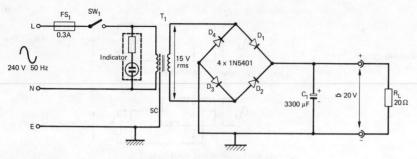

**Figure R2** Unstabilised power supply to give 20 V at 2 A

diodes can be provided relatively cheaply in one encapsulation, and also more importantly the transformer has only to be wound with half the number of secondary turns as the full-wave circuit. The half-wave rectifier is rarely used since it is inefficient and requires relatively large smoothing capacitors. The RIPPLE, that is the a.c. component remaining superimposed on the d.c. output following smoothing, has a frequency of twice the supply frequency for the full-wave circuits and is at the same frequency as the supply for the half-wave rectifier. The amplitude of this ripple depends upon the value of the filter components with respect to the load.

Using a "rule of thumb" the minimum value of the required filter capacitor can be calculated from

$$C_{min} \simeq \frac{1}{2\sqrt{2} \times f_r k_r R_L}$$

where $R_L$ is the load
$k_r$ is the ripple factor
$f_r$ is the ripple frequency

$$k_r = \frac{\text{rms ripple voltage}}{\text{d.c. output voltage}}$$

A typical unstabilised d.c. power supply is shown in fig. R2. The line supply lead is fitted with a fuse. An anti-surge (slow blow) is used because, with a capacitive input filter, a large current may be taken at the instant of switch on. This is because the capacitor is uncharged and acts almost as a short circuit. The mains on/off switch $SW_1$ is a single-pole single-throw type (SPST).

The transformer has an electrostatic screen wound between the primary and secondary coils, which consists of a layer of copper foil extending over the primary winding. The ends of this screen are insulated from each other to prevent it acting as a single shorted secondary. The screen is connected to earth and therefore reduces mains-borne interference as well as providing additional isolation and protection.

As an example assume a 10 V d.c. supply at 500 mA is required from a bridge rectifier and that the rms ripple is to be about 500 mV.

In this case $k_r = 50 \times 10^{-3}$, $R_L = 20\,\Omega$, $f_r = 100$ Hz, therefore

$$C_{min} = 1/2\sqrt{2} \times f_r k_r R_L$$
$$= 1/2\sqrt{2} \times 100 \times 50 \times 10^{-3} \times 20 \text{ farads}$$
$$C_{min} \simeq 3500\ \mu\text{F} \quad (3300\ \mu\text{F is the nearest preferred value}).$$

This, of course, is only a rough approximation since the ripple waveform is not sinusoidal but a complex shape. The ripple voltage will cause a ripple current to flow through the capacitor. A simple calculation will show that the reactance of the 3300 μF at 100 Hz is about 0.5 Ω. Thus the a.c. current flowing through the capacitor is about 1 A. A physically large electrolytic with a ripple current rating well in excess of this must be fitted to prevent the possibility of the capacitor overheating. The ripple current causes $I^2R$ losses in the capacitor. The resistance in this case is the equivalent series resistance (ESR) of the capacitor.

▶ *Power supplies*

## Reliability

A definition of reliability is given in BS 4200 Part 2:

> RELIABILITY is the ability of an item to perform a required function (without failure) under stated conditions for a stated period of time.

Here an item means a component, instrument, or system. Note that a reliability figure cannot be predicted without specifying the operating time and the operating conditions.

To complete the picture, since reliability is concerned with "failure", this also has to be defined:

> FAILURE is the termination of the ability of an item to perform its required function.

Failures may be yet further defined depending on

1) The degree of failure—is the item just out of specification or has it broken down completely?

2) The cause of failure—misuse or inherently weak.

3) The rate or time of failure—sudden or gradual.

### 1) DEGREES OF FAILURE

*a) Partial failure*—failures resulting from deviations in characteristic(s) or parameter(s) beyond the specified limits, but not such as to cause complete lack of the required function.

*b) Complete failure*—failures resulting from deviations in characteristic(s) beyond the specified limits such as to cause complete lack of the required function.

### 2) CAUSES OF FAILURE

*a) Misuse failure*—failures attributable to the application of stresses beyond the stated capabilities of the item.

*b) Inherent weakness failure*—failures attributable to weakness inherent in the item itself when subjected to stresses within the stated capabilities of the items.

### 3) TIME OF FAILURE

*a) Sudden failure*—failures that could not be anticipated by prior examination.

*b) Gradual failure*—failures that could be anticipated by prior examination.

### 4) COMBINATIONS OF FAILURE

*a) Catastrophic failure*—failures that are both sudden and complete.

*b) Degradation failure*—failures that are both gradual and partial.

## Failure rate, MTTF, MTBF

A study of reliability is essentially a study of the failures of components and systems. Field trials of equipment can be used to provide data on system failure rates, but for reliability prediction a designer needs to know with some confidence the failure rate of each type of component that goes to make up the system. We have already defined failure as the inability of a component or piece of equipment to carry out its specified function. In other words, it may be just out of specification limits (a partial failure) or totally non-functioning (a complete failure).

Why should components fail? The answer, of course, is that all man-made items have a finite life. Whatever the article, wear during use, and stresses acting on it, will at some time cause it to fail.

For electronic components the stresses are caused by

*a) Design operating conditions*

Applied voltage and current, power dissipated, and mechanical stress caused by the fixing method.

*b) Environmental conditions*

High or low temperature. Temperature cycling. High humidity. Mechanical vibrations and shocks. High or low pressure. Corrosive atmospheres. Radiation. Dust. Attack by insects or fungi.

Some of these hazards are more damaging than others, for example rapid and large changes in temperature coupled with high humidity could lead, within a short time, to the breakdown of some components.

As far as design is concerned, the life of a component can be greatly improved if it is operated well within the full rated values of current voltage and power. This technique, called DERATING, is used extensively to reduce failure rates. This could be, for example, operating a 1 W resistor at 0.25 W maximum.

The FAILURE RATE of a component can be found by operating large numbers of the component for a long period and noting the number of failures which occur. The initial period of high failure rate, known as BURN-IN or EARLY FAILURE, is followed by a period where the rate of failure levels off to an almost constant value. This period is known as RANDOM FAILURE PERIOD or the USEFUL LIFE, and is the one of most interest since here the failures are entirely random, i.e.

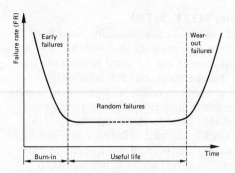

**Figure R3** Failure rate variations with time

**Table R1** *Typical failure rates for common components*

| Components | Type | Failure rate ($\times 10^{-6}$/h) |
|---|---|---|
| Capacitors | Paper | 1 |
| | Polyester | 0.1 |
| | Ceramic | 0.1 |
| | Electrolytic (Al. foil) | 1.5 |
| | Tantalum (soild) | 0.4 |
| Resistors | Carbon composition | 0.05 |
| | Carbon film | 0.2 |
| | Metal film | 0.03 |
| | Oxide film | 0.02 |
| | Wire-wound | 0.1 |
| | Variable | 3 |
| Connections | Soldered | 0.01 |
| | Crimped | 0.02 |
| | Wrapped | 0.001 |
| | Plug and sockets | 0.05 |
| Semi-conductors (Si) | Diodes (signal) | 0.05 |
| | Diodes (regulator) | 0.1 |
| | Rectifiers | 0.5 |
| | Transistor < 1 watt | 0.08 |
| | > 1 watt | 0.8 |
| | Digital—i.c. (plastic DIL) | 0.2 |
| | Linear—i.c. (plastic DIL) | 0.3 |
| Wound components | Audio inductors | 0.5 |
| | R.F. coils | 0.8 |
| | Power transformers (each winding) | 0.4 |
| Switches | (per contact) | 0.1 |
| Lamps & indicators | Filament | 5 |
| | LED | 0.1 |
| Valves | (Thermionic) | 5 |

due to chance alone. Using the failure rate over this period enables predictions to be made of reliability by means of probability theory.

The variation of failure rate (FR) with time is shown graphically in fig. R3. Because of its shape it is often referred to as the *bath-tub curve*.

The term MTTF is normally applied to items that cannot be repaired while MTBF is used for repairable items, i.e. instruments and systems.

The MTBF of a complete system can be calculated by first finding the sum of the failure rates of all components. Consider a very simple circuit using three components, A, B, C. The total failure rate is

$$FR_{(circuit)} = FR_{(A)} + FR_{(B)} + FR_{(C)}$$

Circuit or system failure rate is usually quoted as $\lambda$. Therefore

$$MTBF_{(circuit)} = \frac{1}{\lambda}$$

The best way to see how system MTBF is calculated is to sum the failure rates of components in a circuit, but before we can do that we have to know the failure rates of individual components. The following table, *intended only as a guide*, shows failure rates for commonly used electronic components.

The failure rate for a component depends, of course, on the manufacturing method and its environment in use. Weighting factors must be applied if the environmental conditions are severe as for example in military mobile equipment.

Now, using the figures in Table R1, let us estimate the MTBF for a typical system—say a function generator with its own mains-derived power supply. It uses, say, 30 transistors; 75 metal film resistors; 45 capacitors; 5 potentiometers; 3 switches (30 contacts); a mains transformer (3 windings); 4 rectifiers; 20 small signal diodes; 2 linear ICs; 3 digital ICs; and has 750 soldered joints. The calculation for total system failure rate is made as follows:

| Component | Average failure rate FR ($\times 10^{-6}$/h) | Number used ($n$) | $n$(FR) ($\times 10^{-6}$/h) |
|---|---|---|---|
| Transistors | 0.08 | 30 | 2.4 |
| Resistors (metal film) | 0.03 | 75 | 2.2 |
| Capacitors (polyester) | 0.1 | 45 | 4.5 |
| Potentiometers | 3 | 5 | 15.0 |
| Switches | 0.1 | 30 contacts | 3.0 |
| Transformer | 0.4 | 3 windings | 1.2 |
| Diodes | 0.05 | 20 | 1.0 |
| Rectifiers | 0.5 | 4 | 2.0 |
| Linear ICs | 0.3 | 2 | 0.6 |
| Digital ICs | 0.2 | 3 | 0.6 |
| Soldered joints | 0.01 | 750 | 7.5 |
| | | | 40.0 |

Therefore $\lambda$ is estimated to be $40 \times 10^{-6}$/hour. So the MTBF (symbol $m$) is

$$m = 1/\lambda = 10^6/40 = 25\,000 \text{ hours}$$

It must be pointed out that this is only an example of the way in which MTBF can be calculated. To obtain a reasonable degree of accuracy in the prediction, the actual failure rate of each type of component must be assessed correctly

## Factors affecting equipment reliability

The various stages in the life cycle of an electronic instrument can be separated into four parts (see fig. R4):
a) Design and development
b) Production
c) Storage and transport
d) Operation.

At each stage the overall reliability will be affected by the methods used. If, as should be the case, good reliability is considered an essential part of the manufacturers aim, then a reliability programme has to be established to cover these four stages.

## The exponential law of reliability

If a constant failure rate applies, that is failures are due to chance alone, the relationship between reliability ($R$) and system failure rate ($\lambda$) is given by the formula:

$$R = e^{-\lambda t}$$

where $t$ is the operating time; $\lambda$, the system failure rate, is the sum of all component failure rates; and e is the base of natural logarithms. $R$ is the probability of *zero* failures in the time $t$.

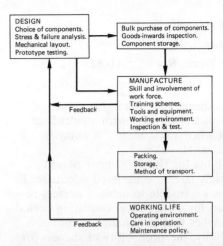

**Figure R4** Factors affecting the reliability of electronic equipment and systems

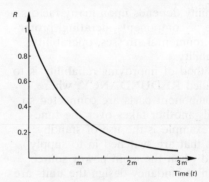

**Figure R5** Plot of $R = e^{-t/m}$

Unreliability $Q = 1 - R = 1 - e^{-\lambda t}$.

Now since MTBF or $m = 1/\lambda$, then $\lambda = 1/m$ and therefore

$$R = e^{-t/m}$$

It is useful to show the graph of $R$ against $t$, with time marked off in intervals of $m$ (fig. R5). This shows that when $t = m$, i.e. the operating time is the same as MTBF, the probability of successful operation has fallen to approximately 0.37 or 37%. Only when the operating time is relatively short compared with MTBF does the reliability become high.

As an example, imagine a naval radar system with an estimated MTBF of 10 000 hours. What is the probability of it working successfully for mission times of 100, 2000, and 5000 hours?

**211**

Using the formula $R = e^{-t/m}$

for $t = 100$ hours, $R = e^{-0.01} = 0.99 \ (99\%)$

$t = 2000$ hours, $R = e^{-0.2} = 0.819 \ (81.9\%)$

$t = 5000$ hours, $R = e^{-0.5} = 0.6065 \ (60.65\%)$

Books of mathematical tables give values of $e^{-x}$, and they can be found with a calculator.

Note that when $x < 0.02$, $e^{-x}$ is approximately equal to $(1 - x)$.

The reliability for the 5000 hour period would certainly not be considered adequate. It is, of course, impossible to achieve perfect reliability, even for a very short operating period, since this implies that the chance of failure is zero. Reliability, the probability of success for a stated period of time under stated conditions, can approach unity (i.e. 0.999) but never equal it.

Good reliability depends upon many factors such as choice of components, derating, protection from environmental stresses, operability, and maintainability.

Another method of improving reliability is to use what is called REDUNDANCY, where subunits or component parts are connected so that if one fails another takes over the function. A good example is the use of standby power sources that are switched in to supply the load if and when the main power source fails. Mostly in redundancy design the units are paralleled. Consider the simple case of two units X and Y wired in parallel so that the whole system does not fail until *both* X and Y have failed. Let the individual units have reliabilities over an operating time of 1000 hours of

$$R_x = 0.85 \quad \text{and} \quad R_y = 0.75$$

Unreliability, or probability of failure, is given by

$$Q = 1 - R$$

Therefore

$$Q_x = 1 - R_x = 1 - 0.85 = 0.15$$

$$Q_y = 1 - R_y = 1 - 0.75 = 0.25$$

The probability of failure of the units in parallel is the product of their unreliabilities:

$$Q_{xy} = Q_x \times Q_y = 0.15 \times 0.25 = 0.0375$$

So the system reliability over 1000 hours is

$$R_{xy} = 1 - Q_{xy} = 1 - 0.0375 = 0.9625$$

Note here the increase in reliability achieved by duplicating the circuits.

Alternatively since $Q_x = 1 - R_x$ and $Q_y = 1 - R_y$, it follows from

$$R_{xy} = 1 - Q_x Q_y$$

that

$$R_{xy} = 1 - (1 - R_x)(1 - R_y)$$
$$= 1 - (1 - R_y - R_x + R_x R_y)$$

i.e. $\quad R_{xy} = R_x + R_y - R_x R_y$

This is covered by the probability theorem:

If $a$ and $b$ are two events which can occur together, the probability of either or both events taking place is $P_a + P_b - P_a P_b$.

Redundancy can be *active* where a standby unit is switched in following a failure, or *passive* where the elements share the load but are each capable of supplying the load or carrying out the function separately. The more units placed in parallel, the greater the overall system reliability, but the cost penalty is obviously high.

It should now be clear that, when units are placed in series such that the failure of one unit means failure for the whole system, then the reliability for the system will be lower than that of the individual units. The system reliability then depends upon the product of the reliabilities:

$$R_{xy} = R_x T_y$$

If $R_x = 0.85$, $R_y = 0.75$, then

$$R_{xy} = 0.85 \times 0.75 = 0.6375$$

▶ *Maintenance*

## Resistor

Before considering the construction of the various types of resistor it is worth looking at the methods used for coding resistors and their preferred values. Many resistors will in future be coded according to BS 1852 [▶ *Colour code*]. The value and the tolerance of the resistor will be printed on the resistor body instead of the now familiar colour bands.

After the value code, a letter will be added to indicate the tolerance:

$F = \pm 1\% \quad G = \pm 2\% \quad J = \pm 5\% \quad K = \pm 10\%$

$M = \pm 20\%$

**Table R2**  BS 2488 *Preferred values (resistors)*

**E24 Series**

10  11  12  13  15  16  18  20  22  24  27  30  33  36  39  43  47  51  56  62  68  75  82  91

**E12 Series**

10    12    15    18    22    27    33    39    47    56    68    82

**E6 Series**

10    15    22    33    47    68

**E96 Series**

100  102  105  107  110  113  115  118 etc.

Thus R33M is a 0.33 Ω ± 20% resistor
    4k7F is a 4700 Ω ± 1% resistor
    6M8M is a 6.8 MΩ ± 20% resistor
    22KK is a 22 000 Ω ± 10% resistor.

*BS 1852 Resistance code*
*Examples*

| Resistance | Marking |
|---|---|
| 0.47 Ω | R47 |
| 1 Ω | 1R0 |
| 6.8 Ω | 6R8 |
| 68 Ω | 68R |
| 100 Ω | 100R |
| 1 kΩ | 1k0 |
| 4.7 kΩ | 4k7 |
| 100 kΩ | 100k |
| 10 MΩ | 10M |

The BS 2488 preferred values of resistance are shown in Table R2. In the E96 series much closer values can be obtained (for ±1% tolerance).

## Construction of fixed resistors

Typical constructions of fixed resistors are shown in fig. R6. The carbon composition type is made by mixing finely ground carbon with a resin binder and an insulating filler. The resulting mixture is compressed, formed into rods, and then fired in a kiln. The ratio of carbon to the insulating filler determines the final value of the resistance. Silver-plated end caps, which have tinned copper leads attached, are then pressed onto the rod. Alternatively some manufacturers mould the carbon around the leads so that the connection is embedded. The latter method is claimed to be more mechanically robust and to reduce the risk of electrical noise resulting from a poor connection. Finally, the whole resistor is either moulded in plastic or

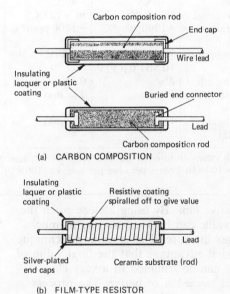

(a)  CARBON COMPOSITION

(b)  FILM-TYPE RESISTOR

**Figure R6**  Construction of fixed resistors

given several coats of insulating lacquer to provide electrical insulation and protection from moisture.

FILM RESISTORS are manufactured by depositing an even film of resistive material onto a high-grade ceramic rod. The resistive material may be pure carbon (carbon film); nickel chromium (metal film); a mixture of metals and glass (metal glaze); or a metal and an insulating oxide (metal oxide). The choice of the ceramic rod is important since this will enhance or degrade the properties of the final resistor. For example, its thermal expansion has to be similar to that of the film material to prevent the film from cracking. Alumina is commonly used. The required resistance value is then obtained by cutting a helical track through the film material, in other words spiralling off part

**213**

**Table R3** *Comparison of common types of general-purpose resistors*

| Resistor type | Carbon composition | Carbon film | Metal oxide | Metal glaze | General-purpose wire-wound |
|---|---|---|---|---|---|
| Range | 10 Ω to 22 MΩ | 10 Ω to 2 MΩ | 10 Ω to 1 MΩ | 10 Ω to 1 MΩ* | 0.25 Ω to 10 kΩ |
| Selection tolerance | ±10% | ±5% | ±2% | ±2% | ±5% |
| Power rating | 250 mW | 250 mW | 500 mW | 500 mW | 2.5 W |
| Load stability | 10% | 2% | 1% | 0.5% | 1% |
| Max. voltage | 150 V | 200 V | 350 V | 250 V | 200 V |
| Insulation resistance | $10^9\,\Omega$ | $10^{10}\,\Omega$ | $10^{10}\,\Omega$ | $10^{10}\,\Omega$ | $10^{10}\,\Omega$ |
| Proof voltage | 500 V | 500 V | 1 kV | 500 V | 500 V |
| Voltage coeff. | 2000 ppm/V | 100 ppm/V† | 10 ppm/V | 10 ppm/V | 1 ppm/V |
| Ambient temp. range | −40°C to +105°C | −40°C to +125°C | −55°C to +150°C | −55°C to +150°C | −55°C to 185°C |
| Temperature coeff. | ±1200 ppm/°C | −1200 ppm/°C | ±250 ppm/°C | ±100 ppm/°C | ±200 ppm/°C |
| Noise | 1 kΩ 2 μV/V 10 MΩ 6 μV/V | 1 μV/V | 0.1 μV/V | 0.1 μV/V | 0.01 μV/V |
| Soldering effect | 2% | 0.5% | 0.15% | 0.15% | 0.05% |
| Shelf life 1 year | 5% max. | 2% max. | 0.1% max. | 0.1% max. | 0.1% max. |
| Damp heat 95% RH | 15% max. | 4% max. | 1% | 1% | 0.1% |

\* Some high values up to 68 M are now available (Mullard VR37 and VR68 range).

† 100 ppm = $100 \times 10^{-6}$ ohms per ohm per volt, i.e. $100 \times 10^{-6} \times 100\% = 0.01\%$.

of the resistive film. By using a close pitch the resistance value can be increased by as much as 100 times or more. This spiralling technique has the added advantage that the final value of the resistor can be trimmed to a very close tolerance of say ±1% or better. The process can be made automatic with the pitch of the diamond cutting wheel being controlled by the output of a bridge circuit which is monitoring the value of the resistor being trimmed.

With film resistors the material used and the thickness of the film will determine the initial resistance value. For example, a metal film of nichrome, approximately 150 Å thick (0.015 μm), will give a resistance of about 125 ohms per square. The term ohms per square is commonly used for the following reason:

$$R = \rho(l/a)$$

where $\rho$ is material resistivity, $l$ = length of material, $a$ = cross-sectional area.

Therefore

$$R = \rho l/dw$$

where $d$ = depth of film, $w$ = width of film.

Therefore

$$R = \rho'(l/w)$$

where $\rho'$ is termed resistivity per unit depth or surface resistivity.

For a square sheet when $l = w$,

$$R = \rho'$$

regardless of the size of the square. Sheet resistance values can therefore be written in ohms per square.

A WIRE-WOUND RESISTOR is manufactured by winding resistance wire onto an insulating former. Common materials used are nickel chromium (Nichrome), copper nickel alloys (Eureka), and alloys of nickel and silver. The wire is produced by drawing the material through a suitable die and then annealing it. The wire must have good uniformity, be ductile, resist corrosion, and have fairly high resistivity. Ductility is important otherwise the wire might crack or break after winding. On the larger resistors, terminations are provided by a band of conducting metal (ferrule) with a lug. Smaller wire-wound resistors may have end caps or a lead brazed onto the ends of the

resistance wire. The connecting lead is then folded and inserted inside the ceramic former. This firmly retains the wire leads and prevents mechanical strain being applied to the actual resistance wire winding. The whole resistor is then covered in some insulating material. Usually, wire-wound resistors are vitreous-enamelled which gives excellent protection against moisture, allows good heat dissipation, and is non-inflammable; other types may be cement-coated which naturally is not impervious to moisture.

In order to achieve relatively high values of resistance, the wire must be thin and many turns must be used, thus the maximum value for wire-wounds is limited to about 100 kΩ. General purpose wire-wounds are used where a relatively large power dissipation is required, but another important application is in providing a precision resistor with a very low temperature coefficient and excellent stability. Tolerances can be ±0.1%, temperature coefficients ±5 ppm/°C, and long-term stability better than 40 ppm/year. Other types of resistor cannot yet approach this performance.

### Resistor Transistor Logic (RTL)

This form of logic using resistors and a transistor invertor was the forerunner of modern digital logic families. It was rapidly superseded by DTL and TTL since it had inferior performance in speed and in drive capability. The fan-out of RTL is typically less than 5.

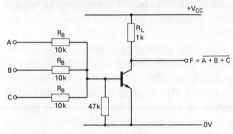

*Figure R7* RTL NOR gate

The basic circuit is a NOR gate shown in fig. R7. Each input resistor is made about ten times greater than the collector load of the transistor to ensure that the transistor saturates. If any one of the three inputs is at logic 1 (high), the transistor receives base current and fully turns on to give a low output. The output is only high when all three inputs are at logic 0.

▶ *TTL*    ▶ *CMOS*

### Safe Operating Area (SOAR)

Any electronic device has specified maximum ratings for voltage, current and power. If one of these is exceeded the device will, in all probability, be damaged. For example a bipolar junction transistor in the common emitter mode will have specified maximum values for

$I_{C(max)}$ collector current
$V_{CE(max)}$ collector/emitter voltage
$P_{tot(max)}$ power dissipation

These three maximum ratings can be drawn onto the output characteristics of the transistor (fig. S1$a$) and the three quantities define the permissible operating area within which the quiescent d.c. operating point must lie.

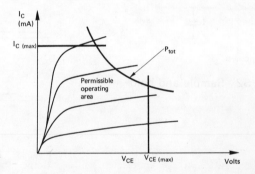

*Figure S1a* Operating area for a bipolar transistor (CE mode)

$V_{CE(max)}$ is set by the maximum reverse voltage that can be sustained across the collector/base junction. If this is too high, avalanche breakdown occurs.

With power transistors another form of breakdown can occur at high voltages and only relatively medium values of current. This is called SECONDARY BREAKDOWN and is caused by signal currents (either increasing or decreasing) tending to concentrate in one region of the base/emitter junction. This concentration leads to an increase in temperature at this point which further increases the current concentration. Local breakdown then occurs at a lower voltage than $V_{CE(max)}$. If this secondary breakdown is included on the output characteristics, the area contained within the four lines is seen to be a safe operating area (fig. S1$b$)

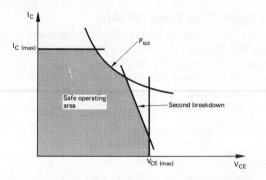

*Figure S1b* Extension to include second breakdown to give SOAR for a power transistor

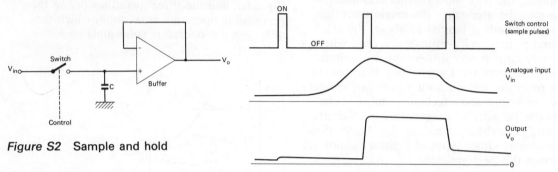

*Figure S2* Sample and hold

polystyrene are best. Also, the leakage current of the switch when in the off position and the input bias current of the buffer must be very low since these too will both cause the capacitor to discharge. These are factors that have to be taken into consideration in determining the hold time for a particular application. For example if the switch, in the off position, has a leakage current of 5 nA and the bias current to the amplifier is 10 nA, a 100 nF storage capacitor will discharge at a rate of 150 mV per second.

## Sample and Hold

As the name implies this is a circuit that periodically measures and stores the amplitude of an analogue input signal. It is widely used in data acquisition systems and pulse code modulators where the input has to be held constant for a fixed time period to allow reasonably error-free analogue-to-digital conversion.

The basic components of the circuit are a fast electronic switch, a capacitor, and a buffer amplifier (fig. S2). The switch, usually a MOS-FET, performs the sampling function. When the control signal is high, the switch closes and the voltage across the capacitor tracks the input signal. At the end of the sample control, as the switch is turned off, the capacitor holds the instantaneous value of the analogue input. The buffer amplifier, which has a very high input impedance, prevents the capacitor from discharging between sampling pulses so that the minimum amount of "droop" occurs. To ensure that the sampled analogue input is stored effectively, the capacitor must be a high-quality low-leakage type; usually a polypropylene or

The speed of circuit response is defined by

*a*) *Aperture time:* which is the delay between the time that the control signal is applied to the switch and the time that the switch actually closes.

*b*) *Acquisition time:* the time required for the capacitor to change from one level of holding voltage to a new value. Typically it is taken as the time required to acquire a new analogue input voltage with an output step of 10 V.

Other important parameters are

*c*) *Dynamic sampling error:* the error introduced into the held output due to a changing analogue input at the time the hold command is given. Error is expressed in mV with a given hold capacitor value and input slew rate.

*d*) *Hold settling time:* the time required for the output to settle within 1 mV of final value after the hold command.

*e*) *Hold step:* The voltage step at the output of the sample and hold when switching from sample mode to hold mode with a steady (d.c.) analogue input voltage.

The LF 398 is an example of a monolithic sample and hold which gives excellent performance. CMOS transmission gates can also be used to create sample and hold circuits.

▶ *CMOS*

## Schmitt Trigger Circuit

This circuit, shown in fig. S3 is used for level detection, reshaping pulses with poor edges, and squaring sine wave signals. The Schmitt is basically a snap-action switch that changes state at a specific trip point. Consider conditions when the input to $Tr_1$ base is at zero. $Tr_2$ is conducting since it has forward bias provided by the potential divider $R_2$, $R_3$ and $R_4$. The voltage at $Tr_2$ base is approximately

$$V_{B2} \simeq V_{CC}R_4/(R_2 + R_3 + R_4)$$

The voltage at $Tr_1$ and $Tr_2$ emitter will be 0.7 V less than $V_{B2}$ and this positive voltage reverse biases $Tr_1$, thereby holding it off. The current flowing through $Tr_2$ is determined by

$$I_{C2} \, \text{mA} \simeq (V_{B2} - 0.7)/R_5 (\text{k}\Omega)$$

and the collector voltage of $Tr_2$ will be

$$V_{C2} = V_{CC} - I_{C2}R_6$$

Usually the circuit is designed so that $Tr_2$ is not saturated, thus allowing faster switching speed.

When the input voltage is increased so that it nearly equals the voltage on $Tr_2$ base, then $Tr_1$ starts to conduct, its collector voltage falls, and $Tr_2$ starts to turn off. Because of the positive feedback between the emitters, $Tr_1$ rapidly turns on and $Tr_2$ off. The output voltage rises to $+V_{CC}$.

A special feature of the Schmitt is that the circuit does not switch back as soon as the input signal is reduced just below the threshold level or trip point, but at a much lower level. The circuit possesses hysteresis or backlash, and this is very useful in eliminating noise superimposed on the input signal.

The reason for hysteresis can be seen by considering that the circuit changes state at the point when the two base voltages are equal. When the threshold level or upper trip point is passed, $Tr_1$ switches on and its collector voltage falls. This means that $Tr_2$ base voltage also falls, so in order for the circuit to switch back to its original state the input voltage to $Tr_1$ base must be reduced to a value equal to the lower voltage on $Tr_2$ base. The effect of the hysteresis of the circuit is shown in fig. S4 where it can be seen that the output switches back only when the input is reduced below the lower trip point.

For the basic transistor circuit (non-saturating design)

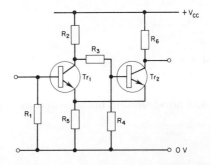

**Figure S3**   Basic Schmitt trigger circuit

Upper trip point $V_{T^+} \simeq V_{CC}R_4/(R_2 + R_3 + R_4)$

Lower trip point $V_{T^-}$

$$\left(V_{CC} + V_{BE}\frac{R_2}{R_5}\right)R_4R_5 \Big/ [R_2R_5 + R_2R_4 + R_5(R_3 + R_4)]$$

Hysteresis $V_H = (V_{T^+} - V_{T^-})$

Because the Schmitt is so useful in reshaping pulses and eliminating noise it is included as a standard gate in most i.c. logic families. The most popular types are

1) TTL standard and low-power Schottky
   - 7413 Dual NAND Schmitt (4 inputs each gate)
   - 7414 Hex inverting Schmitt
   - 74132 Quad NAND Schmitt (2 inputs each gate

For these gates with $V_{CC} = +5$ V

$V_{T^+} = 1.7$ V     $V_{T^-} = 0.9$ V

giving $V_H = 800$ mV

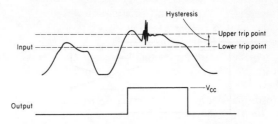

**Figure S4**   Typical waveforms showing the usefulness of the hysteresis in the circuit

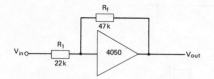

*Figure S5* CMOS non-inverting buffer wired as a Schmitt

2) CMOS $\begin{cases} \text{4584 Hex inverting Schmitt} \\ \text{4093 Quad 2-input NAND Schmitt} \end{cases}$

At $V_{DD} = 10\,\text{V}$, $V_{T^+} = 5.9\,\text{V}$, $V_{T^-} = 3.9\,\text{V}$, giving $V_H = 2\,\text{V}$.

It is also relatively easy to build a Schmitt trigger using one CMOS non-inverting buffer and two resistors (fig. S5). If the feedback resistor is made $47\,\text{k}\Omega$ the upper trip point will be 7.3 V (for $V_{DD} = 10\,\text{V}$) and the lower trip point 2.7 V giving a hysteresis of 4.6 V. While the input to the circuit is low, the feedback resistor reduces the effective level at the gate input so that the circuit does not respond when $V_{in} = V_{DD}/2$ but at the higher voltage of 7.3 V. As soon as the input exceeds 7.3 V the output rapidly switches to a high state aided by the positive feedback. The feedback resistor now keeps the gate input high so that $V_{in}$ has to be reduced to below 2.7 V before the output switches low. By varying the values of the resistors the effective trip points and hysteresis can be set to any required value.

$$V_{T^+} = [0.5 V_{DD}(R_{in} + R_F)]/R_F$$
$$V_{T^-} = V_{DD} - [0.5 V_{DD}(R_{in} + R_F)]/R_F$$

Another Schmitt circuit can be built using a CMOS 4023 3-input NAND i.c. This requires no additional resistors and is faster in operation than the previous circuit. Two of the gates in the i.c. are wired to form a bistable while the third has all its three inputs tied together

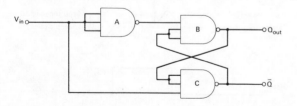

*Figure S6* CMOS 4023 i.c. wired as a Schmitt

as shown in fig. S6. Since a higher input voltage is required by gate A to produce a low output, it follows that a higher input voltage is required to set the bistable than to reset it. The hystersis is approximately equal to $\frac{1}{3} V_{DD}$.

For applications of Schmitts
▶ *Clock pulse generators*  ▶ *Interface circuit*  ▶ *Comparator*

## Schottky Diode and Schottky Transistor

The Schottky diode is formed from a metal to semiconductor rectifying junction and features a low forward threshold voltage of only 300 mV and very fast switching action.

Aluminium, which is a group three element, acts as a p-type impurity when in contact with silicon. Therefore to achieve a low resistance ohmic contact between an aluminium connection pad and an n-type area in a semiconductor device, the region just beneath the contact has to be heavily doped n+ type (fig. S7). Without this heavily doped region, a rectifying junction is formed between the p-type aluminium and the lightly doped high resistivity n-material. This rectifying aluminium to n-silicon junction is the basis of the Schottky diode. This type of diode is a majority carrier device and is sometimes called a hot carrier diode. The carriers are electrons which enter the aluminium when the diode is forward biased. Here the injected electrons become indistinguishable from the electrons in the metal, and when the diode is reverse biased the charge carriers do not have to recross the junction. Hence the switching speed of the Schottky diode is very rapid with values of reverse recovery time approaching 100 pico-seconds. As a discrete component, the diode is mainly used as a microwave detector but it is mainly known because of its use in Schottky TTL [▶ *TTL*].

The main limitation in switching speed of a bipolar transistor is the fact that the transistor saturates giving rise to minority charge storage in the transistor's base region. In a Schottky transistor the transistor is prevented from going into saturation by including a Schottky diode between the base and collector of the transistor. Since the diode has a low forward volt drop (300 mV) when the transistor is turned on, the collector/emitter voltage is clamped to 400 mV. This is sufficient to prevent saturation of the transistor and gives improved switching speed.

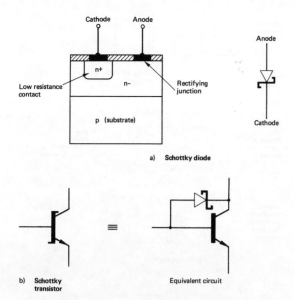

Cathode     Anode

Low resistance contact

n+

n−

Rectifying junction

p (substrate)

Anode

Cathode

a)    Schottky diode

b)    Schottky transistor    ≡    Equivalent circuit

**Figure S7**    Schottky diode and transistor

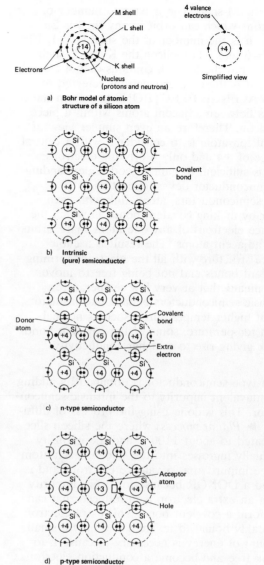

M shell
L shell
(+14)
K shell
Electrons
Nucleus (protons and neutrons)

4 valence electrons
(+4)
Simplified view

a)    Bohr model of atomic structure of a silicon atom

Si +4   Si +4   Si +4
Si +4   Si +4   Si +4
Si +4   Si +4   Si +4

Covalent bond

b)    Intrinsic (pure) semiconductor

Si +4   Si +4   Si +4
Si +4   +5   Si +4
Si +4   Si +4   Si +4

Donor atom

Covalent bond

Extra electron

c)    n-type semiconductor

Si +4   Si +4   Si +4
Si +4   +3   Si +4
Si +4   Si +4   Si +4

Acceptor atom

Hole

d)    p-type semiconductor

**Figure S8**    Silicon atom structure and semiconductors

## Semiconductor Theory

Practically the whole of modern electronic development is due to the exploitation of the special properties of semiconductor materials such as germanium and silicon. In the pure state, referred to as "intrinsic", semiconductors cannot be used by themselves to make useful devices. Materials that are used to make components such as diodes, transistors, FETS and i.c.s require lower values of resistivity than that obtained with intrinsic semiconductors. By the controlled addition of selected impurities the electrical properties of the intrinsic semiconductor are altered to suit the device requirement. This process of introducing impurities into the intrinsic material is called doping, and precise control of impurity content is one of the most important factors in the production of semiconductor devices.

Intrinsic semiconductor (pure silicon or germanium): impurity content typically better than 1 part in $10^{10}$.

Extrinsic semiconductor (material with impurity added): typical doping levels 1 part in $10^8$.

Nearly all modern devices are created from silicon because it has superior performance compared with germanium in terms of lower leakage current and wider operating tempera-

ture range. Therefore the following notes deal with silicon only.

The Bohr model of the atomic structure of a single silicon atom is shown in fig. S8. The nucleus, or central core of the atom, is made up of protons (+ve charge) and neutrons. Electrons orbit around the nucleus in distinct layers or shells. Silicon has a nucleus of +14 and therefore to maintain balance has 14 electrons orbiting in three shells. The shells have been given letters, the K-shell being the nearest to the nucleus, followed by the L-shell and

then the M-shell. The maximum number of electrons which can orbit in any shell is $2n^2$ where $n$ is the number of the shell (K = 1, L = 2, M = 3, etc.). For silicon the K and L-shells are full but the M-shell contains only 4 electrons. It is these outer electrons referred to as the VALENCE ELECTRONS that form the bonds between adjacent atoms within a piece of silicon. Therefore an even simpler view of the silicon atom is to consider it with a central charge of +4 and only 4 orbiting electrons. This is sufficient to gain a good understanding of semiconductor device operation.

In semiconductors, atoms link with each other by making covalent bonds in which the valence electrons of any one atom share orbits with adjacent atoms. The result is a cubic lattice structure with all the electrons forming covalent bonds and not being free to move. This means that at very low temperatures intrinsic semiconductors behave like insulators, but at higher temperatures, such as normal room temperature, some of the covalent bonds break giving rise to a higher conductivity.

**1** n-type semiconductor is obtained by adding a pentavalent impurity to the intrinsic semiconductor. This is done using the process of diffusion (▶ *Planar process*) where the silicon slice is heated to about 1100°C and the impurity gradually ingresses into the silicon. Each atom of the impurity has 5 valence electrons and is called a DONOR because for every impurity atom an extra electron is introduced that cannot form a covalent bond. This extra electron is weakly bound to its atom and only a small amount of energy is necessary to cause it to break free and become a conduction electron. n-type material therefore has an excess of electrons donated by the impurity atoms and these can be used as charge carriers to pass a current. Materials used as donor impurities are
Phosphorus (P)   Arsenic (As)
Antimony (Sb)

**2** p-type semiconductor is made by introducing a trivalent impurity. The impurity atom, called an ACCEPTOR, has only 3 electrons in the valence orbit so that only three covalent bonds can be made with adjacent silicon atoms. The space that is created is called a HOLE and this is a positive charge carrier that can move through the material by accepting elec-

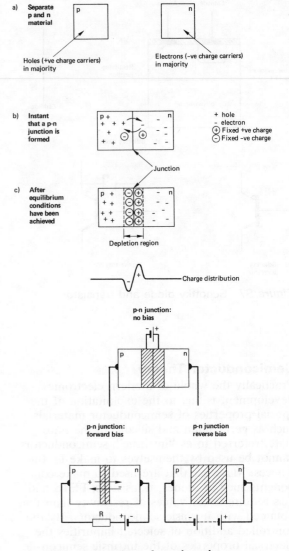

*Figure S9* Formation of p-n junction

trons from other atoms. Materials used as acceptors are
Boron (B)   Aluminium (Al)   Indium (In)
Gallium (Ga)

**3** The fundamental building block of semiconductor devices is the p-n junction. Fig. S9 shows what occurs when a p-n junction is formed. The majority carriers in the p-type are holes while electrons are the majority carriers in the n-type. At the instant the junction is formed, some electrons and holes near the junction recombine. As an electron leaves its

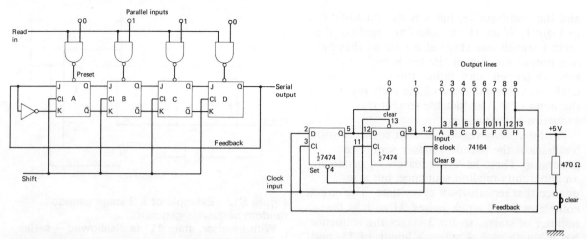

*Figure S11*   Stepping sequence generator; only one output line high at any time

donor atom in the n-type to cross the junction, the donor atom is left with a net positive charge of 1, and as a hole is filled by an electron the acceptor atom in the p-type becomes a fixed negative charge. In this way fixed negative charges are set up in the p-type and fixed positive charges in the n-type in the thin region either side of the junction. Therefore the electron-hole recombination rapidly causes a potential barrier to be set up that opposes the flow of electrons from n to p and the flow of holes from p to n. The thin region either side of the junction becomes empty of charge carriers and for this reason is called the DEPLETION REGION.

The p-n junction has a rectifying action. If a bias voltage greater than about 0.6 V is applied making the p side positive with respect to the n, the action of the potential barrier will be overcome and forward current will flow. This forward current is made up of electrons crossing from n to p and holes moving from p to n.

On the other hand if a bias voltage making the p negative with respect to the n is applied, the effect is to increase the potential barrier to the flow of charge carriers. The depletion region increases and only a very small leakage current flows. This leakage current is due to the thermally generated hole-electron pairs and to any other impurities that are included in the silicon. A typical leakage current in a p-n silicon junction may be less than 10 nA.

The p-n junction as described above is the

basis of the silicon diode and rectifier. Two such junctions, in one crystal, are used to make a transistor and many thousand may occur in an integrated circuit.

▶ *Diode*      ▶ *Transistor*      ▶ *FET*

## Sequence Generator

When a shift register is provided with feedback it can be used to generate a particular binary sequence. In fig. S10 a 4-stage parallel-in/serial-out shift register has its output fed back so that it forms a recirculating store. The required sequence is set up by the parallel inputs and this sequence will appear at the output and be repeated every four shift pulses. Suppose the number 0110 is read in; the sequence with four shift pulses will be

|           | A | B | C | D |
|-----------|---|---|---|---|
| 1st shift | 0 | 0 | 1 | 1 |
| 2nd shift | 1 | 0 | 0 | 1 |
| 3rd shift | 1 | 1 | 0 | 0 |
| 4th shift | 0 | 1 | 1 | 0 |

Another possibility is to use a serial-in/parallel-out shift register to provide a "stepper" type sequential generator, one which has only one output line high (or low) at any time. An example of a 10-state sequential generator is shown in fig. S11. This uses a 7474 dual D bistable and a 74164 TTL shift register. Pressing the Clear button clears the shift register

and the 2nd bistable, but sets the 1st bistable to logic 1. When clock pulses are applied, this logic 1 travels one stage at a time so that only one output line out of the ten is high at any time. After ten clock pulses the logic 1 is fed back from the last stage of the shift register to the input of the 1st bistable so that the sequence is continuous.

An interesting version of shift registers with feedback is the pseudo-random sequence generator. These have the feature of producing an apparently random sequence, but a sequence that repeats itself every time. The circuit completes $2^n - 1$ clock pulses. Here $n$ is the number of stages, so for 3 stages the sequence has a length of 7, 4 stages a length of 15, and so on. For this reason they are also referred to as maximum length sequence generators. Circuits like these are useful in testing audio systems and in data communication system identification and correlation methods. A simple example using 3 D type bistables is shown in fig. S12. The outputs of stages $B$ and $C$ are fed via an exclusive-OR and invertor to the input of stage $A$. The sequence will be as follows:

| Clock | C | B | A | Decimal equivalent |
|-------|---|---|---|--------------------|
| 1 | 0 | 0 | 1 | (1) |
| 2 | 0 | 1 | 1 | (3) |
| 3 | 1 | 1 | 0 | (6) |
| 4 | 1 | 0 | 1 | (5) |
| 5 | 0 | 1 | 0 | (2) |
| 6 | 1 | 0 | 0 | (4) |
| 7 | 0 | 0 | 0 | (0) |

Note that the state 111 is not allowed since this would cause the register to latch up. Additional circuitry may be required to avoid this condition at switch on by initially clearing the register.

If the invertor is removed, the 000 condition will cause latch-up and if this is prevented the sequence will be

| Clock | C | B | A | Decimal equivalent |
|-------|---|---|---|--------------------|
| 1 | 0 | 0 | 1 | (1) |
| 2 | 0 | 1 | 0 | (2) |
| 3 | 1 | 0 | 1 | (5) |
| 4 | 0 | 1 | 1 | (3) |
| 5 | 1 | 1 | 1 | (7) |
| 6 | 1 | 1 | 0 | (6) |
| 7 | 1 | 0 | 0 | (4) |

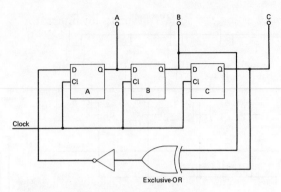

**Figure S12** Example of a 3-stage pseudo random sequence generator.
With invertor, state 111 is disallowed — series is complementary.
Remove invertor to get normal series, then state 000 is disallowed.

As can be seen from the tables the numbers are random but the sequence repeats after every 7th clock pulse. The repetition frequency is

$$f_c/2^{n-1}$$

where $f_c$ is this clock frequency.

By using more registers much longer sequences can be generated as shown in the following table.

*m-sequence (pseudo random) generators*

| Number of stages n | Length | Connection for exclusive-OR feedback |
|--------------------|--------|--------------------------------------|
| 2 | 3 | 1 and 2 |
| 3 | 7 | 2 and 3 |
| 4 | 15 | 3 and 4 |
| 5 | 31 | 3 and 5 |
| 6 | 63 | 5 and 6 |
| 7 | 127 | 6 and 7 |
| 9 | 511 | 5 and 7 |
| 10 | 1023 | 7 and 10 |
| 11 | 2047 | 9 and 11 |
| 15 | 32 767 | 14 and 15 |
| 17 | 131 071 | 14 and 17 |
| 18 | 262 143 | 11 and 18 |

The outputs of pseudo random sequence generators can be converted into noise for testing purposes by

*a*) Passing the serial output to an integrator which has a time constant of less than 1/20 the clock frequency

or *b*) Converting the parallel output into analogue via a DAC.

An example of the use of random sequence generators is in the provision of the code for data scramblers [▶ *Exclusive-OR*].

▶ *Shift register*

## Sequential Logic

A logic circuit that is sequential is one that has a memory and therefore gives an output that depends upon a previously stored bit of information as well as any new data. This distinguishes it from pure combinational logic, which is an arrangement of gates that gives a logic state output dependent solely on the input data existing at any instant in time. The basic building block of any sequential logic circuit is the ▶ *Bistable* and these can be used to make logic circuits such as ▶ *Counters* and ▶ *Shift registers*.

## Shift Register

A shift register is a device used for the temporary storage of digital information which can then be moved (shifted) at a later time. They can be readily constructed using JK master-slave flip-flops to take the form of

*a*) Serial in/Serial out
*b*) Parallel in/Serial out
*c*) Serial in/Parallel out

as shown in fig. S13.

Data stored in the shift register is loaded either in series with shift pulses or in parallel by setting the flip-flops. The data can be shifted to the right one place with every shift pulse.

Large-number shift registers (serial in/serial out) are made in MOS and are the basis of recirculating stores. A bistable can be formed using MOS devices (fig. S14). If the S input is taken high (1), $T_5$ conducts taking $\bar{Q}$ low. This turns off $T_2$, forcing Q to assume logic 1. Similarly if the R input is taken high (1), $T_6$ conducts and Q assumes logic 0 state. Such a bistable forms the basic element for static MOS shift registers. $T_2$, $T_5$ and $T_7$, $T_{10}$ form the two bistables and $T_3$, $T_4$ and $T_8$, $T_9$ are cross-coupling elements. These cross-coupling elements are switched off and on by $\overline{\text{clock 1}}$ and $\overline{\text{clock 2}}$ signals. $T_1$ and $T_6$ are data transfer

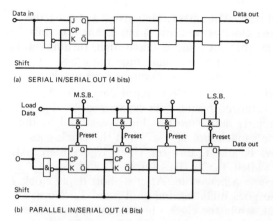

(a) SERIAL IN/SERIAL OUT (4 bits)

(b) PARALLEL IN/SERIAL OUT (4 Bits)

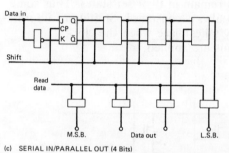

(c) SERIAL IN/PARALLEL OUT (4 Bits)

**Figure S13** Basic shift registers

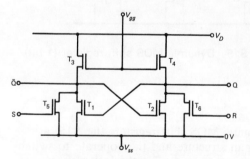

**Figure S14a** MOS bistable

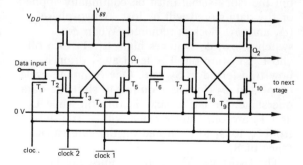

**Figure S14b** MOS static shift register (2 bits shown)

223

switches. The phase relationship between the three clock waveforms is important. To shift data, the clock line is taken high, turning on $T_1$ and $T_6$, and at the same time the cross-coupling elements are switched off by clock 1 and $\overline{\text{clock 2}}$ going low. Input data from $T_1$ to $T_2$ is stored by the gate capacitance of $T_2$, and similarly data from bistable A is stored by the gate capacitance of $T_7$. When the clock goes low, $T_1$ and $T_6$ turn off, $\overline{\text{clock 1}}$ goes high first to switch $T_4$, $T_9$. This forces $T_5$ and $T_{10}$ to assume a new state. After a short delay $\overline{\text{clock 2}}$ also goes high, turning on $T_3$ and $T_8$. Note that, while the clock pulse is not applied, the bistables remain in their set states. Thus some power is always consumed. Shifting of information only takes place when the clock waveforms are applied.

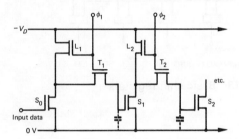

*Figure S15*  Dynamic MOS shift register (1 bit)

Dynamic MOS shift registers (fig. S15) are simpler in structure and they operate to switch load devices on and off with clock pulses. Much less power is consumed from the supply, but the clock signal must be continually applied or the data stored will be lost. A two-phase ($\phi_1$ and $\phi_2$) clock is required. When $\phi_1$ switches low, $\phi_2$ switches high. $L_1$, $T_1$ turn off and $L_2$, $T_2$ turn on. The level at $S_1$ drain is now transferred to $S_2$ gate. One complete cycle of $\phi_1$ and $\phi_2$ clock is required to shift the data one stage. On $\phi_1$, $L_1$ and $T_1$ turn on while $L_2$ and $T_2$ are off. Data applied will be transferred from $S_0$ to $S_1$ to be stored on the gate capacitance of $S_1$.

The two-phase clock signals must not be allowed to overlap since correct storage and transfer of data would not take place.

## Specifications

A full definition of the word SPECIFICATION is: "a detailed description of the required characteristics of a device, equipment, system, product, or process".

As an example, suppose that a signal generator is wanted for general laboratory use. The characteristics required might be as follows:

*a*) Frequency range: 0.1 Hz to 1 MHz continuously variable.

*b*) Output waveform: sine or square.

*c*) Sine wave distortion: less than 0.5% total harmonic distortion.

*d*) Accuracy: better than ±2.5% of dial reading.

*e*) Frequency stability: less than ±0.2% per 24 hours at 23°C.

*f*) Frequency temperature coefficient: 0.01% per °C.

*g*) Maximum output voltage: 10 V peak to peak.

*h*) Output impedance: 60 Ω.

*i*) Output square wave rise and fall time: better than 100 nanosecond.

*j*) Power required: 240 V, 50 Hz or 100 V, 60 Hz.

These are the basic figures that should be contained in the PERFORMANCE SPECIFICATION section of manufacturer's sales leaflets and instruction manuals.

For a particular application, the user has to decide on the performance required and then has to check that the manufacturer's specification can meet those requirements. Inside the factory a TEST SPECIFICATION is issued and used to guide the test department in checking all aspects of the instrument's or component's performance.

The performance specification can be used by the potential purchaser of the equipment to compare and contrast the various instruments or systems offered for sale by several manufacturers. In this way the user can select the best available on both technical and economic considerations.

One of the main purposes of testing, therefore, is to ensure that manufactured items and instruments conform to an agreed specification.

### Standard Specifications

It is naturally important that customers and manufacturers use the same terminology with regard to instrument and component perfor-

mance. It is also important that they can refer to some external standard of measurement and quality. In the UK the British Standards Institution serves this purpose. It started out as the Engineering Standards Committee in 1901 and was incorporated in 1918 as the British Engineering Standards Association. In 1929 it was granted a Royal Charter and took its present name in 1931. It is the recognised body for the preparation and publication of UK national engineering standards, and every year a complete list of these standards is published in the British Standards Yearbook. The specifications important to the electronics and telecommunications industries (including electrical power) are listed in the BSI Sectional List No. SL 26. This can be acquired by writing direct to BSI at 101 Pentonville Road, London N1 9ND. Some of these specifications are

**BS 4727** Glossary of terms used in telecommunication and electronics.

**BS 3939** Graphical symbols for electrical power, telecommunications and electronics diagrams.

**BS 2011** Methods for the environmental testing of electrical components and electronic equipment.

There are, of course, a whole host of other BSI specifications and it would serve no useful purpose to attempt to list them all here. However, a recent addition is worth noting; this is the BS 9000 series which deals with electronic components. The scheme is a single comprehensive system covering specifications and approval procedures for all types of components that can be assessed. For example within the BS 9000 scheme there is

**BS 9110** Fixed resistors of assessed quality: generic data and methods of test. (Generic means "a general class or group of items".)

**BS 9070** Fixed capacitors of assessed quality: generic data and methods of test.

**BS 9300** Semiconductor devices of assessed quality: generic data and methods of test.

The BS 9000 system is made up of

1) Basic specifications comprising the regulations and procedures applicable to all electronic components within the scheme.

2) Specifications covering a specific family or class of components.

    i.e. BS 9300 Semiconductors.

    or  BS 9070 Capacitors.

3) Rules for the preparation of detailed specifications for individual types of components.

    i.e. BS 9301 General-purpose signal diodes.

    or  BS 9340 Low-current thyristors.

Within 3) provision is made for inspection and quality assurance tests to be carried out by the Manufacturer's Quality Assurance Department under the surveillance of an independent supervising inspector.

Every delivery of any components to a factory under the BS 9000 scheme must be accompanied by a Certificate of Conformity. In this way the customer has the advantage of

1) One system of component specification.

2) Components that have been approved for precisely stated performance.

3) An independently controlled inspection system and certified test results.

This reduces costs since it should not be necessary for the customer to test incoming components.

There are also in existence specifications issued by high-volume users of component and equipment. Some of the important ones are

British Ministry of Defence Specifications— prefixed DEF.

American Military Specifications—MIL— STD.

International Specifications—IEC.

These organisations issue specifications covering components, equipments, and methods of testing.

Because of the increasing use of electronics in every area of activity, the various authorities are attempting to combine their specification work in order to reduce the number of standard specifications. For example the BS 9000 system is designed ultimately to replace both the previous DEF and BS component specifications. Also, many BS specifications are taken from International Specifications, and, generally, agreement is sought nowadays.


## Thyristor and Triac

Thyristors (silicon controlled rectifiers) and triacs are semiconductor devices that are now being used extensively in power control circuits. They are particularly suited for a.c. power control applications such as lamp

dimmers, motor speed control, temperature control and invertors; and also are commonly used as overvoltage protection elements in d.c. power supplies.

## Thyristors

The construction of a typical medium power thyristor is shown in fig. T1. It is basically a four-layer p-n-p-n structure. The gate connection to the $p_2$ region enables it to be switched from a non-conducting (forward blocking) state into a low-resistance forward conductive state. Once triggered into forward conduction, the thyristor remains on, unless the current flowing through it is reduced below the holding current value or it is reverse-biased. The characteristics (fig. T2) show that in the reverse direction the thyristor is off and only a small leakage current flows unless the reverse breakdown voltage is exceeded. This is because junctions $J_1$ and $J_3$ have increased depletion regions. When the anode is made positive with respect to the cathode, and with no gate signal applied, junctions $J_1$ and $J_3$ will be forward-biased but $J_2$ will be reversed, and its depletion region increased. Therefore only a small forward leakage current flows and the thyristor is said to be forward blocking or off. It will remain in this high-resistance off state unless the voltage between anode and cathode is made to exceed the forward breakover voltage, or until a pulse of gate power is applied. Triggering by exceeding the forward voltage rating is undesirable and applying a gate signal is the normal preferred method to switch the thyristor on.

The reason why the thyristor switches on and remains on even when the gate signal is removed can be more easily understood by using the two-transistor equivalent circuit of fig. T3. Here the top $p_1n_1p_2$ region is considered as a p-n-p transistor and the bottom $n_1p_2n_2$ region as an n-p-n transistor. Imagine that no gate signal is applied and that a voltage (less than breakdown) is applied making the anode positive with respect to the cathode. The small current that flows in at the anode connection must flow out from the cathode. For $Tr_2$ the emitter current $I$ is the sum of its base currents and its collector current plus any leakage:

$$I = h_{FB1}I + h_{FB2}I + I_{CBO}$$

where $I_{CBO}$ is leakage current.

$$I = I(h_{FB1} + h_{FB2}) + I_{CBO}$$

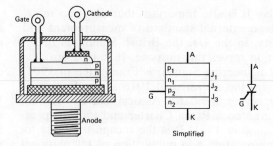

**Figure T1**  **Typical structure of medium power thyristor**

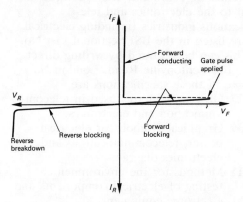

**Figure T2**  **Typical thyristor characteristics**

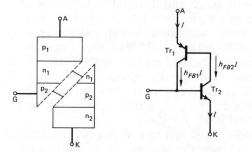

**Figure T3**  **Two-transistor equivalent circuit for thyristor**

Therefore

$$I = I_{CBO}/[1 - (h_{FB1} + h_{FB2})]$$

Note the common base current gain of a transistor, $h_{FB}$, is highly dependent on the value of collector current. While the current flowing through the transistors remains low, the sum of the two current gains ($h_{FB1} + h_{FB2}$) remains less than unity. For example if

$(h_{FB1}+h_{FB2})=0.9,\quad$ then $\quad I=10I_{CBO}$

This will be an insignificant value of current and a high resistance will exist between anode and cathode. If, however, $Tr_2$ is forward-biased by a gate signal, then the increase in current through $Tr_2$ raises the current gain and rapidly $(h_{FB1}+h_{FB2})$ will approach unity. Then the two transistors switch into a conducting state. They are connected in a positive feedback configuration, the collector currents of each transistor supplying the base current to the other, and therefore switch-on is very rapid. The gate signal can be removed and the two transistors remain conducting since the current flowing through them is high enough to ensure that the sum of $h_{FB1}$ and $h_{FB2}$ exceeds unity.

Note that, when switched on, the thyristor will pass large values of current limited only by the external load, with only a small voltage dropped across anode to cathode.

*a) Anode voltage ratings*
*Reverse*

$V_{RWM}$   The maximum continuous peak voltage rating in the reverse direction.

$V_{RRM}$   The repetitive peak reverse voltage. The peak value of reverse voltage transients where the peak lasts for only a short time, i.e. 0.1 ms.

$V_{RSM}$   The non-repetitive peak reverse voltage. Transients which last up to 10 ms but do not occur regularly.

*Forward*

$V_{DWM}$   The crest working off-state voltage applied in the forward direction.

$V_{DRM}$   The repetitive peak off-state voltage applied in the forward direction.

$V_{DSM}$   The non-repetitive peak off-state voltage applied in the forward direction.

$V_T$   The on-state forward drop usually quoted at a specified forward current and junction temperature.

$dv/dt$   The maximum rate of rise of anode–cathode voltage which will not trigger the thyristor.

*b) Current ratings*

$I_{T(AV)}$   Maximum mean current.

$I_{T(r.m.s.)}$   Maximum r.m.s. current.

$I_{TRM}$   Peak current that can be drawn each cycle provided that the mean and r.m.s. current ratings are not exceeded.

$di/dt$   The maximum rate of rise of current when the thyristor is triggered that will not cause unequal distribution resulting in "hot spots" in the junctions of the device.

$I_H$   Holding current.

*c) Gate ratings*

$P_G$   The maximum mean power that will not cause overheating of the gate/cathode junction.

$P_{GM}$   The maximum peak gate power.

$I_{GT}$   The minimum instantaneous trigger current required to initiate turn on at a specified temperature.

$V_{GT}$   The minimum instantaneous trigger voltage required to initiate turn on.

$V_{GD}$   The maximum continuous gate voltage which will not turn on the thyristor.

**Full-wave a.c. power controller**

Fig. T4 shows a typical circuit arrangement for controlling the a.c. power dissipation in the load, which might be for example a stage lighting system. One of the advantages of this circuit is that the actual control element is isolated from the mains supply. This is achieved by the use of the two transformers $T_1$ and $T_2$. $T_1$ provides a 40 V rms a.c. voltage to the bridge rectifier and the 55 V peak amplitude full-wave rectified waveform at the bridge output is limited, on each half cycle, to +10 V by $R_1$ and $DZ_1$. At the beginning of each half cycle, $C_1$ is charged via the resistive network $R_2$, $RV_1$ and $RV_2$ so that the emitter voltage of the unijunction transistor rises positive. The rate of change of this voltage is controlled by the setting of $RV_1$. When the voltage across $C_1$ reaches the peak point of the UJT, the UJT is triggered into conduction and $C_1$ is rapidly discharged through the primary winding of the pulse transformer $T_2$. Positive pulses are induced in the secondary windings and the thyristor which has its anode positive with respect to its cathode will be triggered on. A portion of half cycle power will be applied to the load. On the next half cycle, as $C_1$ is again charged and discharged, the pulse generated will turn on the other thyristor. As the mains goes through zero, the thyristor that had previously been conducting is reverse-biased and therefore turns off.

The amount of power dissipated by the load is controlled by the time position of the trigger

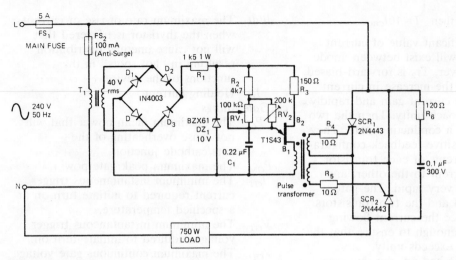

*Figure T4* Full-wave a.c. power controller

pulses relative to the start of each half cycle of the mains waveform. Thus with $RV_1$ at a low value, $C_1$ is charged very rapidly, the trigger pulses occur very early in each half cycle, and maximum power is applied to the load. On the other hand with $RV_1$ at maximum, $C_1$ is charged relatively slowly, the trigger pulses occur very late in each half cycle, and only a small portion of power is applied to the load. $RV_2$ is a preset pot that can be adjusted to give a set value of minimum power.

### Triac

The triac is similar in operation to two thyristors connected in reverse parallel, but with a common gate connection. This means that the device can pass or block current in both directions. Also it can be triggered into conduction in either direction by applying either positive or negative gate signals. Triacs are mostly used in full-wave a.c. control circuits in preference to two thyristors or to a bridge rectifier and thyristor, because simpler heat sinks and more economical trigger circuits can be used.

▶ *Phase control*    ▶ *Burst firing*

### Timer

A fairly large number of monolithic timer circuits are now available but probably the best known are the 555, the 556, and the ZN 1034.

Timing circuits are those which will provide an output change of state after a predetermined time interval. This is, of course, the action of a monostable multivibrator. Discrete circuits can be easily designed to give time delays from a few microseconds up to a few seconds, but for very long delays mechanical devices had often to be used. The 555 timer i.c., first made available in 1972, allows the user to set up quite accurate delays or oscillations from microseconds up to several minutes, while the ZN 1034E can be set to give time delays of up to several months.

**1** The basic operation of the 555 can be understood by referring to fig. T5. For MONO-STABLE OPERATION the external timing component $R_A$ and $C$ are wired as shown. Without a trigger pulse applied, the $\bar{Q}$ output of the flip-flop is high, forcing the discharge transistor to be on and holding the output low. The three internal 5 k$\Omega$ resistors $R_1$, $R_2$ and $R_3$ form a voltage divider chain so that a voltage of $\frac{2}{3}V_{CC}$ appears on the inverting input of comparators 1 and a voltage of $\frac{1}{3}V_{CC}$ on the non-inverting input of comparator 2. The trigger input is connected via an external resistor to $V_{CC}$ so the output of comparator 2 is low. The outputs of the two comparators control the state of the internal flip-flop. With no trigger pulse applied, the $\bar{Q}$ output will be high and this forces the internal discharge transistor to conduct. Pin 7 will be at almost zero volts and the capacitor $C$ will be prevented from charging. At the same time the output will be low.

When a negative-going trigger pulse is applied, the output of comparator 2 goes high momentarily and sets the flip-flop. $\bar{Q}$ output goes low, the discharge transistor turns off, and

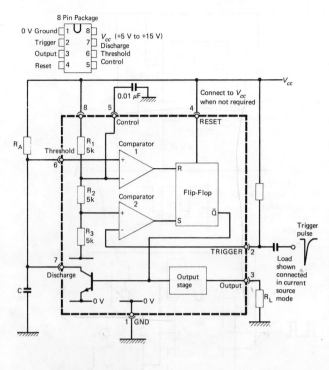

8 Pin Package

| 0 V Ground | 1 | 8 | $V_{cc}$ (+5 V to +15 V) |
| Trigger | 2 | 7 | Discharge |
| Output | 3 | 6 | Threshold |
| Reset | 4 | 5 | Control |

0.01 μF

Connect to $V_{cc}$ when not required

8  5  4

Control  RESET

$R_A$

Threshold  $R_1$ 5k

Comparator 1

R

6

$R_2$ 5k

Comparator 2

Flip-Flop

$\bar{Q}$

S

$R_3$ 5k

7

Discharge

TRIGGER 2

C

Output stage  Output

3

0 V  0 V

1 GND

$V_{cc}$

Trigger pulse

Load shown connected in current source mode

$R_L$

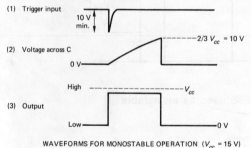

(1) Trigger input

10 V min.

(2) Voltage across C

2/3 $V_{cc}$ = 10 V

0 V

(3) Output

High

$V_{cc}$

Low

0 V

WAVEFORMS FOR MONOSTABLE OPERATION ($V_{cc}$ = 15 V)

*Figure T5*  The 555 timer

the output switches to $V_{CC}$ volts. The external timing capacitor $C$ can now charge via $R_A$ so the voltage across it rises exponentially towards $V_{CC}$. When this voltage just exceeds $\frac{2}{3}V_{CC}$, the output of comparator 1 goes high which resets the internal flip-flop. The discharge transistor conducts and rapidly discharges the timing capacitor, and at the same time the output switches to zero. The width of the output pulse is equal to the time taken for the external capacitor to charge from zero to $\frac{2}{3}V_{CC}$.

$$T = 1.1CR_A$$

$R_A$ can have a value from $1\,k\Omega$ up to $1.3V_{CC}$ MΩ. In other words if a 10 V supply rail is

used $R_A$ can be $1\,k\Omega$ minimum or a maximum of $13\,M\Omega$. In practice, medium values of $R_A$ from $50\,k\Omega$ to $1\,M\Omega$ are used since these tend to give the best results.

The 555 output switches between almost zero (0.4 V) to about one volt below $V_{CC}$ with rise and fall times of 100 nsec. The load can be connected either from the output to ground or from the output to $V_{CC}$. The first connection is known as the current source mode and the latter as the current sink. In both cases up to 200 mA of load current can be accommodated.

Two other input pins are provided. Pin 4, the *reset* terminal, can be used to interrupt the timing and reset the output by application of a negative-going pulse. Pin 5, called the *control*, can be used to modify or modulate the time delay. A voltage applied to pin 5 overrides the d.c. level set up by the internal resistors. In normal timing applications when no modulation is required, pin 5 is usually taken to ground via a $0.01\,\mu F$ capacitor. This prevents any pick up of noise affecting the timing.

One of the important points about the 555 is that the timing is relatively independent of supply voltage changes. This is because the three internal resistors fix the ratio of the threshold and trigger levels at $\frac{2}{3}V_{CC}$ and $\frac{1}{3}V_{CC}$. A typical change of time delay with supply voltage is 0.1% per volt. In addition, the temperature stability of the microcircuit is excellent at 50 ppm per °C. Thus the accuracy and stability of the timing delay depends to a great extent on the quality of the external timing components $R_A$ and $C$. Electrolytic capacitors may have to be used for long time delays but the leakage current must be reasonably low. Also, because the tolerance of electrolytic capacitors is wide (typically $-20\% + 50\%$) part of the timing resistor may have to be a preset to enable the delay to be fairly accurately set.

An example of a 555 used as a simple 10 second timer is shown in fig. T6. Pressing the start button momentarily takes pin 2, the trigger input, down to 0 V. The output will switch high and the LED will come on. $C$ then charges from 0 V towards $+V_{CC}$. After 10 sec the voltage across $C$ reaches 6 V ($\frac{2}{3}V_{CC}$) and the 555 is reset, the output switching back to the low state.

**2**  The 555 can also be wired up to operate as an ASTABLE MULTIVIBRATOR, the output being a train of positive pulses with the width and frequency determined by external timing

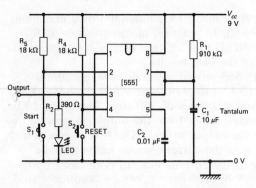

**Figure T6    10-second timer using a 555**

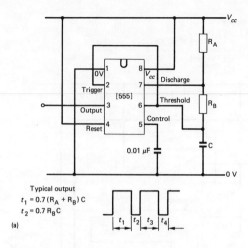

Typical output
$t_1 = 0.7 (R_A + R_B) C$
$t_2 = 0.7 R_B C$

(a)

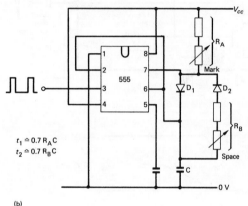

$t_1 \simeq 0.7 R_A C$
$t_2 \simeq 0.7 R_B C$

(b)

**Figure T7    555 used as an astable**

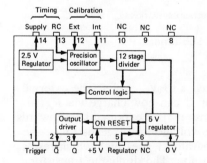

**Figure T8a    The ZN 1034 timer**

components (fig. T7). Pins 2 and 6 are connected together which allows the capacitor to charge and discharge between the threshold and trigger levels. At switch-on, $C$ charges via $R_A$ and $R_B$ towards $V_{CC}$. When the voltage across $C$ reaches $\frac{2}{3}V_{CC}$, the output changes state and $C$ is discharged via $R_B$ towards $0\,V$. When the voltage across $C$ falls to $\frac{1}{3}V_{CC}$, the circuit again changes state, the internal discharge transistor turns off, and $C$ charges via $R_A$ and $R_B$ towards $V_{CC}$. Thus a continuous train of pulses appears at the output. To get an almost symmetrical output waveform, $R_B$ should be made very large with respect to $R_A$, say 50 times, and then the frequency $f$ will be primarily determined by $R_B$ and $C$.

$$f \simeq 1/1.4 R_B C \quad \text{when} \quad R_B \gg R_A$$

To achieve some control over the duty cycle, that is the ratio of mark-to-space of the output waveform, a circuit such as shown in fig. T7b is used. Here diode $D_1$ conducts when $C$ is charging, and $D_2$ conducts when $C$ is discharging.

Gated oscillators can be created by using pin 4, the reset, as a control. While pin 4 is held at $0\,V$, the oscillator is inhibited, but if a positive voltage is applied then the circuit is allowed to oscillate. A suitable control signal can be the output of another 555 wired as a monostable.

Many other useful circuits can be made using the 555 or the 556 timers such as temperature- or light-controlled oscillators. The 556 is in effect two 555 devices within one common package age.

**3** The other popular i.c. timer, the ZN 1034, is shown in fig. T8a. This has an internal oscil-

lator, the frequency of which can be controlled by an external resistor and capacitor. The oscillator output is connected inside the i.c. to a 12-stage divider so the oscillator has to complete $(4096 - 1)$ cycles before the timing

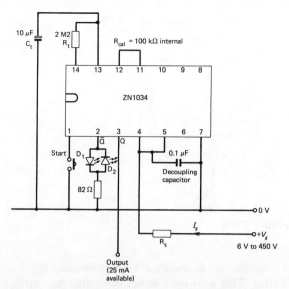

*Figure T8b* One-week timer using a ZN1034

$D_1$ and $D_2$ form the timing-state indicator

$D_1$ ON: timing in progress

$D_2$ ON: timing over

$I_S$ in mA is calculated by adding the device quiescent current of 7 mA to the required on-state output current.

$$R_S = \frac{V_S - 5}{I_S} \text{ k}\Omega$$

period ends. Thus very long time delays can be achieved by using only medium values of timing components. The time delays of weeks or months, if required, can be repeated with high accuracy as long as good-quality components are used for the oscillator timing.

The total timing period $T$ is related to $R_t$ and $C_t$ by the formula

$$T = 4095 \, KC_t R_t$$

where $K$ is a constant determined by the value of the calibration resistance. An internal calibration resistor of 100 kΩ is provided by connecting pins 11 and 12 and this gives $K = 0.668$. This method is recommended for best temperature stability. Thus, if $C_t = 100 \, \mu\text{F}$ and $R_t = 2 \, \text{M}\Omega$, and $K = 0.668$, then $T = 7$ days.

A circuit for this is shown in fig. T8*b*. Two complementary outputs are provided: $\bar{Q}$ on pin 2 is normally at low voltage but rises to about +3 V at the end of the timing period; $Q$ on pin 3 will be high at 3 V after the start button

is pressed and will fall to less than 0.4 V (low) at the end of the timing period. The outputs can each sink or source up to 25 mA and can, therefore, be used to drive relays via a transistor switch, or thyristors, or small signal lamps. The calibration can be checked by measuring the time period of the oscillator on pin 13. However, a high impedance meter must be used with an impedance of at least $10R_t$, otherwise the oscillator period will be changed. For this circuit the oscillator will have a periodic time equal to the total time period, 7 days, divided by $4095K$, i.e.

Oscillator time period = 7 days/2736

▶ *Monostable*    ▶ *Gated oscillator*

## Touch Switch

The availability of high-input-impedance devices such as FETs and CMOS logic have made possible the design of simple and reliable touch switch systems. The principle of operation (fig. T9) depends on one of:

*a*) The finger directly bridging two electrodes. Skin resistance is typically less than a few hundred thousand ohms.

*b*) The finger capacitively bridging two electrodes or body capacitance (250 pF) being used to alter some circuit condition.

*c*) With the body acting as an aerial, the finger being used to inject some 50 Hz (60 Hz) mains hum onto the pick-up plate.

In order to create reliable and safe touch switches the following points must be observed.

*a*) All switch contacts and circuits *must be effectively isolated from the a.c. mains supply.*

*b*) The sensing circuit should be provided with a good earth.

The circuit should be designed so that it operates with the minimum amount of sensitivity to prevent spurious operation.

*d*) The output signal should be conditioned, i.e. debounced before being applied to logic, otherwise a hesitant input could cause several output pulses. A bistable action is usually best, or a monostable with a relatively long time delay.

*Examples*

**1** *Touch-operated bistable using VMOS power FETs* (fig. T10*a*). This uses the direct bridging method. Briefly touching the On contact switches $T_1$ off, its drain voltage rises, forcing $T_2$ to conduct, and the circuit latches in this

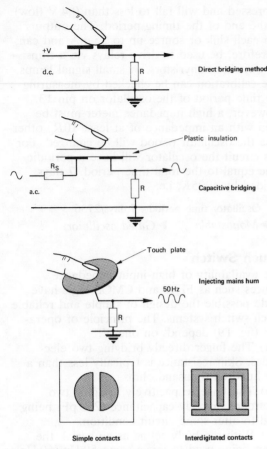

Direct bridging method

Plastic insulation

Capacitive bridging

Touch plate

Injecting mains hum

50Hz

Simple contacts    Interdigitated contacts

**Figure T9   Touch switches**

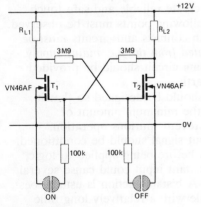

**Figure T10a**

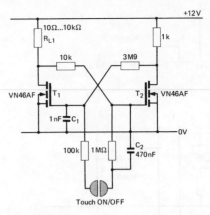

**Figure T10b**

state. Briefly touching the Off contact will reverse the condition as $T_2$ is turned off and $T_1$ is forced to conduct. A disadvantage is that when power is applied either FET will conduct giving an indeterminate output. An improved circuit is shown in the next diagram.

**2**  *Single-touch-operated VMOS bistable* (fig. T10$b$). This also uses direct bridging. At switch on, because $C_1$ is across $T_1$ gate, $T_2$ will conduct and $T_1$ will be off. By briefly touching the contacts, the potential on $C_2$ is transferred to $C_1$ causing $T_1$ to begin conduction. The positive feedback forces $T_1$ to switch on and $T_2$ is turned off. This is a stable condition and the circuit will therefore latch in this state. When the contacts are again touched, the potential on $C_1$ is transferred to $C_2$ and the circuit switches back to its initial state with $T_2$ on and $T_2$ off. If the finger is left on the contact the circuit will oscillate giving a low frequency square wave output.

**3**  *Touch control using direct bridging method and a 4013 dual D bistable* (fig. T10$c$). The first half of the i.c. is wired as a monostable. Briefly touching the contacts forces the bistable into the set condition so that the $Q$ output switches high. The *RC* network delays the reset until $C_1$ has charged to about $\frac{1}{2}V_{DD}$. The bistable then resets automatically so that the output pulse from the $Q$ output is about 1 sec in duration. The second half of the i.c. is connected as a divide-by-two which is forced to change state on the trailing edge of the monostable output. In this way the touch switch can be used for on/off control. A simple *RC* circuit to the reset of the bistable ensures that the output is low when power is first switched on.

**4**  The last circuit is one based on inducing some 50 Hz (or 60 Hz) hum by touching the pickup plate (fig. T10$d$). The resistors $R_1$ and $R_2$ are arranged so that a small amount of a.c. will cause the 4011B gate to switch giving a

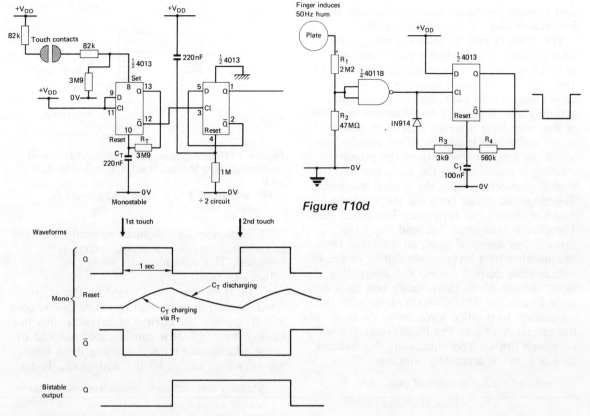

**Figure T10d**

**Figure T10c**

50 Hz square wave as the clock input to the retriggerable monostable. The output of the mono will switch high, and remain high, until the input is removed. While the 50 Hz square wave is present, $C_1$ is discharged every time the NAND gate output goes low and this prevents the monostable from resetting. When the input is removed, $C_1$ can charge via $R_4$ and the monostable resets bringing the Q output low. The second half of the 4013 can be used as a divide-by-two to give on/off control as shown in the previous circuit.

▶CMOS     ▶VMOS power FET

## Transformer

This is an electro-magnetic component used to change one alternating voltage level to another. For example a step-down transformer would be used in an instrument's power supply to "transform" the 240 V a.c. mains (or 110 V a.c.) to say 15 V a.c. which is a more suitable value for supplying, after rectification, the low-voltage d.c. required by a solid state instrument. The transformer, consisting of two separate coils,

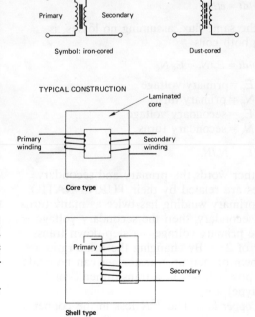

**Figure T11a   Transformer**

also provides essential isolation between the instrument and the a.c. mains supply.

The primary and secondary windings, each several turns of varnished copper wire, are linked together by a magnetic circuit (fig. T11$a$). The mutual inductance between the magnetically coupled coils provides the transformer action. Radio frequency transformers with coils of fine wire have a ferrite dust core, while audio and power tranformers are usually iron cored. In a power transformer the primary winding is connected to the a.c. input and the load is connected across the second winding. The magnetic circuit coupling the windings is made of special low reluctance laminated steel. The core is laminated, i.e. made up from several thin sheets of steel, all insulated from one another by a layer of varnish or oxide, to reduce eddy current losses. The alternating input voltage across the primary sets up a magnetic flux, all of which, ideally, links with the secondary. In practice some losses do occur but the efficiency of a power transformer is usually very high (98%). The alternating flux induces an e.m.f. in the secondary winding.

Induced e.m.f. = Number of turns

× Rate of change of flux

$$e = N \frac{d\Phi}{dt}$$

$d\Phi/dt = e/N$

Since the same flux, assuming no losses, is linking both coils

$d\Phi/dt = E_p/N_p = E_s/N_s$

where $E_p$ = primary voltage
$N_p$ = primary turns
$E_s$ = secondary voltage
$N_s$ = secondary turns

$E_p/E_s = N_p/N_s$

In other words the primary and secondary voltages are related by their TURNS RATIO. If the primary winding has twice as many turns as the secondary, then the secondary voltage is half the primary voltage—a step-down transformer of 2 : 1. By changing the turns ratio a step-down or step-up transformer can be made.

The power losses in a transformer (iron-cored type) are

a) *Copper loss*. The $I^2R$ loss in the copper windings caused by the current flowing in them.

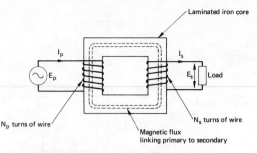

Laminated iron core

$I_p$   $E_p$   $E_s$   $I_s$   Load

$N_p$ turns of wire   $N_s$ turns of wire

Magnetic flux linking primary to secondary

*Figure T11b*   Suppose turns ratio = 10 : 1, and assuming zero losses, then if $E_p$ = 240 V, $E_s$ = 24 V.
For a load of 1 A, $I_s$ = 1 A and $I_p$ = 0·1 A.

b) *Hysteresis loss*. A small amount of energy is lost each time a.c. primary current is reversed. This is reduced by using low reluctance steel.

c) *Eddy current loss*. Caused by the input e.m.f. setting up a current in the magnetic core which opposes the change of magnetic flux producing the e.m.f. It is considerably reduced by using a laminated core. Assuming these losses are very low, i.e. an ideal transformer, then

Primary volt-amperes = secondary volt-amperes

$$E_p I_p = E_s I_s$$

Since $E_p/E_s = N_p/N_s$ then $I_p/I_s = N_s/N_p$.

Thus in the transformer the current that can be supplied by the secondary is $NI_p$, where $N$ is the turns ratio.

When a transformer is off load, since the power losses are usually very small, the primary current will also be very small. Appreciable primary current only flows when the secondary winding is loaded.

▶ *Autotransformer*   ▶ *Power supply*
▶ *Rectifier circuits*

**Transistor**
A bipolar junction transistor consists of two p-n junctions formed close together in one single piece of silicon. This gives either an n-p-n or p-n-p configuration and the symbols with simplified view of construction are shown in fig. T12. An alloy junction type of transistor has to be constructed individually but the ▶ *Planar process* enables up to a thousand transistors to be manufactured on one slice of silicon.

The operation of the n-p-n is as follows (for p-n-p reverse the supplies and note that majority charge carriers are holes):

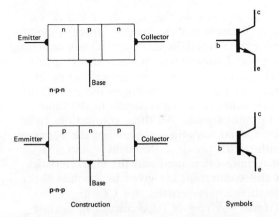

Figure T12   Bipolar junction transistors

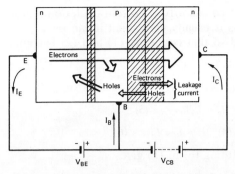

*Figure T13*   Flow of charge carriers in a correctly biased n-p-n transistor

*a) No bias voltages applied*   Two depletion regions are formed at the collector/base and base/emitter junctions.

*b) Correct bias for operation*   This is given by a reverse voltage at the collector/base junction with the collector more positive than the base, and a forward bias of approximately 0.6 V to the base/emitter junction. The collector/base depletion region becomes wider but the base/emitter depletion region is much reduced. As a result the emitter injects electrons into the base region and simultaneously the base injects holes into the emitter. However, during manufacture, the emitter is more heavily doped than the base giving a high electron to hole ratio. This ratio is referred to as emitter efficiency $\gamma$. Because of this only a few of the electrons that have been injected from the emitter into the base recombine with holes; the majority diffuse across the base region. The action of diffusion is one of the spreading of charge carriers from the region of high concentration near the base/emitter junction to areas of lower concentration. When the electrons reach the collector/base depletion region, they come under the influence of the positive field set up by $V_{CB}$ and are swept up into the collector. Because of the high emitter efficiency the majority (99% or greater) of the electrons injected from the emitter reach the collector to make up the collector current $I_C$. The small number that recombine with holes in the base region go to make up the base current $I_B$. The full picture of the flow of charges in a correctly biased n-p-n transistor is shown in fig. T13 and this includes the small leakage current across the collector/base junction.

The proportion of emitter current that reaches the collector is called $\alpha$, or more usually the common base current gain $h_{FB}$. This is one of the *h*-parameters

$h_{FB}$ the large signal gain or d.c. common base current gain is

$$\left. \frac{\Delta I_C}{\Delta I_E} \right|_{V_{CB}\text{ held constant}}$$

$h_{fb}$ the small signal gain or a.c. common base current gain is

$$\left. \frac{\delta i_c}{\delta i_e} \right|_{V_{CB}\text{ held constant}}$$

A typical value for $h_{FB}$ lies between 0.98 and 0.995.

A more useful measure of the amplifying properties of the transistor is the ratio of collector current to base current. This is called $\beta$ or more usually the common emitter current gain $h_{FE}$.

$h_{FE}$ the large signal gain or d.c. common emitter current gain is

$$\left. \frac{\Delta I_C}{\Delta I_B} \right|_{V_{CE}\text{ held constant}}$$

$h_{fe}$ the small signal gain or a.c. common base current gain is

$$\left. \frac{\delta i_c}{\delta i_b} \right|_{V_{CE}\text{ held constant}}$$

The collector current flowing in the n-p-n transistor can be represented by two formulae:

$$I_C = h_{FB}I_E + I_{CBO} \qquad (1)$$
$$I_C = (I_E - I_B) + I_{CBO} \qquad (2)$$

where $I_{CBO}$ is the leakage current flowing

across the collector/base current depletion region. Since this is usually very small it can be neglected so that equations (1) and (2) become

$$I_C = h_{FB}I_E \qquad (3)$$

$$I_C = I_E - I_B \qquad (4)$$

From (3) $\quad I_E = I_C/h_{FB} \qquad (5)$

and from (4) $\quad I_E = I_C + I_B \qquad (6)$

Substituting (6) in (5) gives

$$I_C + I_B = I_C/h_{FB}$$

$$\therefore \quad I_B = I_C\left(\frac{1}{h_{FB}} - 1\right) = I_C\left(\frac{1 - h_{FB}}{h_{FB}}\right)$$

$$\therefore \quad I_C/I_B = h_{FB}/(1 - h_{FB})$$

Now $h_{FE} = I_c/I_B \quad \therefore \quad h_{FE} = h_{FB}/(1 - h_{FB})$
Similarly $\quad h_{fe} = h_{fb}/(1 - h_{fb})$

The importance of this relationship is that only a small change in $h_{FB}$ results in a very large change in $h_{FE}$. Suppose a transistor has an $h_{FB}$ of 0.99 then the value of $h_{FE}$ is

$$h_{FE} = 0.99/(1 - 0.99) = 99$$

If another transistor, of the same type, has an $h_{FB}$ of 0.995, only 0.5% higher than the first, then its value of $h_{FE}$ will be

$$h_{FE} = 0.995/(1 - 0.995) = 199$$

an increase in $h_{FE}$ of 101%

It is therefore difficult for any manufacturer to produce transistors without a wide spread on the $h_{FE}$ parameter. For example a 2N2222 transistor at $I_C = 150$ mA has a range of $h_{FE}$ values from 100 to 300. $h_{FE}$ is a parameter that varies with collector current $I_C$ and also, to a lesser extent, with collector/emitter voltage $V_{CE}$. In any design the spread on $h_{FE}$ must be taken into account, otherwise bias levels and gain in a circuit will vary considerably between units and whenever a transistor is replaced.

Transistors are classified according to intended application in roughly the following areas:

Audio and general purpose
Low-noise types
Audio power amplifiers (>1 W)
High-speed switching
Power and high voltage switching
Video and r.f. amplifiers
U.h.f. and microwave.

The emphasis on the data sheet will be on the application but in whatever group a particular transistor falls there are certain common ratings and parameters that are always given.

The transistor can be connected in one of three configurations with either the base, emitter, or collector being common to the input and output signals. All three connections have their special advantages but, since the common emitter (CE) provides the highest gain, it is much more often used and the characteristics for this connection are given in the data sheet. The three characteristics (for CE) are

**1** *Input* A plot of base current $I_B$ against base/emitter voltage $V_{BE}$ at a fixed value of collector/emitter voltage $V_{CE}$.

The inverse of the slope of this graph gives the input impedance $h_{ie}$:

$$h_{ie} = \frac{\delta v_{be}}{\delta i_b}\bigg|_{v_{ce}=0}$$

**2** *Output* A set of graphs of collector current $I_C$ against collector/emitter voltage $V_{CE}$ for various values of base current $I_b$.

The slope of one of these gives the output admittance $h_{oe}$:

$$h_{oe} = \frac{\delta i_c}{\delta v_{ce}}\bigg|_{i_b=0}$$

**3** *Transfer* A plot of collector current $I_C$ against base current $I_B$ for a fixed value of collector/emitter voltage $V_{CE}$.

The slope of this graph gives $h_{fe}$:

$$h_{fe} = \frac{\delta i_c}{\delta i_b}\bigg|_{v_{ce}=0}$$

**Transistor as an amplifier**
A simple example using a CE connection will show how a transistor produces signal gain (fig. T14). The transistor must be provided with a d.c. bias current and in this circuit a relatively simple bias arrangement is shown for clarity. $R_1$ provides a d.c. base bias current $I_B$ which sets up a d.c. collector current $I_C$. In this way the operating voltage at the collector, referred to as the quiescent point, will be $(V_{CC} - I_C R_L)$ and should be adjusted to be about half the supply voltage. By setting the operating point at about $\frac{1}{2}V_{CC}$ the a.c. signal voltage developed at the collector can be allowed to swing either side of the operating point without too much distortion.

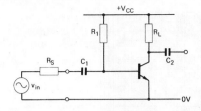

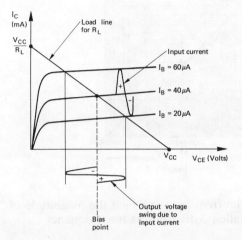

Figure T14  The transistor as an amplifier

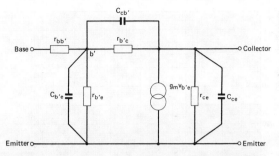

**Figure T15**  Hybrid-$\pi$ model for the bipolar junction transistor

Even more simply, since

$$\frac{h_{fe}}{h_{ie}} = \frac{\delta i_c/\delta i_b}{\delta v_{be}/\delta i_b}\frac{\delta i_c}{\delta v_{be}} = g_m$$

we can write  $A_v = -g_m R_L$

The $g_m$ for nearly all transistors has a value of 40 mS per milliamp of $I_C$.

This parameter $g_m$ is used in the hybrid-$\pi$ model (fig. T15). This useful model corresponds fairly closely to the physical operation of the transistor by taking into account parasitic resistances and internal junction capacitances. It is valid for signal frequencies up to $\frac{1}{3}f_T$. The effective base node is represented by the point $b'$ and the external base connection (b) is shown connected to $b'$ via a resistor $r_{bb'}$.

The component parts are as follows;

$r_{bb'}$  Base spreading resistance; typical value 50 $\Omega$ to 400 $\Omega$.

$$r_{bb'} = h_{ie} - r_{b'e}$$

$r_{b'e}$  Resistance of the conducting base/emitter junction; typical value 1 k$\Omega$ to 5 k$\Omega$.

$$r_{b'e} = h_{fe}/g_m$$

$r_{b'c}$  Feedback resistance; typical value 1 M$\Omega$.

$$r_{b'c} = r_{b'e}/h_{re}$$

$r_{ce}$  Output resistance; typical value 50 k$\Omega$.

$$r_{ce} = 1/[h_{oe} - (1 + h_{fe})(1/r_{b'c})]$$

$g_m$  Forward transfer conductance; 40 mS per mA of $I_C$.

$C_{b'e}$  Capacitance of the base/emitter junction; typical value 100 pF.

$$C_{b'e} \simeq g_m/2\pi f_T$$

Having set the d.c. bias conditions, imagine that an a.c. input signal is applied. As it goes positive a small increase in base current will occur and this will be amplified by the transistor ($i_c = h_{fe}i_b$). Because the common emitter current gain is relatively large, the change in collector current will be much greater than the small change in base current. As the collector current rises, the collector voltage falls because the change in current develops a change in voltage across $R_L$. Note that the output is inverted. On the negative half-cycle of the input the base current is slightly decreased, leading to a larger fall in collector current and a rise in output voltage.

Voltage gain of single CE stage is

$$A_v = \frac{-h_{fe}R_L}{(1 + h_{oe}R_L)(h_{ie} + R_s)}$$

Here the negative sign indicates inversion between output and input.

If $h_{oe}R_L \leqslant 0.1$ and $R_s \ll h_{ie}$

$$A_v = \frac{-h_{fe}}{h_{ie}}R_L$$

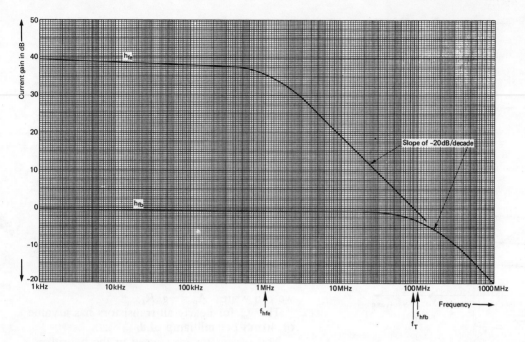

*Figure T16* Frequency response of current gains in a bipolar transistor

$C_{cb'}$ Depletion capacitance of the collector/base junction; typical value 5 pF.

$C_{cb'}$ forms a feedback path between collector and $b'$ and degrades the h.f. performance of a transistor amplifier. It is amplified by the Miller effect and appears as a much larger capacitance in parallel with $C_{b'e}$.

Thus, for an amplifier with resistive load.

$$C_{in} = C_{b'e} + C_{cb'}(1 - A_v R_L)$$

Here $A_v = -g_m R_L$

The base spreading resistor in parallel with $r_{b'e}$ form a low pass filter with $C_{in}$, giving an upper high frequency point of

$$f_h \simeq \frac{1}{2\pi C_{in}}\left(g_{b'e} + \frac{1}{r_{bb'} + R_s}\right)$$

**Frequency limitations of the bipolar transistor**

As the operating frequency of the input signal to a transistor is increased, so its value of current gain will fall. This is due to the fact that when carriers are injected into the base region from the emitter they have to diffuse across the base before being swept up by the collector field. This process of diffusion is relatively slow.

The three frequencies used to specify transistors are

$f_{h_{fb}}$ The frequency at which the magnitude of $h_{fb}$ has fallen 3 dB from its low-frequency value.

$f_{h_{fe}}$ The frequency at which the magnitude of $h_{fe}$ has fallen 3 dB from its low-frequency value.

$f_T$ The transition frequency, the frequency at which the magnitude of $h_{fe}$ falls to unity.

The relationship between these is shown in fig. T16.

Note that $|h_{fb}| = \dfrac{h_{fbo}}{[1 + (f/f_{h_{fb}})^2]^{1/2}}$

and $|h_{fe}| = \dfrac{h_{feo}}{[1 + (f/f_{h_{fe}})^2]^{1/2}}$

$h_{fbo}$ is the low-frequency value of $h_{fb}$
$h_{feo}$ is the low frequency value of $h_{fe}$
$f$ is the signal frequency.

$$h_{feo} = \frac{h_{fbo}}{1 - h_{fbo}} \quad \text{and} \quad f_{h_{fe}} = f_{h_{fb}}(1 - h_{fbo})$$

$f_T$ is also referred to as the GAIN BAND-WIDTH product:

$$f_T = f_{h_{fe}} h_{feo}$$

Therefore it follows that for frequencies above $f_{h_{fe}}$, if $f_T$ is known, then $|h_{fe}|$ can be found.

*Example* If $f_T = 500$ MHZ, then at 50 MHz, $|h_{fe}| = 500/50 = 10$.

▶ *Amplifier*   ▶ *Planar process*

238

## Transistor Transistor Logic (TTL)

A widely used logic family that is available with many functions. TTL combines fast speed with moderate power consumption and reasonable levels of noise immunity. It was developed as a successor to DTL and is continually being updated. The latest version called Advanced Low Power Schottky TTL (ALS) has a propagation delay time of typically 4 ns per gate and a dissipation of only 1 mW per gate.

*TTL types*

| | |
|---|---|
| Standard | 74 series |
| Schottky | 74S |
| Low power Schottky | 74LS |
| Advanced low power Schottky | 74ALS |
| High speed | 74H⎫ being |
| Lower power | 74L⎭phased out |

*Specification* (main points)

Power supply: $V_{CC}$ +5 V ±250 mV, absolute maximum 7 V.

$V_{IL}$  input voltage (logic 0) guaranteed to appear as a low state input voltage 800 mV.

$V_{IH}$  input voltage (logic 1) guaranteed to appear as a high state input voltage 2.0 V.

Absolute maximum input voltage 5.5 V.

$V_{OL}$  low state (logic 0) output voltage 0.4 V max. (0.5 V for 74LS).

$V_{OH}$  high state (logic 1) output voltage 2.4 V min. (2.7 V for 74LS).

Noise margin: typically 1 V, worst case 400 mV

$I_{IL}$  input current with input low ($V_{IL}$ 0.8 V), 1.6 mA.

$I_{IH}$  input current with input high ($V_{IH}$ 2 V), 40 $\mu$A max. at $V_{in}$ = 2.4 V.

$I_{sink}$  output current from standard NAND totem pole type output in logic 0 state ($V_{OL}$ 0.4 V), 16 mA max.

$I_{source}$  output current from standard NAND totem pole type output in logic 1 state ($V_{OH}$ 2.4 V), −400 $\mu$A.

(Reference book: Texas Instruments, *Designing with TTL*, McGraw Hill.)

## TTL standard NAND gate

The circuit of one of the two gates inside the 7420 is shown in fig. T17. To ease the understanding of the circuit operation it can be split into three sections:

$Tr_1$  A multi-emitter transistor, performing the AND function on inputs A, B, C, D.

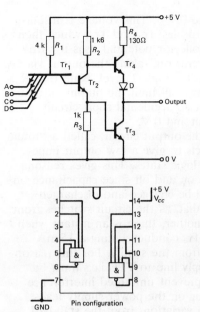

*Figure T17*  TTL NAND gate ($\frac{1}{3}$ 7420)

$Tr_2$  A phase splitter.

$Tr_3$ and $Tr_4$  An output stage having a low output impedance in both logic states (known as the totem pole stage).

Assume that all inputs A, B, C and D are high. Because the base of $Tr_1$ is connected to $+V_{CC}$ via $R_1$ all the emitter/base junctions will be reverse-biased. The base/collector junction of $Tr_1$ will be forward-biased and current will flow via $R_1$ into $Tr_2$ and $Tr_3$, turning them both on. The output of the gate will be at logic 0. Under these conditions, the voltage on $Tr_1$ base will be approximately 2.1 V (i.e. $V_{BE3}$ + $V_{BE2}$ + $V_{BC1}$). Therefore, as long as the input levels on A, B, C and D are all greater than 2 V, all the emitter/base junctions of $Tr_1$ are reverse-biased. Since $Tr_2$ is saturated, its collector voltage falls to 0.8 V (i.e. $V_{BE3}$ + $V_{CE(sat)2}$) and therefore $Tr_4$ and $D_1$ are non-conducting because the output is only at approximately 0.2 V. With $Tr_4$ off, there is no connection between the output and $V_{CC}$. The diode is provided to ensure that $Tr_4$ remains off while $Tr_3$ is on. $Tr_3$ acts as a *current sink* and up to 10 standard inputs of 1.6 mA each can be taken by $Tr_3$ without the output level rising above 0.4 V.

If any one input A, B, C or D is taken to a logic 0, i.e. 0.2 V, that particular emitter/base junction of $Tr_1$ will be forward biased, and $Tr_1$

base will fall to 0.9 V. This voltage is insufficient to forward-bias $Tr_2$ and $Tr_3$, which then turn off. $Tr_2$ collector potential rises towards $V_{CC}$ and $Tr_4$ turns on, taking the output via $D_1$ to +3.3 V which is logic 1 level. Note that $R_4$, a 130 Ω resistor, will limit the output current in the event of an accidental short circuit between output and 0 V.

Effectively the output circuit, called a "totem pole stage", acts to give a low output impedance in both logic states. This gives reasonably fast switch on and off since capacitance on the output will be charged and discharged rapidly. Note that, as the circuit switches from one state to another, there is an instant when both $Tr_3$ and $Tr_4$ conduct, taking a 15 mA current pulse from the supply. For this reason the power supply line to TTL i.c.s must be decoupled to prevent unwanted interference pulses occurring on the power supply leads.

Some of the variations from the standard NAND gate circuit in TTL are shown in fig. T18. The 7401, a quad 2-input NAND, has no collector load and is termed "open collector". Several of these gates can be wired in parallel to give the wired-OR facility. The collector load resistor has to be provided externally. With this external resistor at 2 kΩ a reasonable fan-out is still available but, since the circuit has a high logic 1 output impedance, the operating speed is reduced.

The 7440 is an example of a NAND buffer, a gate that has higher drive capability. The Darlington emitter follower gives the gate a fan-out of 30.

A NOR gate (7402) is also available as are hundreds of other special circuits such as AND-OR-NOT gates, Schmitt triggers, monostables, bistables, etc. For a complete list consult manufacturers data sheets.

▶ Decoupling
▶ Open collector gate

## Schottky TTL
The main limitation in switching speed of standard TTL is the fact that the transistors saturate. Schottky TTL overcomes this by including Schottky barrier diodes between base and collector of the transistors. These metal-to-semiconductor rectifying junctions have a low forward voltage drop (300 mV) so that, when the transistor is turned on, the collector voltage is held at about 400 mV. This is sufficient to prevent saturation of the transistor. The

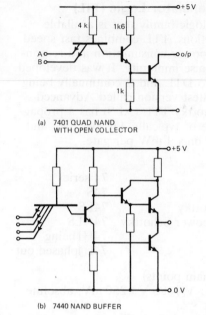

(a) 7401 QUAD NAND WITH OPEN COLLECTOR

(b) 7440 NAND BUFFER

**Figure T18** Variations in TTL gates

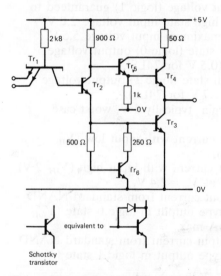

**Figure T19** Schottky TTL NAND gate

Schottky NAND circuit is shown in fig. T19.

The operation is almost identical to standard TTL except for the additions of $Tr_5$ and $Tr_6$. $Tr_5$ increases output drive in logic 1 state by acting as a Darlington pull-up, and $Tr_6$ is included to decrease still further the switching delays. $Tr_6$, called an "active turn-off" element, operates as a non-linear load during the switch on and switch off of $Tr_2$ and $Tr_3$. When $Tr_2$ turns off for example, $Tr_6$ goes rapidly into a

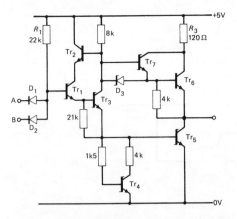

*Figure T20*  Low-power Schottky NAND

high resistance state assisting turn off. Schottky
TTL with a propagation delay of approximately
3 ns and a power dissipation of 20 mW per
gate competes with ECL in high-speed logic
systems.

An important development in TTL is low-
power Schottky (54LS/74LS). The NAND gate
is shown in fig. T20. This gate has a propagation delay of typically 7 ns, but consumes only
2 mW per gate. It is intended as a replacement
for the 54/74 series, since apart from its other
improvements in performance, it requires only
one fifth of the power of standard TTL.

The threshold level is 1.5 V, set by the three
base/emitter volt drops ($Tr_1$, $Tr_3$, $Tr_5$) from 0 V
minus the input diode drop ($D_1$ or $D_2$). Note
that the multi-emitter transistor of other TTL
circuits is replaced by a simple diode AND
circuit. Low-power Schottky TTL is actually an
updated DTL circuit. Fast switching speed and
high fan-out are achieved by using Schottky
barrier diodes across transistors and by clever
circuit design. $Tr_7$ is a Darlington pull-up
giving fast switching at the output from 0 to 1.
$Tr_4$ acts as an active turn off, and $Tr_2$ is an
emitter follower supplying the collector current
for $Tr_1$. With both inputs going high, $Tr_2$
supplies an initial current pulse into $Tr_1$ to aid
switching at the output from 1 to 0. $Tr_3$ and
$Tr_5$ turn on rapidly and the output goes low
(400 mV). At this point $Tr_3$ collector goes low
(1.1 V), thus reducing the collector current to
$Tr_1$ via $Tr_2$ to a lower value during the static
state. Diode $D_3$ also assists by supplying a
rapid discharge path for $Tr_6$. $R_3$, the 120 Ω
collector load of $Tr_6$, limits the output current
to a safe value of about 30 mA in the event of

an output short circuit. The fan-out is typically
20 and the noise margin is better than earlier
types of TTL. Table T1 shows comparisions.

**Table T1**  *Comparison table for TTL types (typical
values)*

| Type | Propagation delay | Fan-out | Worst case d.c. noise margin 1 | 0 | Power dissipation per gate |
|---|---|---|---|---|---|
| TTL 74 | 10 ns | 10 | 0.4 V | 0.4 V | 10 mW |
| Schottky 74S | 3 ns | 10 (low) 20 (high) | 0.7 V | 0.3 V | 20 mW |
| Low-power Schottky 74LS | 7 ns | 20 | 0.7 V | 0.3 V | 2 mW |

▶ *Digital circuit*      *CMOS*

### Tri-State Logic

In situations where the outputs of logic gates
have to drive common signal lines, a facility
has to be provided that enables gates to be
disconnected from the common line. Early
types of logic such as RTL and DTL allowed
common connections to be easily made since
several gate outputs could be wired together
(wired-OR connection) without damaging the
gates or degrading the speed of operation.
However, connecting gate outputs together
when using TTL or CMOS logic is not allowed
since the totem pole type output stages in
these i.c.s have a low output impedance in
both logic states. Therefore if one gate on the
common output line is in the low state while
others are in the high state, a damaging overload current would flow. A compromise can be
achieved by using open-collector gates and
providing an external pull-up resistor to the
common connection — but this method restricts
the maximum operating speed of the system.
Tri-state overcomes this problem by having
three possible output conditions:

   Low state   logic 0
   High state   logic 1
   High impedance state   effectively open
                                       circuit

In addition to the normal data input, a
tri-state gate has an enable input that either
allows the output to assume its correct state
(0 or 1) or effectively open circuits the gate

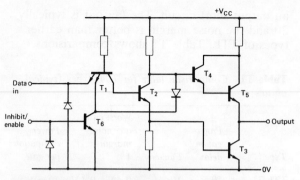

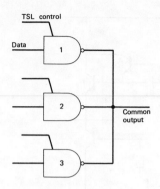

**Figure T21** Tri-state buffer gate
When a logic 1 is applied to the base of $T_6$, transistors $T_2$, $T_3$, $T_4$ and $T_5$ will all be off and the output will be in a high impedance state

output (fig. T21). This means that several such gates can be used to drive a common signal line with *only one* tri-state gate being enabled at any one time. In other words one gate is switched in while the others are switched out. In complex systems such as microcomputers tri-state is particularly useful in reducing the number of interconnections. Bidirectional bus systems, using 8 or 16 lines, are used with tri-state gates to allow data to be presented to the input of the system or to allow output data to be sent to the output ports. A simple arrangement for one common line is shown in the diagram.

▶ *Open collector gate*

## Tunnel Diode

This is a p-n junction diode which is made using very high doping levels. This results in an ultra thin depletion region and an effect known as tunnelling. In the forward bias region, the current increases rapidly for only low values of bias voltage and then decreases as shown (fig. 22a) to a valley point before rising again as forward voltage is further increased. A negative resistance region is exhibited in the forward direction as the tunnel current falls, and this type of characteristic allows the tunnel diode to be used as a fast switch or a high-frequency oscillator. An example of an oscillator is shown in fig. T22b. The diode is biased so that it operates on the negative resistance part of the characteristic and its negative resistance then makes up for the losses in the tuned circuit.

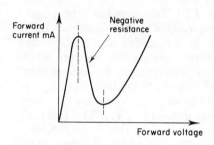

**Figure T22a** Tunnel diode characteristics

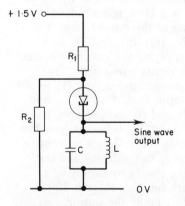

**Figure T22b** Typical tunnel diode oscillator

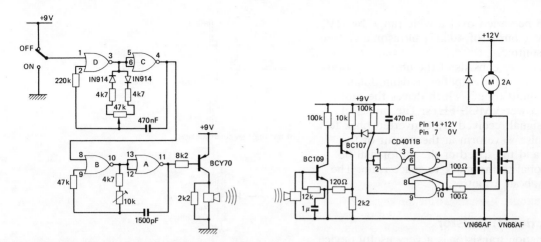

*Figure U1* Application of ultrasonics in the remote control of a d.c. motor

## Ultrasonics

Ultrasonics is the application and use of sound waves with frequencies above the human audio range, in other words upwards from about 20 kHz. Applications include
*Industrial*
Cleaning
Mixing/emulsions
Cutting brittle materials
Welding
Flaw detection in materials
*Specialised*
Echo sounding (sonar)
Medical diagnosis
Surgery
*General purpose*
Remote control
Burglar alarms
Liquid-level meters
Anti-collision devices
Object detection

Sound travels as a compressional wave and must have a medium through which to travel. The velocity of a sound wave is independent of frequency but does vary with the type of medium. It is about ten times faster in solids than it is in air. (Velocity of sound in air 331 m/s.)

Ultrasonic waves are produced using electro-mechanical resonators. The smaller transducers depend for their operation on either the piezo-electric or magneto-strictive effect. The former type is most commonly used for the general-purpose applications with a typical transducer operating frequency of 40 kHz. The piezo-electric effect is exhibited in materials such as quartz and ceramics (lead zirconate-titanate) in which charge dipoles within the ceramic are aligned by a process called polarisation. The material is heated to a temperature above its Curie point and then, as it cools through the Curie point, an electric field is applied. This forces the dipoles to align themselves in the direction of the applied field and this alignment remains when the material cools to normal temperatures and the field is removed. The material is now piezo-electric in that it develops a voltage across it if it is mechanically strained in the polarization direction. Conversely if an electric field is applied along the polarisation direction, the material will change its length. In this way a piezo-electric material converts mechanical energy to electrical energy and vice versa. An alternating voltage will force it to vibrate and maximum mechanical changes will take place when the applied signal is at the same frequency as the natural resonant frequency of the transducer. The mechanical changes will be transmitted as an ultrasonic wave.

An application using these transducers for remote control of the speed of a d.c. motor is shown in fig. U1. The transmitter is pulse width modulated. A CMOS 4001B quad NOR i.c. is wired as the oscillator section. Two gates, A and B, form the 40 kHz square wave drive to the ultrasonic transmitter, and this oscillator is gated on and off by the square wave generated by the other two gates C and D. This runs at about 40 Hz and the mark-to-space

ratio can be varied over a wide range by $RV_1$. In this way bursts of 40 kHz ultrasonic energy are transmitted.

The receiver consists of the ultrasonic transducer, a high gain amplifier, a demodulator, and a Schmitt trigger which drives the motor via two power VMOS FETs. The bursts of 40 kHz signal received are converted back to a 40 Hz pulse waveform at the output of the Schmitt and the speed of the motor will be proportional to the mark to space ratio. The range is about 5 metres.

## Unijunction Transistor

A unijunction transistor is a very useful device for making simple oscillator circuits that have quite good stability. It is constructed (fig. U2) of either a bar or a cube of lightly doped n-type silicon. Ohmic contacts are made at each end and these are called base 2 and base 1. The resistance between the two base connections is typically 6 k$\Omega$ for the T1S43. Somewhere near the centre of this bar a p-type region is formed during manufacture, and the connection to this is called the emitter. An equivalent circuit can be drawn as shown to consist of two resistors and a diode. The diode is the pn junction formed between the emitter and the bar. It is important to realise that the cathode connection of the equivalent diode is internal to the UJT and cannot be reached. Imagine the device connected across a d.c. supply with $B_2$ positive with respect to $B_1$ and $V_E$ equal to 0 V. The voltage between $B_2$ and $B_1$ is called $V_{BB}$ the *interbase voltage*. $RB_2$ and $RB_1$ form a potential divider, so the voltage across $RB_1$ (internal to the UJT) is a value dependent on the ratio of these resistances. In fact the voltage across $RB_1$ is called $\eta V_{BB}$, where $\eta$, known as the *Intrinsic Stand-Off Ratio*, depends upon the geometry of the device and is determined by the physical position of the p-region relative to $B_2$ and $B_1$. $\eta$ has values between 0.55 to 0.82 for T1S43 devices.

If $V_E$ is now gradually increased in value, a voltage between emitter and $B_1$ will be reached when the diode is almost forward biased. This voltage is called $V_p$ and from the equivalent circuit we can see that

$$V_p = \eta V_{BB} + V_D$$

Once $V_E$ exceeds $V_p$ the diode conducts, and

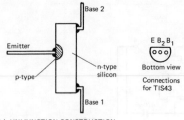

(a) UNIJUNCTION CONSTRUCTION

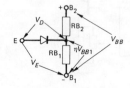

(b) EQUIVALENT CIRCUIT

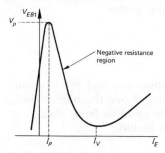

(c) TYPICAL CHARACTERISTICS

*Figure U2* Unijunction transistor

it injects holes into the $B_1$ region. The resistance of the $B_1$ region, that is $RB_1$, falls. However, $RB_2$ remains almost unchanged so the voltage across $RB_1$ also falls in value. This increases the forward bias across the diode causing it to conduct more, and therefore further decreasing the value of $RB_1$. This is a positive feedback effect and so very rapidly the resistance of the $B_1$ region falls to a very low value. Obviously in any circuit application the current injected at the emitter must be limited or the device will be destroyed. The characteristics for the device are shown in fig. U2, where the negative resistance region can clearly be seen. This negative resistance characteristic enables the UJT to be used in simple relaxation oscillators.

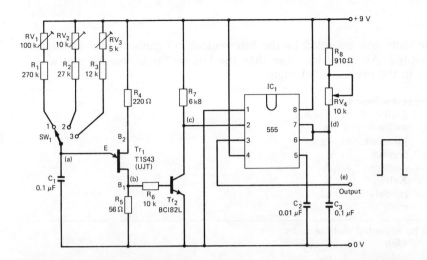

*Figure U3* **Pulse generator**

Now consider the operation of the oscillator section of the circuit in fig. U3. At switch-on $C_1$ is uncharged so $V_{EB1}$ is zero. A small current of just over 1 mA flows through the UJT from $B_2$ to $B_1$. $C_1$ now charges exponentially via $RV_1$ and $R_1$ towards +9 V. When the voltage across $C_1$ reaches the trigger voltage $V_p (V_p = \eta V_{BB} + V_D)$ then the unijunction switches to its low resistance state and $C_1$ is rapidly discharged through $R_5$. This causes a short-duration positive pulse to be developed across $R_5$. When the capacitor is discharged, the emitter current falls to a value which is insufficient to maintain conduction between the emitter and base 1, and so the unijunction switches back to its high-resistance state. $C_1$ is again free to charge and the cycle repeats itself. Thus a sawtooth type waveform is generated at the emitter and short duration positive pulses appear at $B_1$. The frequency of the oscillations depend upon the time constant in the emitter circuit, and $V_p$ the UJT trigger voltage. With $SW_1$ in position (1) the time constant is $C(R_1 + RV_1)$. For the T1S43, since $\eta$ can have a value of between 0.55 and 0.82, the trigger voltage $V_p$ will also vary from one device to another.

Since $V_p = \eta V_{BB} + V_D$ the values of $V_p$ for this circuit lie between 5.3 V and 7.9 V.

The actual formula for the frequency is

$$f = 1 \Big/ CR \log_e \left( \frac{1}{1 - \eta} \right)$$

where $CR$ is the time constant of the emitter circuit.

Note that $V_{BB}$ does not appear in this formula so that variations in supply voltage should not affect the frequency. However, because of the large variations in $\eta$ the frequency cannot be accurately set by fixed values of timing components. For each switch position, a preset is included to allow the three frequencies of 30 Hz, 300 Hz and 600 Hz to be set on test.

$R_4$ is included as a temperature-compensation component. The interbase resistance of the UJT will fall with increasing temperature and this will affect the trigger point. With $R_4$ in circuit any change in interbase resistance results in a volt drop across $R_4$ which tends to balance the changes in $V_p$ and $I_p$.

▶ *Programmable device*

## Unipolar Device
This refers to a semiconductor device in which the current flow is made up of only one type of charge carrier, either electrons or holes but not both. Thus all field effect transistors are unipolar devices, which distinguishes them from bipolar transistors.

▶ *FET*

# Units: the SI System

International system of metric units now accepted by the International Organisation of Standardisation (ISO) and generally adopted. All countries other than the United States use the metric system of measurement or are in the process of changing.

1   The S.I. System is founded on **seven base units.**

| Physical quantity | Unit | Symbol |
|---|---|---|
| Length | metre | m |
| Mass | kilogramme | kg |
| Time | second | s |
| Electric current | ampere | A |
| Temperature | Kelvin | K |
| Luminous intensity | candela | cd |
| Amount of substance | mole | mol |

2   **Two supplementary units** can be regarded as base units

| Physical quantity | Unit | Symbol |
|---|---|---|
| Plane angle | radian | rad |
| Solid angle | steradian | sr |

3   **Derived units** are expressed algebraically in terms of base units or supplementary units, for example, the S.I. unit for velocity is metre per second (m/s). For many of the S.I. derived units special names and symbol exist.

| Physical quantity | Unit | Symbol | Expressed in terms of other units |
|---|---|---|---|
| Frequency | Hertz | Hz | $1\,Hz = 1\,s^{-1}$ |
| Force | Newton | N | $1\,N = 1\,kg\,m/s^2$ |
| Pressure and stress | Pascal | Pa | $1\,Pa = 1\,N/m^2$ |
| Work, energy, quantity of heat | Joule | J | $1\,J = 1\,Nm$ |
| Power | Watt | W | $1\,W = 1\,J/s$ |
| Quantity of electricity | Coulomb | C | $1\,C = 1\,As$ |
| Electrical potential | Volt | V | $1\,V = 1\,W/A$ |
| Electric capacitance | Farad | F | $1\,F = 1\,As/V$ |
| Electric resistance | Ohm | $\Omega$ | $1\,\Omega = 1\,V/A$ |
| Electric conductance | Siemens | S | $1\,S = 1\,\Omega^{-1}$ |
| Magnetic flux | Weber | Wb | $1\,Wb = 1\,Vs$ |
| Magnetic flux density | Tesla | T | $1\,T = 1\,WB/m^2$ |
| Inductance | Henry | H | $1\,H = 1\,Vs/A$ |
| Luminous flux | lumen | lm | $1\,lm = 1\,cd\,sr$ |
| Illuminance | lux | lx | $1\,lx = 1\,lm/m^2$ |

4   **Units in general use**

| Physical quantity | Unit | Symbol | Definition |
|---|---|---|---|
| Time | minute | min | $1\,min = 60\,s$ |
|  | hour | h | $1\,h = 60\,m$ |
|  | day | d | $1\,d = 24\,h$ |
| Plane angle | degree | ° | $1° = (\pi/180)\,rad$ |
|  | minute | ′ | $1′ = (1/60)°$ |
|  | second | ″ | $1″ = (1/60)$ |
| Volume | litre | l | $1\,l = 1000\,cc$ |
| Mass | tonne | t | $1\,t = 10^3\,kg$ |
| Pressure of fluid or gas | bar | bar | $1\,bar = 10^5\,Pa$ |
| Temperature | ° Celsius | °C | $1°C = 273.15\,K$ |

The unit of the temperature interval for both the Kelvin and the Celsius scales is the same. In English-speaking countries the general term for the Celsius scale is Centigrade. This is not recommended because in some countries the right angle is divided into 100 parts known as ° Centigrade.

## 5 Multiples and sub-multiples of units

| Factor | | Prefix | Symbol |
|---|---|---|---|
| $1\ 000\ 000\ 000\ 000 = 10^{12}$ | | tera | T |
| $1\ 000\ 000\ 000 = 10^{9}$ | | giga | G |
| $1\ 000\ 000 = 10^{6}$ | | mega | M |
| $1\ 000 = 10^{3}$ | | kilo | k |
| $100 = 10^{2}$ | | hecto | h |
| $10 = 10^{1}$ | | deca | da |
| $0.1 = 10^{-1}$ | | deci | d |
| $0.01 = 10^{-2}$ | | centi | c |
| $0.001 = 10^{-3}$ | | milli | m |
| $0.000\ 001 = 10^{-6}$ | | micro | $\mu$ |
| $0.000\ 000\ 001 = 10^{-9}$ | | nano | n |
| $0.000\ 000\ 000\ 001 = 10^{-12}$ | | pico | p |
| $0.000\ 000\ 000\ 000\ 001 = 10^{-15}$ | | femto | f |
| $0.000\ 000\ 000\ 000\ 000\ 001 = 10^{-18}$ | | atto | a |

## VMOS Power FET

These are enhancement mode MOSFETs that can be used in a wide variety of power, interface, and h.f. applications. Devices are available with

*a*) Current handling capabilities of 10 A or greater $(I_D)$

*b*) Breakdown voltages up to 500 V $(BV_{DSS})$

*c*) Drain/source on resistance values of less than $1\ \Omega$ $(R_{DS(on)})$.

In addition the VMOS power FET is capable of very fast switching action. Some can switch 1 A of drain current in less than 4 ns.

The n-channel type is the most common and will be described here, but some complementary p-channel devices such as the ITT BD512 and BS 250 do exist.

The structure of a typical n-channel VMOS is shown in fig. V1. It has a heavily doped $n^+$ substrate and a high resistivity $n^-$ epitaxial layer. A lightly doped $p^-$ region called the body and an $n^+$ region for the source are diffused into this epitaxial layer. Following this, a V-groove is etched through the $n^+$ and $p^-$ regions into the epitaxial layer. Insulating oxide is grown over the groove and aluminium deposited over the oxide to form the gate. Connections are made to the top $p^-$ and $n^+$ regions for the source and to the bottom of the substrate for the drain.

Operation is similar to a conventional MOS-FET in which a conducting channel is induced from source to drain by a controlling voltage on the gate, but unlike the conventional MOS-FET the VMOS uses vertical, rather than lateral, current flow. In other words electrons flow down from source to drain in a VMOS.

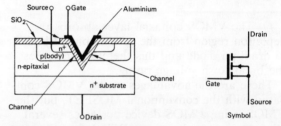

*Figure V1*  n-channel VMOS

With both gate and drain positive with respect to the source, the gate induces an n-type channel in both surfaces of the p-type source in the regions facing the groove. Current then flows as electrons are attracted from the source to the drain.

There are several advantages resulting from this type of construction:

*a*) The lengths of the induced channels are determined by the diffusion process not by masking, and are therefore more controllable. The lengths are relatively short (1.5 $\mu$m) giving greater current density (ratio of channel width to length determines current density).

*b*) Since each V-groove has two channels, the current density is doubled.

*c*) The substrate forms the drain contact, and this enables heat to be removed efficiently especially as the drain can be mounted directly on the header.

*d*) Since drain metallisation is not required on the top of the chip, the chip area can be reduced and this gives lower capacitance.

*e*) As with a bipolar transistor, the substrate is heavily doped, therefore the saturation resistance of the VMOS is low.

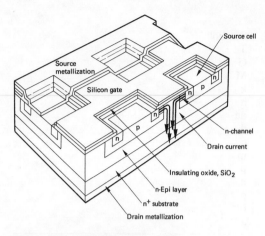
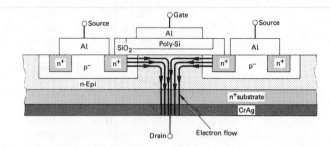

*Figure V2* TMOS power FET

*f*) The VMOS epitaxial layer absorbs the depletion region from the reverse biased body to drain pn diode and this therefore increases the breakdown voltage rating.

These are the advantages of the VMOS compared with the conventional MOSFET, but the VMOS being a MOS device possesses several advantages over the power BJT. It has very high input impedance, requiring input currents of less than 100 nA. This means the device can be directly interfaced with CMOS logic or other high-output impedance circuits. The VMOS being a unipolar device exhibits negligible charge storage effects and this gives it its fast switching speed.

Finally the temperature coefficient of the VMOS drain-to-source on voltage is positive, the VMOS current decreasing as temperature rises. There are then fewer problems of thermal runaway than in BJTs. The latter point can cause problems with high voltage supplies when the drain current is held relatively constant, since a rise of $V_{DS(on)}$ caused by the temperature will cause a rise of $V_{DS}$ and increase dissipation. Other versions of VMOS, both of which conduct current vertically, are

**1** TMOS — manufactured by Motorola. It is a planar, non V-groove structure, and is virtually many power MOSFETS all connected in parallel with a common drain (see fig. V2). Several source cells, square in shape, are each isolated from the common drain by a p-well. When the gate is made positive with respect to the source, the p-regions around the periphery of each cell are inverted (changed from p to n type) and electrons can then flow from source

cells down to the drain. The structure gives very low values of $R_{DS(on)}$ and high current capability. For example an MTM 1225 TMOS power FET has

$I_D$ 12 A $BV_{DSS}$ 100 V
$R_{DS(on)}$ 0.25 Ω at 6 A

**2** HEXFET — manufactured by International Rectifier. Very similar structure to TMOS except that the source cells are hexagonal in shape. It is claimed that up to half a million cells per square inch are used. Each cell is separated from the common drain by a p-region around the periphery of each cell. Again, drain current flows when the gate is made positive with respect to the source.

For applications    ▶ *Alarm circuit*
    ▶ *Ultrasonics*    ▶ *Touch switch*

**Voltage regulator using VMOS FETs** (fig. V3) VMOS devices can be used to create simple low-cost linear regulators. The VMOS has a high value of $g_{fs}$ (typically 200 mS), a high input impedance, and can pass relatively high values of drain current. Another and important advantage, shown in this example, is the fact that VMOS devices can be connected in parallel to provide more output current. They have a positive temperature coefficient of $V_{DS}$ and therefore tend to share the load current equally.

The operation is similar to the conventional linear regulator [▶ *Power supply*] except for the fact that the VMOS FETS require virtually no gate current. $Tr_1$ is the error amplifier with its emitter tied to the 6.2 V reference. When the unregulated input is switched on, $R_1$ provides

**Table V1** *Data for a representative selection of VMOS devices*

| Type no. | n/p | $V_{DS}$ | $I_D$ | $P_{tot}$ (25°C case) | $R_{DS(on)}$ | $g_{fs}$ (min.) | $V_{GS(threshold)}$ |
|----------|-----|----------|-------|-----------------------|--------------|-----------------|---------------------|
| VN46AF | n | 40 V | 2 A | 15 W | 3 Ω | 150 mS | 1.7 V |
| VN66AF | n | 60 V | 2 A | 15 W | 3 Ω | 150 mS | 1.7 V |
| VN88AF | n | 80 V | 2 A | 15 W | 4 Ω | 150 mS | 1.7 V |
| BD512 | p | −60 V | −1.5 A | 10 W | 6.5 Ω | 150 mS | 2.2 V |
| BD522 | n | 60 V | 1.5 A | 10 W | 2.5 Ω | 270 mS | 1.6 V |

All the above are To 202$^m$ case

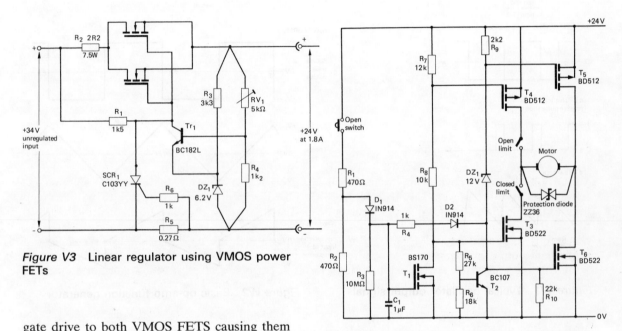

*Figure V3* Linear regulator using VMOS power FETs

*Figure V4* Automatic door control using VMOS power FETs

gate drive to both VMOS FETS causing them to conduct. The output voltage rises and stabilises at +24 V. This level can be adjusted by $RV_1$. If more load current is taken, the output voltage falls slightly and $Tr_1$ conducts less. Its collector voltage rises to adjust the gate voltage at the VMOS FETs to a value sufficient to provide the extra current demanded. The voltage at $Tr_1$ collector is about 26 V for a load current of 100 mA and increases to approximately 31 V when the load is 1.8 A.

Both the VMOS FETs must be mounted on a suitable heatsink with a thermal resistance of better than 10°C/W. To protect the devices against an overload, such as a short circuit, a foldback current limit using a C103YY thyristor is used. This trips at about 2 A and switches both VMOS FETs off since the gate drive falls to less than 1 V (below $V_{GS}$ threshold).

The regulator has an output resistance of about 60 mΩ and provides +24 V with loads up to 1.8 A. Larger currents can be provided by using more parallel VMOS FETs.

**Automatic door control** (fig. V4)
Here complementary VMOS FETs are used to switch power to a d.c. motor which drives the door. $T_1$, the BS170 n-channel VMOS FET, is wired as a monostable. When the Open switch is operated, $T_1$ is triggered on . This causes $T_2$ to turn off reverse biasing $D_2$. $C_1$ therefore discharges slowly via $R_3$ and $T_1$ remains on for approximately 20 sec. While $T_1$ is on, $T_3$ and $T_5$ will be off and $T_4$ and $T_6$ will be on. The motor is driven to the open position as current

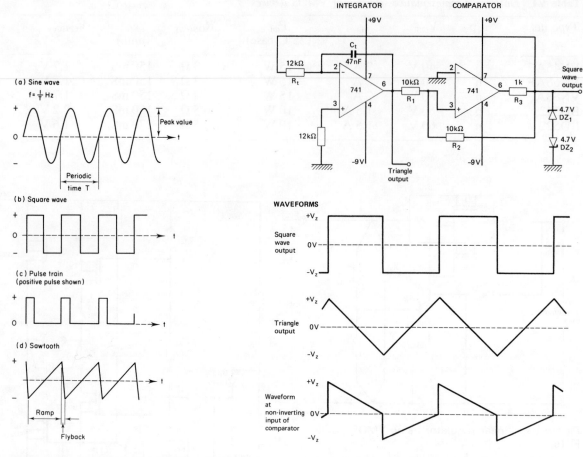

(a) Sine wave

$f = \frac{1}{T}$ Hz

Peak value

Periodic time $T$

(b) Square wave

(c) Pulse train (positive pulse shown)

(d) Sawtooth

Ramp

Flyback

**Figure W1** Typical oscillator output signals

INTEGRATOR   COMPARATOR

+9V   +9V

$C_t$   47 nF

12 kΩ   $R_t$   2   7   741   6   10 kΩ   $R_1$   2   7   741   6   1k   $R_3$   Square wave output

3   4   3   4

12 kΩ   10 kΩ   $R_2$   4.7 V $DZ_1$

−9V   Triangle output   −9V   4.7 V $DZ_2$

**WAVEFORMS**

Square wave output   +$V_z$   0V   −$V_z$

Triangle output   +$V_z$   0V   −$V_z$

Waveform at non-inverting input of comparator   +$V_z$   0V   −$V_z$

**Figure W2** Basic op-amp function generator

is supplied from the +24 V rail via $T_4$ and $T_6$. The limit switch will open to remove drive from the motor when the door reaches the end of its travel. When the monostable switches back, $T_1$ turns off forcing $T_2$ to conduct so that $C_1$ is rapidly discharged. With $T_1$ off, both $T_4$ and $T_6$ also turn off, but $T_3$ and $T_5$ are forced to conduct. The motor is then driven in the opposite direction to close the door.

## Waveforms and Waveform Generators

A wide variety of signal shapes occur in electronics. The basic waveforms are Sinusoidal, Square, Triangle, Sawtooth, Pulse. For any regularly recurring waveform the frequency in Hertz is given by

$f = 1/T$ Hz

where $T$ is the *periodic time* measured in seconds.

The sine wave is basic to all other waveshapes. Any regularly recurring signal can be shown to be made up of the sum of a particular set of sinusoidal waves. For example, a square wave consists of a fundamental sine wave component (one that has the same frequency as the square wave) plus sine waves at various amplitudes of all the odd harmonics of that fundamental, i.e. $f_3$, $f_5$, $f_7$, etc. [▶ *Fourier analysis*].

Sine waves are generated by oscillators such as the ▶ *Colpitts* and ▶ *Wien bridge*, which consist of a frequency determining network, an amplifier, and positive feedback. Square waves can be generated either by passing a sine wave through a Schmitt trigger or by using an asta-

**250**

ble multivibrator. Both types are described elsewhere [▶ *Oscillator* ▶ *Clock pulse generator*].

This leaves the FUNCTION GENERATOR as the waveform-producing circuit to be discussed in this section. A function generator produces simultaneous square and triangle wave outputs with often a waveform shaper included to convert the triangle into a sine wave of medium distortion. The classic circuit for a triangle/square wave function generator consists of an integrator and comparator wired together in a positive feedback loop (fig. W2).

The operation can be described by considering that the comparator output has just switched high to $+V_{o(sat)}$. A reference voltage level of $+V_z$ is then applied to the input of the integrator causing the integrator output to ramp from $+V_z$ towards $-V_z$ at a rate determined by values of $R_t$ and $C_t$. The integrator output is compared by the circuit formed by the other op-amp and $R_1$ and $R_2$. With $R_1 = R_2$, as shown, the output of the comparator will switch from $+V_{o(sat)}$ to $-V_{o(sat)}$ at the point when the integrator output reaches a voltage of $-V_z$. This occurs when the noninverting input of the comparator op-amp is just negative of zero volts. A voltage level of $-V_z$ is now applied to the input of the integrator which causes its output to ramp in a positive direction from $-V_z$ to $+V_z$. When this ramp reaches $+V_z$, the comparator is forced to switch from $-V_{o(sat)}$ to $+V_{o(sat)}$ and the cycle repeats. In this way continuous square and triangle waves are produced. With $R_1 = R_2$ the amplitudes of the two waveforms will be almost identical at $\pm V_z$. The circuit will oscillate without the reference diodes by connecting $R_2$ directly to the comparator output; but $R_1$ must then be a lower value than $R_2$ or the circuit will latch in one saturated output state. Without the reference, and with $R_1$ less than $R_2$, the triangle will be at a smaller amplitude than the square wave and the frequency will vary with voltage supply. With $R_3$ and $DZ_1$ and $DZ_2$ in circuit, and $R_1 = R_2$, the frequency is set by the integrator time constant $C_t R_t$.

Since the inverting input terminal of the integrator can be considered as a virtual earth, the capacitor is charged by an almost constant current. This current is given by

$$I_{charge} = V_z / R_t$$

Now  $I\,dt = C\,dV$      (since $Q = CV$)
and   $dV = 2V_z$
since the output moves from $-V_z$ to $+V_z$ and vice versa.

$$\therefore \quad t_1 = t_2 = dt = \frac{C\,dV}{I} = \frac{C2V_z}{V_z/R_t} = 2C_t R_t$$

Periodic time = $t_1 + t_2 = 4C_t R_t$

$$\therefore \quad f = 1/4C_t R_t$$

With the values given, $f = 440\,\text{Hz}$.

Using this basic circuit, and with careful choice of op-amps, triangle/square waves can be generated from fractions of a Hertz up to several hundred kiloHertz. For very low frequency operation the design of the integrator becomes critical, since the input bias current must then be low compared to the charging current (use a BIFET type or MOSFET such as a 3130), while at high frequencies it is the switching speed of the comparator that is important. An op-amp (or better still a comparator i.c.) with fast slew rate is required.

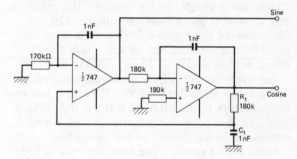

**Figure W3**   Quadrature (sine/cosine) oscillator
  $f_0 = 1/2\pi C_t R_t$

Two integrators wired together in a feedback loop will cause simultaneous generation of sine and cosine waves — a so-called QUADRATURE oscillator. An example for a 1 kHz design is shown in fig. W3.

**1**   One very useful feature that can be added to basic waveform generators is the ability to control the frequency of oscillations by some externally applied voltage or current a VOLTAGE CONTROLLED OSCILLATOR (VCO) is shown in fig. W4, and it can be seen to be a modification to the previously described triangle/square wave generator. The controlling voltage $V_{in}$ is applied to the integrator which then produces a negative going ramp. The level of this ramp is compared by the second op-

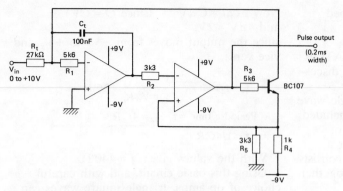

**Figure W4** Voltage-controlled oscillator

amp which has its non-inverting reference input tied to a negative voltage set by the potential divider $R_4$ and $R_5$. When the ramp reaches a level that is just more negative than this reference voltage (approximately $-6.8\,\text{V}$), the comparator op-amp switches positive and forces the transistor $\text{Tr}_1$ to conduct. $C_t$ is now forced to charge fairly rapidly in the opposite direction and a pulse is given from the output. The frequency of these pulses is determined by the input voltage and for the values given, the range is from 40 Hz at 1 V to 320 Hz at 8 V.

An even simpler VCO can be constructed using a 555 timer (or CMOS 555) and a few of the invertors in a CMOS 4049B i.c. (fig. W5). The 555 is used as a Schmitt with pins 6 and 2 connected to the output of the integrator formed round gate C. The triangle wave is fed to the 555 input and, without any applied voltage to $V_C$, the triangle amplitude will be set by the upper and lower trip points of the internal components in the 555. These limits are $\frac{2}{3}V_{CC}$ and $\frac{1}{3}V_{CC}$. The frequency of the square waves at pin 3 can then be varied by applying a voltage to pin 5 of the 555. This overrides the internal voltages at which the Schmitt trigger action takes place and frequency control over a wide range is possible. One drawback is that the triangle wave amplitude will vary with applied control voltage.

**2** The final circuit (fig. W6) shows a CURRENT CONTROLLED OSCILLATOR (ICO) using a 3080 operational transductance amplifier [▶ *Op-amp*] wired in a feedback loop with the limits of the triangle wave set between $\frac{1}{3}$ and $\frac{2}{3}V_{CC}$. The 3080 is connected as a non-inverting current-programmable integrator with the rate of charge of the timing capacitor

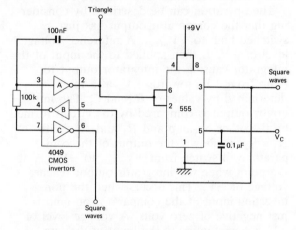

**Figure W5** VCO using a 555 and CMOS invertors

$C_t$ set by $I_{ABC}$. When the output of the 555 switches high, the capacitor is charged linearly from $\frac{1}{3}V_{CC}$ to $+\frac{2}{3}V_{CC}$. As soon as this level is exceeded, the 555 output switches low and the capacitor voltage will ramp in the opposite direction. The frequency is set by the current $I_{ABC}$ which can be varied from a few microamps up to 1 mA maximum. To avoid non-linearity an op-amp buffer is used to drive any external load. A simple modification of connecting a resistor of medium value (330 $\Omega$) from the integrator output to the Discharge pin of the 555 (pin 7) converts the output into a sawtooth wave. With this resistor in circuit the discharge transistor conducts to rapidly discharge $C_t$ when the positive going ramp exceeds $+\frac{2}{3}V_{CC}$.

**3** There are several alternative methods of producing triangle or sawtooth waves. One useful method is to use a digital counter and to convert its outputs via an R-2R ladder

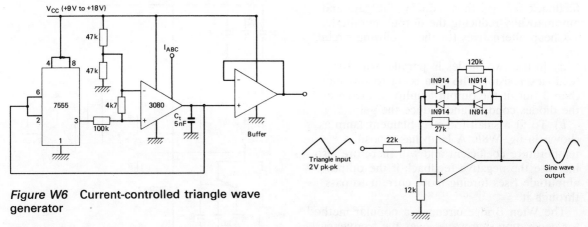

**Figure W6** Current-controlled triangle wave generator

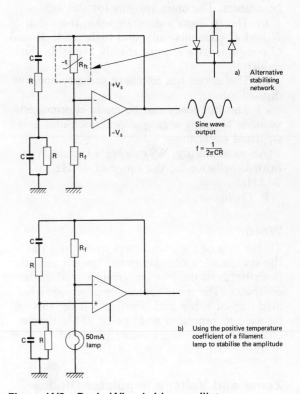

**Figure W7** Triangle to sine wave convertor

network. The resulting ramp waveform then consists of several small and equal interval steps. [▶ *Counters and dividers*].

The triangle output from a function generator can be shaped into a sine wave using a circuit with a non-linear transfer function. A simple circuit (fig. W7) consists of diodes wired across the feedback resistor of an inverting amplifier. The diodes conduct on the peaks of the triangle wave input and therefore reduce the gain giving a "rounding off" effect. For good results the input wave should swing equally about zero, i.e. have very low d.c. component, and the amplitude must be accurately adjusted. If these two conditions are met, distortion in the resulting sine wave output may be less than 1%.

## Wien Bridge Oscillator

Many oscillator circuits consist of an amplifier with positive feedback via a frequency determining network. In the Wien bridge oscillator the frequency determining network is a *CR* phase shift network (fig. W8). At one particular frequency the phase shift from the output to the non-inverting input will be zero giving positive feedback. The frequency is given by

$$f = 1/2\pi RC$$

It can also be shown by analysis that the overall gain of the amplifier need only be 3 to maintain oscillations, as this just makes up for the losses in the network. Therefore a negative feedback path is provided to stabilise the amplitude of the output sinewave. By fitting a

**Figure W8** Basic Wien bridge oscillator

device with a non-linear characteristic within the negative feedback loop, the gain can be held to just over 3 and the output will then be a sinewave with low harmonic distortion. In many circuits a sealed thermistor type R53 is used for this purpose. Its resistance falls if the output amplitude increases. This increases the

**253**

feedback voltage, thus reducing the gain and automatically reducing the output amplitude.

Cheap alternatives for the stabilising circuit are

*a*) to use two diodes in parallel with the feedback resistor. The gain is set to just more than 3 but, if the output amplitude increases, the diodes conduct and reduce the gain.

*b*) To fit a small (50 mA) filament lamp as shown in fig. W8*b*. A lamp has a positive temperature coefficient and will therefore also increase the negative feedback if the output amplitude rises forcing more current to pass through it.

The Wien Bridge circuit is a popular method for generating sine waves over the frequency range of 1 Hz to 1 MHz and it is the standard circuit used in audio and low frequency generators. The main reasons for this are

*a*) The frequency depends upon the value of *R* and *C* elements only, and high-grade *R* and *C* components are more readily available than inductors.

*b*) The circuit has excellent stability and low distortion.

*c*) The frequency can be made continuously variable by using a ganged potentiometer and switched capacitors.

An example (fig. W9) shows a typical Wien bridge oscillator for the range of 15 Hz to 22 kHz.

▶ *Oscillator*

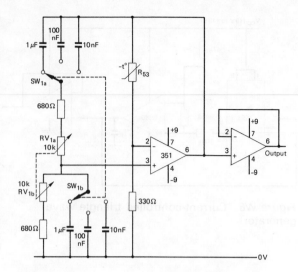

*Figure W9* Wien bridge oscillator 15 Hz to 22 kHz; distortion <0.1%

## Word

This is one of the parameters that is used in the specification of a computer system and it describes how the bits are grouped within the computer. The most commonly used word is made up of 8 bits and is called a byte; this can be used to represent any one of 255 different characters or messages. More powerful machines use a data word of 16 bits.

## Zener and Voltage Regulator Diodes

These are silicon p-n junction devices specifically designed to be used in the reverse breakdown region as voltage regulators and reference elements. The doping level is higher than in ordinary signal diodes and this higher impurity concentration results in a relatively thin depletion region. Because of this, for only small reverse voltages very high field strengths exist in the depletion region, leading to a low value of reverse breakdown voltage. By controlling the doping level, diodes of varying reverse breakdown voltage can be produced. The typical range is from 2.7 V up to 100 V or higher, with power ratings from 500 mW up to 75 W.

The device behaves like a normal silicon diode in the forward direction but in the reverse mode has a very high resistance until the breakdown voltage is reached. At breakdown the resistance falls and a rapid increase in reverse current takes place. As long as the power dissipation of the zener is not exceeded, this reverse current does not damage the diode. Two mechanisms account for this rapid rise in current at breakdown:

*a*) *Zener effect* Where the high field strengths caused by the reverse voltage in the narrow depletion region pull electrons out of the covalent bonds of silicon atoms. Hole-electron pairs then created make up the reverse current. This effect occurs mostly in low-voltage diodes with breakdown voltages of less than 6 V and is characterised by a "soft", rather than an abrupt change in current at breakdown. Consequently the slope resistance of a true zener diode is rather high (100 Ω at 5 mA being typical). Zeners also have a negative temperature coefficient.

*b*) *Avalanche effect* This predominates in diodes with breakdown voltages in excess of 6 V. The change in current is very rapid at breakdown giving a sharp knee to the charac-

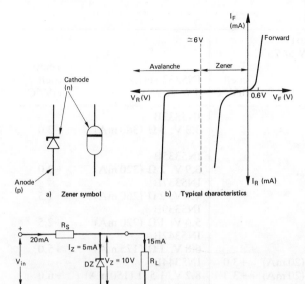

a) Zener symbol     b) Typical characteristics

c)   Simple voltage regulator

*Figure Z1*  Zener diode

teristics. The temperature coefficient is positive.

In practical devices both effects occur and therefore diodes with breakdown voltages of between 4 V and 6 V exhibit low values of temperature coefficient. Specially selected units are available with very low drift of breakdown voltage with temperature.

The parameters usually quoted for a zener diode at 25°C ambient and zener current of 5 mA are

$V_Z$  Breakdown voltage (tolerance ±5% or ±10%).

Slope or a.c. resistance $r_z = \delta V_z/\delta I_z$ ohms.

Temperature coefficient: change of $V_z$ with temperature quoted as mV/°C or as a percentage

$P_Z$  Maximum power rating.

Fig. Z1 shows a zener being used as a simple voltage regulator supplying a load with 15 mA at a constant voltage of 10 V. As $I_Z$ is chosen to be 5 mA the value for the series resistor given by

$$R_s = (V_{in} - V_Z)/(I_Z + I_L) = (30 - 10)/20 \text{ mA} = 1 \text{ k}\Omega$$

The zener operates to keep the voltage at the output relatively constant irrespective of changes in input voltage and in load current. For example if the load current is halved, falling to 7.5 mA, then an extra 7.5 mA must

flow in the zener. The change in output voltage assuming a fixed value of $r_z$ (here 25 Ω) is

$$\Delta V_z = \Delta I_z r_z = 187.5 \text{ mV} \qquad \text{i.e. } 1.87\%$$

Similarly with the load reconnected, if the input voltage increased by 10 V from 30 V to 40 V the change in output voltage will be

$$\Delta V_z = \Delta I_z r_z = (\Delta V_{in}/R_s)r_z = 250 \text{ mV (a rise of } 2.5\%)$$

The maximum power rating of the zener is determined by the worst-case conditions. In this case assuming a rise in input voltage up to 40 V max. with the load disconnected:

$$P_{Z(T)} = I_Z(V_Z + \Delta V_Z)$$
$$= 30 \times 10^{-3}(10.437) = 313.13 \text{ mW}$$

Therefore a 500 mW device would be used.

*Zener reference diodes*

| | $P_Z$ | $r_Z$ at 7.5 mA | $V_Z$ at 7.5 mA | Temp. coeff. |
|---|---|---|---|---|
| IN821 ⎤ | 400 mW | 15 Ω | 6.2 V ±5% | 0.01%/°C |
| IN827 ⎦ | | | | 0.001%/°C |

▶ *Clipping circuit*    ▶ *Power supply*
▶ *Programmable device*

## Zero Crossing Detector

Used when it is required to detect the point at which an a.c. input signal goes through zero. The circuit (fig. Z2) is usually based on a high-speed comparator which has its reference input

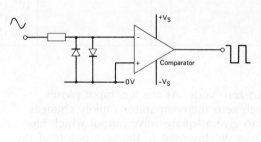

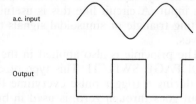

*Figure Z2*  Principle of zero crossing detector

# Table Z1 *Zener diode data*

| $V_z$ | BZY88 500 mW at 25°C $r_z(\Omega)$ | Temp. coeff. mV/°C | BZY85 1.3 W at 25°C $r_z(\Omega)$ | Temp. coeff. mV/°C | BZX61 1.3 W at 25°C $r_z(\Omega)$ | Temp. coeff. mV/°C | IN5333 series 5 W at 25°C | Temp. coeff. mV/°C |
|---|---|---|---|---|---|---|---|---|
| 2.7 | 120 (5 mA) | −1.8 | 20 (80 mA) | −2.2 | | | IN5333B | |
| 3.0 | 120 | −1.8 | 20 (80 mA) | −2.4 | | | 3.3 V, 3 Ω (380 mA) | −2.0 |
| 3.3 | 110 | −1.8 | 20 (80 mA) | −2.6 | | | | |
| 3.6 | 105 | −1.8 | 15 (60 mA) | −2.9 | | | IN5335B | |
| 3.9 | 100 | −1.4 | 15 (60 mA) | −2.9 | | | 3.9 V, 2 Ω (320 mA) | −2.0 |
| 4.3 | 90 | −1.0 | 13 (50 mA) | −2.2 | | | IN5337B | |
| 4.7 | 85 | +0.3 | 13 (45 mA) | +1.9 | | | 4.7 V, 2 Ω (260 mA) | +1.5 |
| 5.1 | 75 | +1.0 | 10 (45 mA) | +2.0 | | | IN5339B | |
| 5.6 | 55 | +1.5 | 7 (45 mA) | +2.5 | | | 5.6 V, 1 Ω (220 mA) | +2.5 |
| 6.2 | 27 | +2.0 | 4 (35 mA) | +3..4 | | | IN5342B | |
| 6.8 | 15 | +2.7 | 3.5 (35 mA) | +4.1 | | | 6.8 V, 1 Ω (175 mA) | +5.0 |
| 7.5 | 15 | +3.7 | | | 5 (20 mA) | +3.0 | IN5344B | |
| 8.2 | 20 | +4.5 | | | 7.5 (20 mA) | +3.3 | 8.2 V, 1.5 Ω (150 mA) | +6.0 |
| | | | | | | | IN5346B | |
| 9.1 | 25 | +6.0 | | | 8 (20 mA) | +4.6 | 9.1 V, 2 Ω (150 mA) | +6.0 |
| | | | | | | | IN5347B | |
| 10 | 25 | +7.0 | | | 8.5 (20 mA) | +5.0 | 10 V, 2 Ω (125 mA) | +7.0 |
| 11 | 35 | +8.0 | | | 9 (20 mA) | +5.5 | IN5349B | |
| 12 | 35 | +9.0 | | | 9 (20 mA) | +6.0 | 12 V, 2.5 Ω (100 mA) | +10 |
| 13 | 35 | +10.5 | | | 10 (20 mA) | +6.5 | IN5352B | |
| 15 | 40 | +12.5 | | | 14 (20 mA) | +9.0 | 15 V, 2.5 Ω (75 mA) | +13 |
| 16 | | | | | 16 (10 mA) | +9.6 | | |
| 18 | | | | | 20 (10 mA) | 10.8 | | |
| 20 | | | | | 22 (10 mA) | 12.0 | | |
| 22 | | | | | 23 (20 mA) | 13.2 | IN5359B | |
| 24 | | | | | 25 (10 mA) | 14.4 | 24 V, 3.5 Ω (50 mA) | +20 |
| 27 | | | | | 35 (10 mA) | 16.2 | | |
| 30 | | | | | 40 (10 mA) | 21.0 | | |
| 33 | | | | | 45 (10 mA) | 23.1 | | |
| 36 | | | | | 50 (10 mA) | 25.2 | | |
| 39 | | | | | 60 (5 mA) | 27.3 | | |
| 47 | | | | | 80 (5 mA) | 37.6 | | |
| 56 | | | | | 105 (5 mA) | 44.8 | | |
| 68 | | | | | 120 (5 mA) | 54.4 | | |
| 72 | | | | | 130 (5 mA) | 57.6 | | |

tied to zero volts. As the a.c. input passes through zero the comparator rapidly changes state to give a square wave output which has its edges synchronised to the zero points of the a.c. input. A circuit like this is useful in converting triangle or sinusoidal signals to square waves.

The principle is also applied in the ZERO VOLTAGE SWITCH. This type of circuit, producing a trigger pulse everytime the a.c. input goes through zero, is used in burst firing a.c. power control. In this method power is applied to the load in complete half-cycles rather than in portions of half-cycles as in phase control. By using zero voltage switching the triac or thyristor is only switched on just after the a.c. mains has gone through zero and in this way only low levels of radio frequency interference are generated.

An example (fig. Z3) shows the use of a zero voltage switch i.c. type L121 in the on/off control of a 1 kW heater. The thermistor controls the temperature at which the triac is triggered on and the i.c. produces a 200 $\mu$s pulse to

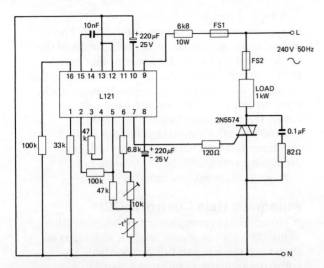

*Figure Z3* Use of zero voltage switch i.c. in heater control

switch the triac when the a.c. mains next goes through zero. Temperature range is from 15°C to 45°C.

A.C. power controllers using zero voltage switching can only be applied to loads, such as heaters, that have relatively slow response times. Phase control is the method used in lamp dimmers and a.c. motor speed controllers.

▶ *Burst firing*    ▶ *Thyristor*

## Automatic Frequency Control (AFC)

Generally this can be taken to mean any circuit that maintains the frequency of an oscillator at an almost fixed value. More specifically, AFC is a circuit technique used in VHF and UHF communication receivers to keep the receiver accurately tuned to the frequency of the selected carrier wave input. In VHF and UHF receivers, the i.f. (intermediate frequency) is only a low value in comparison to the carrier frequency of the signal being received. Therefore any drift in the frequency of the local oscillator will have an immediate effect on receiver performance. AFC is an arrangement that feeds back a signal to control the local oscillator so that the output frequency from the MIXER is in line with the i.f. The basic block diagram of a system is shown in fig. (A)1. Here the output of the i.f. amplifier is fed to a discriminator circuit which itself is tuned to the same value as the intermediate frequency. Any drift in frequency from the mixer, via the i.f. amplifier, results in a d.c. voltage output from the discriminator. This voltage is used to control the frequency of the local oscillator. Normally a varactor diode is used to alter the tuning of the frequency-determining network of the local oscillator. In this way, the intermediate carrier wave from the mixer is prevented from drifting. AFC is switched out when the receiver is retuned to another station.

A typical circuit arrangement for AFC in an f.m. radio receiver is shown in fig. (A)2. The output from the ratio detector will consist of the audio signal plus a d.c. component that varies with de-tuning. When the receiver is on tune, this voltage is zero. The d.c. component is fed back, after further filtering, to control the voltage across $D_1$, a varactor diode. This diode is connected to the tuned circuit of the local oscillator and will therefore force the local oscillator to maintain the receiver on tune.

▶ *Radio systems and circuits*

## Automatic Gain Control (AGC)

A circuit used in communication receivers that adjusts the gain of the r.f. and i.f. amplifiers so as to keep the input to the demodulator at approximately the same level irrespective of changes in signal strength at the aerial. The signal at the aerial will vary between one signal source and another and, in cases where the receiver is tuned to a distant signal source, fading will also occur. These changes would naturally affect the output; in the case of a radio receiver the volume would vary considerably.

Basically AGC is a feedback system where a signal derived from the demodulator and proportional to the rectified r.f. carrier is fed back to adjust the gain of the r.f. and i.f. amplifiers [see fig. (A)3]. As the received signal strength

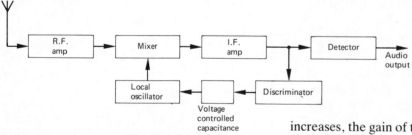

Figure (A)1   Block diagram of AFC

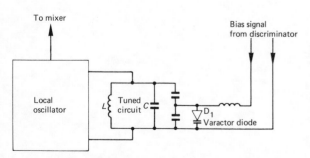

Figure (A)2   Using a varactor diode for AFC

increases, the gain of the amplifiers is reduced. Delayed AGC is a variant that does not apply a reduction in gain until a certain level of signal strength is exceeded.

A typical circuit for an a.m. radio receiver is shown in fig. (A)4. From the demodulator output the rectified carrier is smoothed by the filter $R_1C_2$. This filter must have a time constant that eliminates audio signals (modulation) from the AGC voltage but not so long that it prevents the AGC from being able to cope with fading. This negative voltage is applied to the base of the i.f. amplifier (n-p-n transistor) and will reduce the i.f. gain as the signal strength increases.

▶ *Radio systems and circuits*

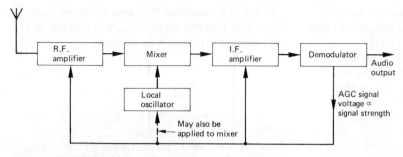

*Figure (A)3* Automatic gain control in a superhet receiver

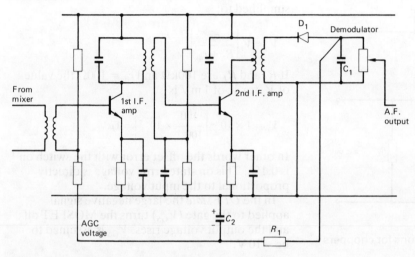

*Figure (A)4* Circuit showing how AGC is applied to the i.f. amp.

## Chopper

The name usually given to an electronic switch unit used to convert a d.c. voltage signal into a.c. The switch literally "chops" the d.c. to produce a square wave with an amplitude (pk-pk) proportional to the value of the d.c. voltage and a frequency set by the switch drive frequency [fig. (C)1].

The switch element is normally an enhancement mode MOSFET since these devices give the best performance in terms of low switch offset, switching speed and high off resistance.

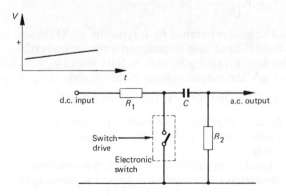

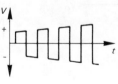

*Figure (C)1* Principle of chopper circuit (shunt switch shown)

MOSFETs can be used in the series or shunt mode [fig. (C)2] but are normally combined using a series-shunt configuration [fig. (C)3].

Taking the shunt MOS chopper as an example, the magnitude of errors can be estimated as follows. An equivalent circuit for steady state conditions is as shown in fig. (C)4.

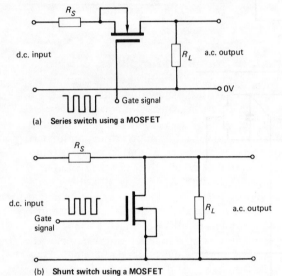

(a)  Series switch using a MOSFET

(b)  Shunt switch using a MOSFET

*Figure (C)2*  Switch configurations for choppers

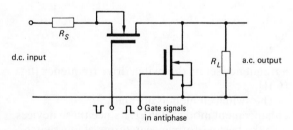

*Figure (C)3*  Series shunt configuration

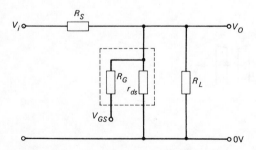

*Figure (C)4*  Equivalent circuit for the shunt chopper under steady state conditions

In the *ON condition* the output voltage should be zero. A finite value of $r_{ds\,(on)}$ will prevent this. The output voltage $V_o$ is given by:

$$V_o = V_i \left[ \dfrac{\dfrac{r_{ds}R_L}{r_{ds}+R_L}}{R_S + \dfrac{r_{ds}R_L}{r_{ds}+R_L}} \right]$$

Since $R_L$ is normally large compared with $r_{ds}$ [typical values of $r_{ds\,(on)}$ for a MOSFET lie between 100 to 300$\Omega$], the above formula can be simplified to:

$$V_o = V_i \left[ \dfrac{r_{ds}}{R_S + r_{ds}} \right]$$

If $R_S$ and $R_L$ are 100k$\Omega$ and $r_{ds} = 100\Omega$ the value of $V_o$ for $V_i$ of 1 mV is

$$V_o = 1 \times 10^{-3} \left[ \dfrac{100}{100.1} \times 10^{-3} \right] = 1\mu V$$

In other words the offset error with the switch on is 0.1%. This on state error voltage is directly proportional to the input voltage.

In the *OFF state* the large negative signal applied to the gate ($V_{GS}$) turns the MOSFET off and the output voltage rises. $V_{GS}$ is assumed to be $-10$ V.

$$V_o = V_i \left[ \dfrac{\dfrac{r_{ds}R_L}{r_{ds}+R_L}}{R_S + \dfrac{r_{ds}R_L}{r_{ds}+R_L}} \right] + V_{GS} \left[ \dfrac{\dfrac{r_{ds}R_L}{r_{ds}+R_L}}{R_G + \dfrac{r_{ds}R_L}{r_{ds}+R_L}} \right]$$

$r_{ds\,(off)}$ is used in this case and with a MOSFET has values of nearly $10^9\Omega$. Therefore the equation above simplifies to

$$V_o = V_i \left[ \dfrac{R_L}{R_S + R_L} \right] + V_{GS} \left[ \dfrac{R_L}{R_G + R_L} \right]$$

The gate resistance $R_G$ is typically $10^{12}\Omega$ (in a MOSFET the gate is insulated from the body of the device). Using $R_L = R_S = 100\,$k$\Omega$ and $V_i = 1\,$mV the output voltage in the off state is

$$V_o = 1 \times 10^{-3}[0.5] - 10\,[10^5/10^{12}] \simeq 0.5\,\text{mV}$$

The error is given by the second term of this equation and is 1 $\mu$V or 0.2% of the output voltage.

Apart from steady state errors, the switching action itself causes spikes to appear at the output.

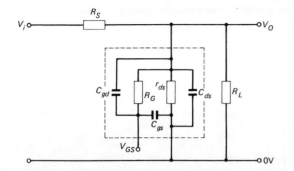

Figure (C)5   A.C. equivalent circuit of shunt MOSFET chopper

These result from feedthrough via $C_{gd}$ to $R_L$ [see fig. (C)5]. On each edge of the switch drive signal, a small spike appears at the output. The amplitude of these spikes can be reduced by using a switch with slow rise and fall times, a MOSFET with a low value of $C_{gd}$, or a series-shunt combination.

## Chopper Amplifiers

One of the problems of amplifying small d.c. signals (less than 1 mV) is the drift inherent in the d.c. amplifier itself. In an op-amp circuit, wired as a d.c. amplifier, the drift is due to changes in $V_{io}$ (input offset voltage) and $I_{io}$ (input offset current) with temperature. The offsets themselves can be nulled out, but the drifts with temperature may well mask the signal being amplified and will cause large errors.

An alternative method of d.c. amplification is to chop the d.c. signal to produce a square wave and then to amplify this a.c. signal with an a.c. amplifier [fig. (C)6]. The output of the amplifier is then synchronously demodulated so that the output of the system is an amplified version of the original d.c. input.

Since the amplifier is a.c. coupled, it cannot add any drift error to mask the d.c. signal being amplified. The bandwidth of the system is, however, limited by the chopper drive frequency.
    ▶ *Chopper*   ▶ *Differential amplifier*   ▶ *Op-amp*

## Chopper Stabilised Amplifier

This is a type of d.c. amplifier which combines low drift with wide bandwidth. The circuit, usually in i.c. form, consists of two amplifiers. The main amplifier is a directly coupled op-amp [fig. (C)7] connected as an inverting amplifier with voltage gain set by $R_1$ and $R_2$. Any drift at the input of this circuit (from point x) is fed to a chopper amplifier. The bandwidth of the chopper is restricted so that it is only effective at frequencies associated with drift, i.e. less than 0.05 Hz, and its output is connected to the non-inverting input of the main op-amp. In this way any drift is automatically nulled out. The bandwidth of the main circuit is not limited by the chopper drive frequency since the chopper amplifier has the sole task of dealing with drift. Thus, wide bandwidth coupled with low drift performance is achieved. Typical bandwidths might be 10 MHz with drift lower than 1 μV/°C.
    ▶ *Chopper*   ▶ *Chopper amplifier*   ▶ *Op-amp*

*Figure (C)6*   Chopper amplifier

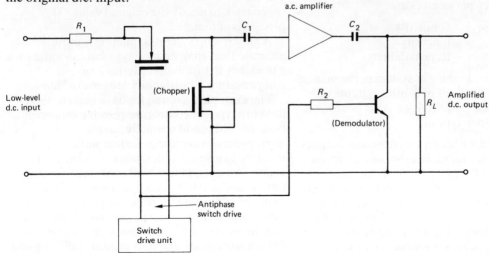

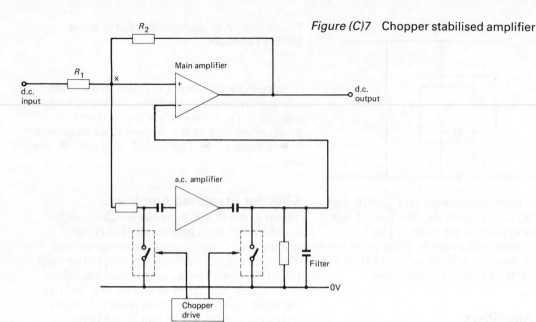

*Figure (C)7  Chopper stabilised amplifier*

$R_2$

$R_1$

d.c. input

x

Main amplifier

+

−

d.c. output

a.c. amplifier

Filter

0V

Chopper drive

## Control Systems

Any control system consists of a group of elements or sub-systems (amplifiers, comparators, transducers and power devices) connected together so that a controlling action can be carried out on an output variable. This output variable might be position, flow rate, light level, temperature and so on; with the overall action of the system being to set the required output according to a command input and then to maintain it constant.

Certain parameters will obviously be of importance in defining the performance of a system. These key parameters are:

Speed of response    Sensitivity
Accuracy                    Linearity
Stability                     Repeatability

However before looking at system performance the various classifications of control systems will be given. The most basic division is between *open loop* and *closed loop* types.

In an OPEN LOOP control system, the output is set by a reference input, but the control action is independent of the effect produced at the output. Open loop systems have no feedback. An example of an open loop system for speed control is shown in fig. (C)8. The command input set by the input potentiometer is amplified by the rest of the system and drives the motor at a fixed

speed. If increased friction caused the load to be slowed down, the open loop system would be unable to adjust the drive to the motor in order to compensate.

The system can be converted to CLOSED LOOP by using a transducer (sensor) to sense the speed of the load and to feed back this signal for comparison with the input reference level [fig. (C)9]. The feedback path (negative) closes the loop and enables the system to respond to changing conditions at the output. If the output speed falls, the feedback signal decreases, causing an increase in the error signal to the controller. The motor is then driven harder to bring the load speed back to nearly the original value. This process is automatic.

A closed loop system is inherently more accurate than an open loop type but has more of a tendency to instability. Methods of compensating for instability are covered later.

The examples given so far have been analog, but both types of control are possible using digital systems. These will normally have a microprocessor with appropriate software in memory to give great flexibility over the control action. An open loop digital controller will use output devices such as stepper motors which can easily be controlled for speed or position by the rate or number of digital pulses outputted by the system. For closed loop control, the sensor signal (which is often analog) may require buffering and

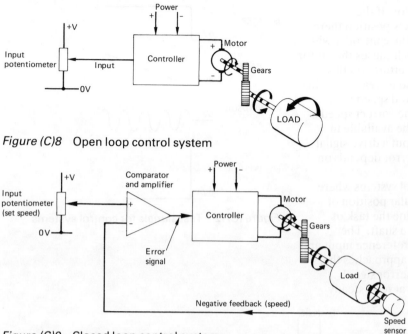

Figure (C)8   Open loop control system

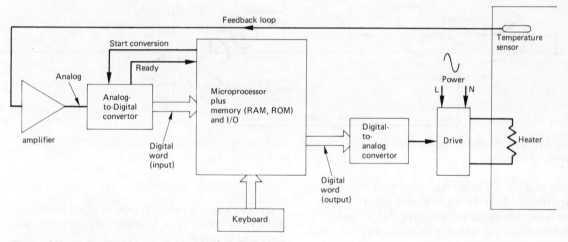

Figure (C)9   Closed loop control system

Figure (C)10   Typical block diagram of digital
closed loop control system

conversion via an ADC before being suitable as
an input to the microprocessor system. A typical
digital system block diagram for an oven
temperature controller (closed loop) is shown in
fig. (C)10. The stability problem for this sort of
system can be worse than the analog type
because, during the time between samples of the
ADC, the system switches from closed loop to
open loop. If the sample rate is too low, the
output will easily overshoot the desired value and
may well oscillate.

The two main classifications of closed loop
control systems are:

a) Regulators
b) Servomechanisms (servos)

*Regulators* include all those systems designed
to control speed, temperature, flow and so on.
With a regulator there must be a finite error signal
always present, even under steady state
conditions, in order to maintain the required
output value. The speed control system in fig.

(C)8 is an example of a regulator. If the potentiometer is moved to a new position there will initially be a large error. The controller will output an increased signal which causes the motor to accelerate to its new speed setting. As this happens, the sensor output also increases causing the error to reduce as the desired speed is approached. But even when the correct speed is reached, an error signal must be available in order for the controller to output a drive signal to the motor. The size of this error depends on the gain set in the controller.

*Servos* are positioned control systems where the output is the linear or angular position of some part of a machine. Imagine the task of setting the angular position of a shaft. The required position is set by the reference input and the shaft is then rotated. As it approaches the final steady state position, the error signal reduces, and when the shaft is at rest (steady

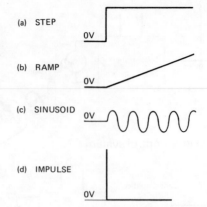

Figure (C)12   Test signals for control systems

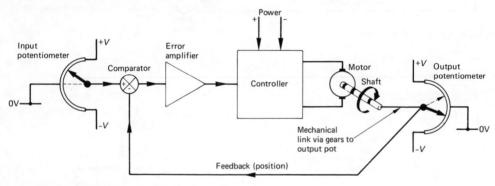

Figure (C)11   Position control system (servomechanism)

state) and ideally in perfect alignment with the required position the error is zero [fig. (C)11]. The basic difference between regulators and servomechanisms is that with regulators there will be a steady state error signal.

The basic position control system shown in fig. (C)11 can be altered using different gears and drives to give linear movement, and systems can be combined to give XY or XYZ position control. Part of the requirement for such systems is fast response together with high sensitivity and accuracy. To achieve these desired characteristics the gain of the error amplifier and controller must be set to a high value. This naturally can lead to instability. Additional stabilising circuits must then be added to the system.

The performance of the system can be checked using standard test signals [fig. (C)12]:

 a)  *The step*—a rapid change of input from one fixed value to another. Used for testing response time and accuracy.

 b)  *The ramp*—a linear rate of change of the input. Used for testing the linearity and lag.

 c)  *The sinusoid*—a sine wave input. Used for testing the bandwidth.

 d)  *The impulse*—a short-duration pulse. Used for transient testing.

Imagine a step input applied to a system. A typical response might be as shown in fig. (C)13a. Here the output overshoots the required value and takes a few oscillations before finally settling down. In fact there are four types of response that any system can give to a step input. These are [fig. (C)13b]:

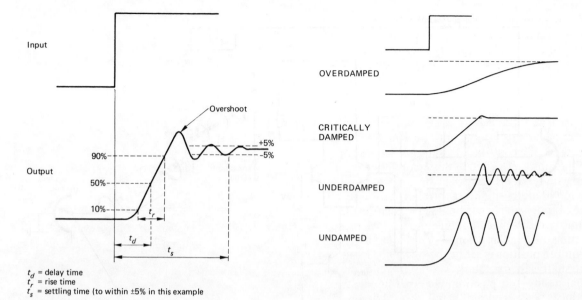

Input

Output

90%--
50%--
10%--

Overshoot

+5%
-5%

$t_r$

$t_d$

$t_s$

$t_d$ = delay time
$t_r$ = rise time
$t_s$ = settling time (to within ±5% in this example)

*Figure (C)13a*   Typical response of a control system (servo) to a step input

OVERDAMPED

CRITICALLY DAMPED

UNDERDAMPED

UNDAMPED

*Figure (C)13b*   Types of response that a system could give to a step input

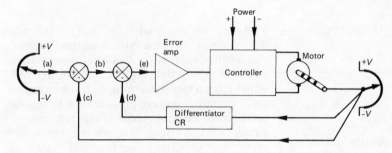

Power
+        −

+V

(a)        (b)        (e)

Error amp

Motor

Controller

+V

−V

(c)        (d)

Differentiator CR

−V

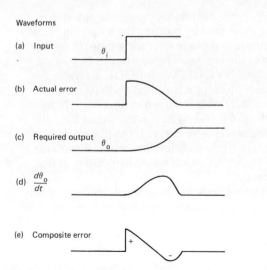

Waveforms

(a)   Input
$\theta_i$

(b)   Actual error

(c)   Required output
$\theta_o$

(d)   $\dfrac{d\theta_o}{dt}$

(e)   Composite error
+
−

*Figure (C)14a*   System block diagram using output derivative feedback for damping

*a*)  Overdamped—a sluggish response with poor accuracy.

*b*)  Critically damped—fast response with good accuracy.

*c*)  Underdamped—fast response but with a large overshoot and damped oscillations.

*d*)  Undamped—oscillatory and unstable.

*Instability* occurs in a system because of the high gain necessary for accurate control and because of lag networks within the system. Since the feedback path is negative, any further lag networks (mechanical or electrical) tend to make the feedback positive. If the frequency at which this occurs is such that the loop gain is greater than unity, then oscillations will set up in the system's output.

One standard method of stabilising a control system is to feed back an additional signal which is proportional to the rate at which the output is changing [fig. (C)14*a*]. This signal is then mixed

265

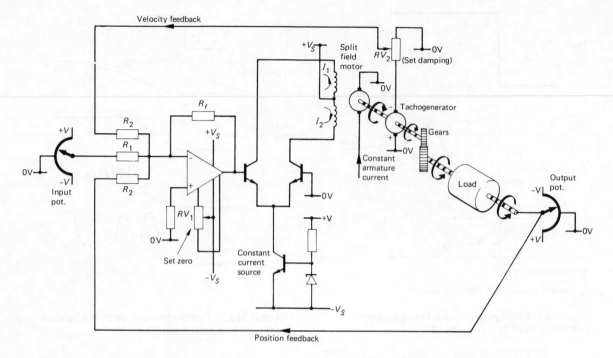

*Figure (C)14b*  Circuit diagram of servo system
using stabilising feedback

*Figure (C)15*  Nyquist plot for a control system
which is conditionally stable

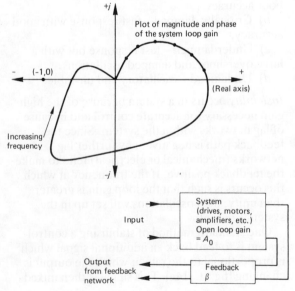

with the error and retards or "brakes" the output
if it moves at too fast a rate. A method such as
this is called *velocity feedback* or *output derivative
feedback*. The stabilising signal can be derived
either from a tachogenerator (a transducer that
gives an output voltage proportional to angular
rotation) or from an electrical differentiator
circuit (CR). A circuit diagram of a simple servo
system that uses stabilising feedback is shown in
fig. (C)14*b*.

The degree of stability present in any system
can be estimated by plotting the loop gain
magnitude and phase over the whole frequency
range of the system. The results can be plotted
either as a NYQUIST or a BODE diagram. A
typical Nyquist plot for a system that is
conditional stable is shown in fig. (C)15. Note
that the point $(-1,0)$ lies on the left of the
frequency response curve. If the plot enclosed the
point $(-1,0)$ then oscillations would occur at the
output since feedback would be positive at that
frequency.

From this, indications of the degree of stability
are:

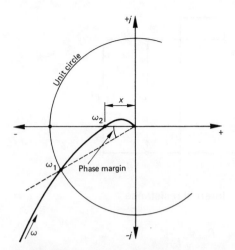

*Figure (C)16*   Gain and phase margin in a closed loop system

a) *Phase Margin*: specified at $\omega_1$ the *gain crossover frequency*, the frequency at which the loop gain equals 1. Phase margin is the angle in degrees by which the phase lag can be increased without a change in gain to give a total phase shift of 180°.

b) *Gain Margin*: specified at $\omega_2$ the *phase crossover frequency*, the frequency at which the phase shift is 180°. Gain margin is the amount by which the loop gain can be increased without change in phase shift so that the gain equals 1.

These points are shown in fig. (C)16, where

Phase margin $= 35°$
Gain margin $= 1/0.3 = 10.46$ dB
(Gain margin is normally expressed in dB.
$G_m = 20 \log 1/x$.)
► *Sample and hold* ► *Transducers*

## Inductor/Inductance

An inductor (abbreviation of inductive resistor) is a passive device which consists of a coil of wire wound on an insulating former (fig. (I)1). Inductors have a variety of uses that include tuned circuits and tuned amplifiers, some oscillators (see Hartley and Colpitts Oscillators) and filters. The coil or winding is used to introduce inductance into a circuit, which is the property of a component to oppose a change in current flow. As the current flowing in a coil changes, the magnetic flux set up by the current also changes and this induces an opposing e.m.f. in the coil.

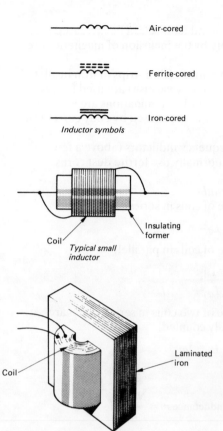

*Inductor symbols*

*Typical small inductor*

*A.F. Inductor (choke)*

*Figure (I)1*   The inductor

Assume the current in a coil changes uniformly by $\Delta I$ amps in $\Delta t$ seconds. Then the induced e.m.f. is given by

$$e = -\text{ constant} \times \frac{\Delta I}{\Delta t}$$

The constant is the inductance of the coil (symbol of inductance is $L$).

$$\therefore e = -L \frac{dI}{dt}$$

The unit of the inductance is the Henry (H). For a coil the inductance value depends upon two main factors:

a) the number of turns
b) the material of the core.

For an air-cored inductor (simply a coil of $n$ turns of wire wound on an insulating former),

$$L \propto n^2$$

The value of inductance can be increased substantially by the inclusion of magnetic core material.

Iron laminations (thin strips of iron insulated from each other by varnish) are used for low-frequency coils. The laminations are used to reduce power losses caused by eddy currents in the core.

High-frequency inductors (above a few kilohertz) normally use ferrite dust cores.

*Useful formula*
Inductance of coils in series:

$$L = L_1 + L_2 + \ldots + L_n$$

Inductance of coils in parallel:

$$\frac{1}{L} = \frac{1}{L_1} + \frac{1}{L_2} + \ldots + \frac{1}{L_n}$$

Inductance of two coils in series which are magnetically coupled:

$$L = L_1 + L_2 \pm 2m$$

Here

Mutual Inductance $m = \dfrac{N_2\Phi_2}{I_1}$ H

where $N_2\Phi_2$ is the flux linkages with a second coil when a current $I_1$ flows in the first coil.

Coupling coefficient $k = \dfrac{m}{\sqrt{(L_1 L_2)}}$

Inductive reactance $= X_L = \omega L = 2\pi f L \ \Omega$

Current in a pure inductor lags voltage across the inductor by 90°.

For a coil which possesses some resistance:

Impedance $Z = R + jX_L$

$|Z| = \sqrt{[R^2 + X_L^2]}$

Phase angle $= \phi = \tan^{-1}\dfrac{X_L}{R}$

## Internal Resistance

Any source of e.m.f. such as a battery, power supply, or generator will have some finite resistance which limits the power that the circuit can supply. When the source is loaded, the current taken causes a voltage drop across the internal resistance resulting in a lower output voltage [fig. (I)2]:

$$V_L = E - I_L R_{int}$$

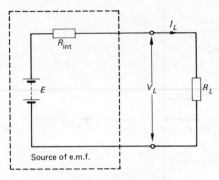

*Figure (I)2* Internal resistance

By definition $\quad R_{int} = \dfrac{V_{oc}}{I_{sc}}$

where $V_{oc}$ = open circuit output voltage ($R_L$ removed)

$I_{sc}$ = short circuit output current ($R_L$ made a short circuit).

However it is not practical to measure $R_{int}$ by placing a short circuit across the output terminals of a generator or battery. The internal resistance can be measured by loading the output with a known resistance $R_L$.

Then since $\quad V_L = V_{oc} - I_L R_{int}$

and $\quad I_L = V_L / R_L$

$$R_{int} = R_L\left(\frac{V_{oc} - V_L}{V_L}\right)$$

▶ *Network theorems*

## Network Theorems

There are often situations in electronic design, testing and fault diagnosis where network analysis of some kind is required. The following theorems are the most useful:

### Kirchhoff's Laws

**1**  At any junction in a circuit, the sum of the instantaneous (or phasor) values of the currents is zero [fig. (N)1].

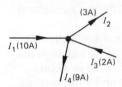

$I_1 + I_2 + I_3 + I_4 = 0$
In this case
$10 + (-3) + 2 + (-9) = 0$  *Figure (N)1* Kirchhoff's First Law

**2** In any closed network, the sum of the instantaneous (or phasor) values of the potential differences across the circuit elements is equal to the sum of the electromotive forces [fig. (N)2].

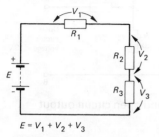

$$E = V_1 + V_2 + V_3$$

*Figure (N)2*  Kirchhoff's Second Law

*Example*  For the circuit of fig. (N)3 find the voltage across $R_2$.

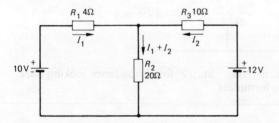

*Figure (N)3*  Circuit for illustration of Kirchhoff's Laws

For the first loop:

$4I_1 + 20(I_1 + I_2) = 10$
∴  $24I_1 + 20I_2 = 10$        (1)

For the second loop:

$10I_2 + 20(I_1 + I_2) = 12$
∴  $20I_1 + 30I_2 = 12$        (2)

Multiplying equation (1) by 1.5 gives

$36I_1 + 30I_2 = 15$        (3)

Subtracting equation (2) from (3) gives   $16I_1 = 3$

∴  $I_1 = 0.1875\text{ A}$

Substituting this value into equation (1) gives

$20I_2 = 10 - 4.5$
∴  $I_2 = 0.275\text{ A}$
∴  Current through $R_2 = I_1 + I_2 = 0.4625\text{ A}$
∴  Voltage across $R_2 = (I_1 + I_2)R_2 = 9.25\text{ V}$

**Superposition Theorem**
Useful for networks where there is more than one source of e.m.f. The effect of each e.m.f. is considered separately and then finally superimposed. The theorem states:

The current in each branch of a network is the sum of currents in that branch due to each e.m.f. acting alone, when all other e.m.f.s have been replaced by their internal resistances (or impedances).

*Example*  For the circuit of fig. (N)3 (as used for Kirchhoff example), find the voltage across $R_2$.

First replace the 12 V source with a short circuit; this gives fig. (N)4a.

$$\text{Current through } R_1 = \frac{10}{4 + 20//10} = 0.9376\text{ A}$$

∴  Current in $R_2$ due to 10 V e.m.f. acting alone
   = 0.3125 A

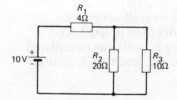

*Figure (N)4a*  First modification for analysis using superposition theorem

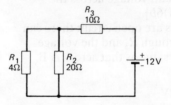

*Figure (N)4b*  Second modification for analysis using superposition theorem

Then replace the 10 V source with a short circuit. This gives fig. (N)4b.

$$\text{Current through } R_3 = \frac{12}{10 + 20//4} = 0.9\text{ A}$$

∴  Current in $R_2$ due to 12 V e.m.f. acting alone
   = 0.15 A
∴  Total current in $R_2$ from both sources of e.m.f.
   = 0.3125 + 0.15 = 0.4625 A
∴  Voltage across $R_2 = 9.25\text{ V}$

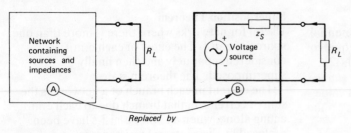

*Figure (N)5* Thévenin's theorem

## Thévenin's theorem

Any two-terminal network containing resistances (or impedances) and voltage sources and/or current sources can be replaced by a single voltage source in series with a single resistor (or impedance). The e.m.f. of the voltage source is the open circuit e.m.f. at the network terminals, and the series resistance (or impedance) is the resistance (or impedance) between the network terminals when all sources are replaced by their internal impedances [fig. (N)5].

This very useful theorem allows any one component in a network to be isolated. The whole of the remaining network can be replaced by a single source of e.m.f. and a single resistor (or impedance).

*Example*  For the circuit of fig. (N)6*a* find the current flowing in $R_L$.

First, disconnect $R_L$ from the network and calculate the open circuit voltage across the terminals A B [fig. (N)6*b*].

Since A B terminals are open circuit there can be no current flow through $R_3$ and the voltage across $R_2$ must be the same as that across A B.

$$\therefore \quad V_{AB} = \frac{V_{in} R_2}{R_1 + R_2} = 18\,\text{V}$$

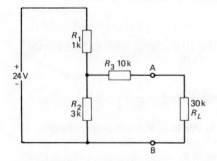

*Figure (N)6a*  Circuit for analysis using Thévenin's theorem

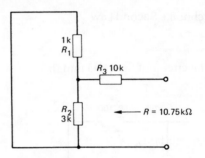

*Figure (N)6b*  Step 1: find open circuit output voltage

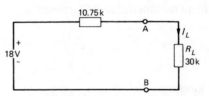

*Figure (N)6c*  Step 2: find impedance looking back into terminals

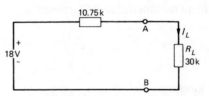

*Figure (N)6d*  Final Thévenin equivalent circuit

This voltage will be the value of the source of e.m.f. in the Thévenin equivalent circuit.

Next, imagine the 24 V d.c. source short circuited and determine the resistance looking back into terminals A B (fig. (N)6*c*):

$$R_T = 10\text{k} + 3\text{k}//1\text{k} = 10.75\text{k}$$

The Thévenin equivalent circuit can now be drawn. With $R_L$ connected across terminals A B [fig. (N)6*d*],

$$\text{Voltage across } R_L = \frac{18 \times 30}{40.75} = 13.251\,\text{V}$$

$$\therefore \quad \underline{I_L = 0.442\,\text{mA}}$$

## Norton's Theorem

Any two-terminal network containing resistance (or impedances) and voltage sources and/or current sources can be replaced by a single current source in parallel with a single resistance (or impedance). The current source has a value equal to the short circuit output current from the network, and the parallel resistance (or impedance) is the resistance (or impedance) seen looking back into the network when all sources are replaced by their internal impedances [fig. (N)7].

This theorem is the dual of Thévenin's theorem.

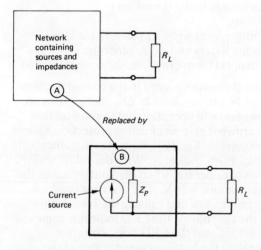

Figure (N)7   Illustration of Norton's theorem

*Example*   In fig. (N)8a (same as circuit for Thévenin's example), find the current in the 30 kΩ load.

First, find the short circuit output current [fig. (N)8b]. Here the 30 kΩ load has been replaced by a short circuit.

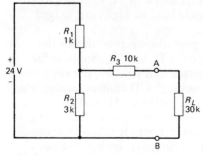

Figure (N)8a   Circuit for analysis using Norton's theorem

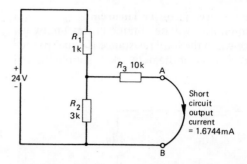

Figure (N)8b   Step 1: find short circuit output current

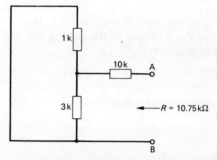

Figure (N)8c   Step 2: find value of parallel resistance

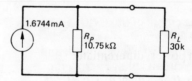

Figure (N)8d   Final Norton equivalent circuit

$$I_1 = \frac{24}{1k + 3k//10k} = 7.2558\,\text{mA}$$

$$\therefore\quad I_{sc} = 1.6744\,\text{mA}$$

This is the value of the Norton current generator.

Next, replace the voltage source with a short circuit and determine the resistance looking back into terminals A B [fig. (N)8c]:

$$R_n = 10.75k$$

The Norton equivalent circuit can now be drawn [fig. (N)8d].

When the 30 kΩ load is connected the current flowing in the load is

$$I_L = \frac{I_N \times R_n//R_L}{R_L} = \underline{0.442\,\text{mA}}$$

271

## Maximum Power Transfer Theorem

Maximum power will be obtained from a network or source when the load resistance is equal to the output resistance of the network or source [fig. (N)9].

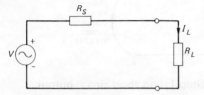

*Figure (N)9*   Maximum power transfer

In general, for maximum power transfer, the load impedance $Z_L/\theta$ should be the conjugate of the source impedance $\overline{Z_S/\theta}$, i.e.

$$Z_L/\theta = \overline{Z_S/\theta}$$

*Proof of theorem using resistors*
From fig. (N)9,

$$I_L = \frac{V}{R_S + R_L}$$

Power in the load $= I_L{}^2 R_L$

$$\therefore \quad P_L = \frac{V^2 R_L}{(R_S + R_L)^2}$$

To find maximum power, differentiate:

$$\therefore \quad \frac{dP_L}{dR_L} = V^2 \left[ \frac{(R_S + R_L)^2 - R_L \cdot 2(R_S + R_L)}{(R_S + R_L)^4} \right]$$

For maximum $\dfrac{dP_L}{dR_L} = 0$

and the expression can only be zero if

$$(R_S + R_L)^2 = 2R_L (R_S + R_L)$$

i.e. when $R_L = R_S$.
    ► *Equivalent circuits*  ► *Internal resistance*

## Protection Circuits

Some form of protection is essential in most electronic systems, either to limit the current (or voltage) supplied to components and therefore prevent damage by overheating, or to ensure that supplies or clock signals are maintained in the event of a power failure.

Any electronic device has maximum permissible ratings for current, voltage and power. A transistor for example, operating as a switch in common emitter mode, will have maximum limits specified for the following main values: $I_{c\,max}$, $V_{CE\,max}$ and $P_{tot}$. Exceeding any of these for more than a few milliseconds will invariably lead to the transistor's failure.

Active devices like transistors, FETs and i.c.s are particularly vulnerable to overloads, whereas components like resistors can withstand overloads for relatively long periods. This means that the amount and type of protection required depends a lot on the circuit and its application.

Protection circuits include the following:

Fuses
Current limits
Overvoltage limits (crowbars)
Isolators
Standby power supplies (battery back-up)
Standby timers and clock generators
Interrupts (for microprocessor-based systems).

The first three are covered in the named sections (► *Fuse* ► *Current limit* ► *Crowbar*). Fuses are relatively slow in operation but in combination with a crowbar give an effective protection against an overvoltage. Fig. (P)1 shows this together with a "blown-fuse" indicator. With the values shown, the crowbar operates when the voltage across the load rises above $+10\,V$. The thyristor conducts, taking point x low and causing the fuse to blow. Once the fuse blows, the LED indicator comes on, provided that the LED "on" current is greater than the holding current value of the thyristor.

Power output stages are particularly susceptible to damage by voltage surges at the output and by overcurrent caused by a short circuit. Fig (P)2 shows typical protection circuits used for a class B (or AB) complementary power output. Diodes $D_3$ and $D_4$ are used to protect the darlingtons against voltage surges caused by the back e.m.f. generated by the inductive load: they conduct if the voltage at $Tr_3$ emitter or $Tr_6$ emitter exceeds $0.5\,V$. They should both be high current type diodes.

Overcurrent protection is provided by the current limit circuits $Tr_1$ and $Tr_4$. Suppose the output were shorted to zero volts; the voltage developed across the $0.47\Omega$ emitter resistor on positive half cycles will force $Tr_1$ and $D_1$ to conduct, and base current drive to the darlington of $Tr_2$ and $Tr_3$ will be restricted. With the values shown, the output current is limited to a peak value of $5.5\,A$. Similarly $Tr_4$ will limit the current through $Tr_5$ and $Tr_6$ on negative half cycles.

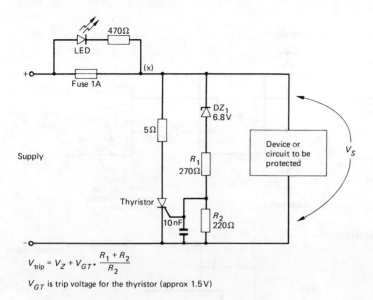

$$V_{trip} = V_Z + V_{GT} \cdot \frac{R_1 + R_2}{R_2}$$

$V_{GT}$ is trip voltage for the thyristor (approx 1.5 V)

*Figure (P)1* **Protection against an overvoltage**

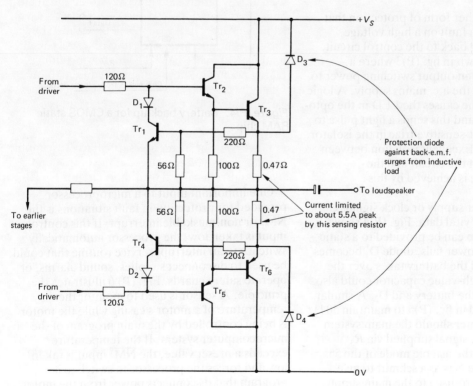

*Figure (P)2* **Protection circuits for power output stage**

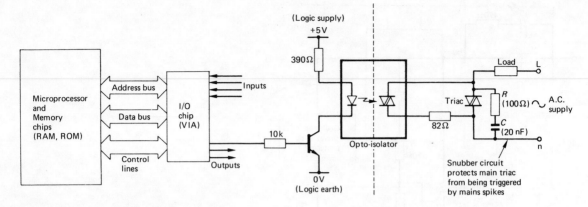

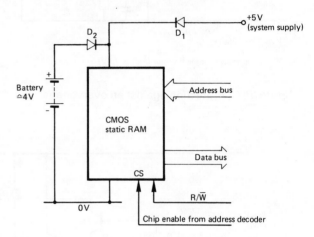

*Figure (P)3*  Illustration of isolation protection

*Figure (P)4*  Battery back-up for a CMOS static RAM chip

*Isolators* are another form of protection that effectively prevent a fault on a high voltage output from feeding back to the control circuit. The principle is shown in fig. (P)3 where a microprocessor has an output switching power to a load connected to the a.c. mains supply. A logic 1 on the I/O chip line causes the LED in the opto-isolator to conduct and this sends a light pulse to switch both the light-sensitive triac in the isolator and the main triac. Excellent isolation between the mains portion of the circuit and the microprocessor side is achieved by this arrangement.

Failure of a power supply or clock signals can often lead to loss of vital data. Fig. (P)4 shows how battery back-up can be provided to a static RAM chip. If the power fails, diode $D_1$ becomes reversed biased and the battery takes over the supply via $D_2$. A high-value capacitor could also be used in place of the battery and $D_2$. A similar arrangement is used in fig. (P)5 to maintain 50 Hz clock signals to a timer should the main system supply fail. The a.c. signal supplied via $R_3$ normally overrides the astable mode of the 555 chip and it therefore acts as a schmitt to give timing signals synchronised to the main supply. If the power fails, the battery back-up comes into operation and the 555 then operates as a square wave generator to give approximately 50 Hz output signals.

One important input to a microprocessor provided for protection in fault situations is the $\overline{\text{NMI}}$ or non-maskable interrupt. If this control input is taken low, the processor automatically switches to an interrupt service routine that could be used to disconnect supplies, sound alarms, or operate safety guards. Fig. (P) 6 illustrates the principle. A sensor is used to monitor the temperature of a motor's casing while the motor is being controlled by the main program of the microcomputer system. If the temperature exceeds a preset value, the $\overline{\text{NMI}}$ input is taken low and forces the processor to switch to a program that disconnects power from the motor and then operates a flashing alarm.

▶ *Current limit*  ▶ *Crowbar*
▶ *Fuse*  ▶ *Opto-devices*
▶ *Relays*  ▶ *Suppression*

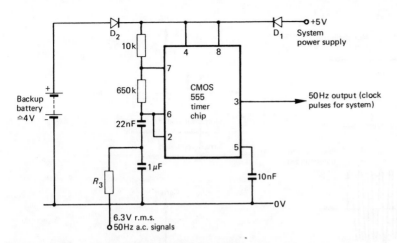

*Figure (P)5*  Protection of timing signals

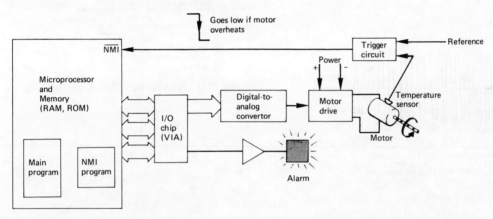

*Figure (P)6*  Using the nonmaskable interrupt of a microprocessor for protection

## Radio Systems and Circuits

Radio waves—electromagnetic waves that can travel through space—range in frequency from about 50 kHz up to 300 GHz. The wavelength is given by

$$\lambda = c/f$$

where $c$ = velocity of electromagnetic wave in free space $300 \times 10^6$ m/s.

Thus a 3 MHz radio wave has a wavelength of 100 m.

The various frequency classifications used in communication engineering are given on page 125.

The frequency used at the transmitter and radiated from the aerial is the *carrier wave frequency*. The signal that is superimposed onto this carrier by a process called *modulation* is the information.

The most common forms of radio communication use either *amplitude modulation* (a.m.) or *frequency modulation* (f.m.) [see fig. (R)1]. The block diagrams of a basic a.m. transmitter and receiver are shown in fig. (R)2. The r.f. oscillator which gives the carrier frequency is crystal-controlled. The audio signal, the information to be transmitted, is then superimposed onto the carrier by the modulator circuit. A final r.f. power amplifier (tuned class B) then feeds the modulated carrier to the aerial system. The simplest type of receiver, called a tuned radio frequency receiver (t.r.f. or "straight" set) consists of a receiving aerial and an *LC* resonant circuit tuned to the carrier wave frequency. The received carrier signal together

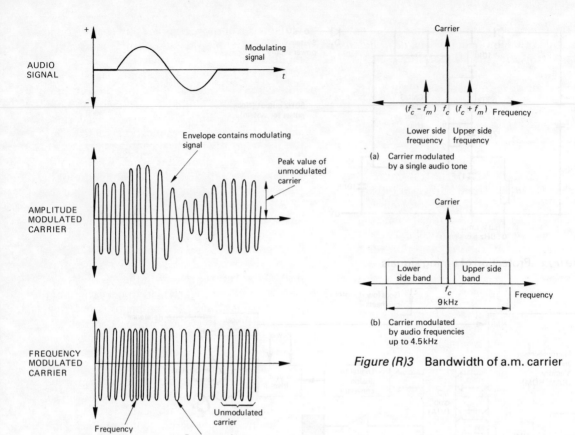

AUDIO
SIGNAL

Modulating
signal

Carrier

$(f_c - f_m)$  $f_c$  $(f_c + f_m)$  Frequency

Lower side   Upper side
frequency    frequency

(a)   Carrier modulated
      by a single audio tone

Envelope contains modulating
signal

Peak value of
unmodulated
carrier

AMPLITUDE
MODULATED
CARRIER

Carrier

Lower
side band

Upper side
band

$f_c$
9 kHz

Frequency

(b)   Carrier modulated
      by audio frequencies
      up to 4.5 kHz

FREQUENCY
MODULATED
CARRIER

Unmodulated
carrier

Frequency
of carrier
increased

Frequency of
carrier decreased

Figure (R)3   Bandwidth of a.m. carrier

Figure (R)1

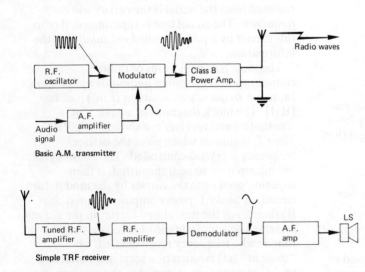

Radio waves

R.F.
oscillator

Modulator

Class B
Power Amp.

Audio
signal

A.F.
amplifier

Basic A.M. transmitter

Tuned R.F.
amplifier

R.F.
amplifier

Demodulator

A.F.
amp

LS

Simple TRF receiver

Figure (R)2

with the sidebands is amplified in a tuned r.f. stage and then demodulated. The *demodulator* or *detector* strips the audio signal from the r.f. carrier.

In an a.m. wave the signals present will be

the carrier $f_c$
upper sideband $f_c + f_m$
lower sideband $f_c - f_m$

When $f_m$ is audio signal in the range 50 Hz to 4.5 kHz, the bandwidth of the transmitted wave takes up 9 kHz [fig. (R)3]. However only one sideband is required to enable information to be transmitted and received. Transmission of the carrier itself represents a power loss and efficiency is improved if the carrier power is reduced. The various types of transmission for a.m. are:

SSBRC    Single sideband with reduced carrier. A limited amount of carrier power makes receiver tuning easier.
SSBSC    Single sideband with suppressed carrier. No carrier power is transmitted and it must be reinserted at the receiver. With some systems a pilot signal may be transmitted to assist tuning.
DSBSC    Double sideband suppressed carrier.

The block diagram of an SSBSC transmitter is shown in fig. (R)4. A balanced modulator (in inset) is used, with the diode bridge circuit suppressing the carrier frequency. Because of the requirement in SSBSC systems to reintroduce the carrier at the receiver, which makes the design

of the receiver more complex and costly, sound broadcast systems do not use single-sideband modulation.

The t.r.f. receiver mentioned earlier has severe drawbacks and is not suitable for a.m. reception over more than a limited frequency band. For reasonable reception, especially for weak signals, several stages of r.f. amplification would be required and all of these stages would have to be tuned and ganged together. Thus, when another input signal is selected, all r.f. amplifier must be changed by exactly the same amount so that they are all tuned to the new frequency.

The difficulties of alignment and tracking over a wide frequency range are almost insurmountable. This problem does not exist in the *superheterodyne* receiver which is therefore the standard arrangement. In the superheterodyne [fig. (R)5 for block diagram] the problem of retuning is overcome by converting the wanted signal frequency into a constant frequency signal (called the intermediate frequency i.f.). Basically the wanted signal from the aerial is mixed with the signal from an oscillator (called the local oscillator) to give the intermediate frequency. This new signal will still have all the modulation present.

When two signals of different frequencies are mixed together in a nonlinear device or circuit, new frequencies will be produced. Among these are the sum $(f_0 + f_c)$ and the difference frequencies $(f_0 - f_c)$. By using a tuned circuit, the difference $(f_0 - f_c)$ is selected out from the mixer to give the i.f. Typically in a.m. broadcast receivers, the i.f. is chosen to be 470 kHz.

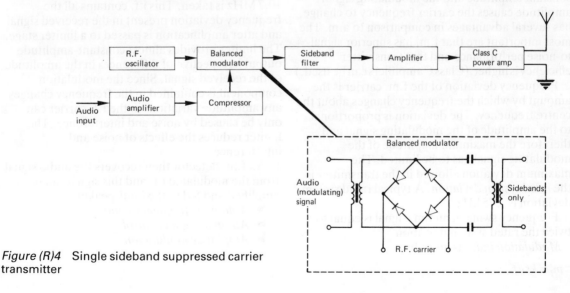

*Figure (R)4*    Single sideband suppressed carrier transmitter

Suppose that the required signal at the aerial has a frequency of 1 MHz; then to give an i.f. of 470 kHz, the local oscillator must be running at 1470 kHz. If the receiver is then retuned to a signal with frequency of say 600 kHz, the local oscillator frequency must then be changed to 1070 kHz. The local oscillator and r.f. tuned circuit at the aerial are ganged together so that the resulting difference frequency is always 470 kHz (the i.f. may in fact be from 450 kHz to 470 kHz depending on the design). The advantage of this system is that the i.f. amplifiers, which provide most of the selectivity and gain for the superheterodyne are fixed and do not have to be retuned. One disadvantage is that there will always be a second signal frequency (unwanted) which, when mixed with the local oscillator signal, produces the i.f. In the last example, this second channel or image frequency is 1540 kHz since:

$$f_0 - f_c = (1070 - 600) \text{ kHz} = 470 \text{ kHz}$$
$$\text{and} \quad f_c - f_0 = (1540 - 1070) \text{ kHz} = 470 \text{ kHz}.$$

The tuned r.f. stage at the aerial has the task of eliminating any second channel signals.

The typical circuit of a combined mixer/oscillator for an a.m. radio receiver is shown in fig. (R)6. This will be followed by two stages of i.f. amplification. The signal is then demodulated and AGC taken off to be fed back leaving the audio signal. The a.f. amplifier is usually a standard class AB complementary push-pull which then delivers audio power to the loudspeaker.

Frequency modulation, where the carrier has constant amplitude and the modulating signal amplitude causes the carrier frequency to change, has several advantages in comparison to a.m. The most important are that f.m. has superior signal-to-noise performance and that transmitter efficiency is higher. (Class C amplifiers can be used.)

Frequency deviation of the f.m. carrier is the amount by which the frequency changes about the centre frequency. The deviation is proportional to the amplitude of the modulating signal and therefore the maximum amplitude of the modulating signal has to be limited. The maximum deviation allowed in the transmitter is the *rated system deviation*. A typical rated system deviation is ±75 kHz.

Frequency swing of an f.m. signal is equal to twice the rated system deviation.

*Modulation index* is given by

$$m_f = k f_d / f_m$$

where $k$

$$= \frac{\text{modulating signal voltage}}{\text{maximum allowable modulating signal voltage}}$$

$$f_d = \text{rated system deviation}$$
$$f_m = \text{modulation frequency.}$$

and *deviation ratio* by

$$D = f_d / f_{m(max)}$$

For an f.m. signal, the sidebands produced are as shown in fig. (R)7. This is for a 10 MHz carrier modulated by a 5 kHz audio tone. The first-order side frequencies are $(f_c \pm f_m)$, the second-order are $(f_c \pm 2f_m)$, the third-order $(f_c \pm 3f_m)$ and so on.

Thus the upper side frequencies are:

1.0005 MHz
1.0010 MHz
1.0015 MHz
1.0020 MHz
and so on

The f.m. system inherently requires a wider bandwidth than a.m.:

$$\text{Bandwidth} = 2(k f_c + f_m)$$

Typically with a rated system deviation of ±75 kHz, a maximum modulating signal of 15 kHz and deviation ratio of 5, the required bandwidth is 180 kHz.

The block diagram of a basic f.m. receiver is shown in fig. (R)8. Again the superheterodyne principle is used. The r.f. tuner, with a bandwidth of about 200 kHz, is used to select the required v.h.f. signal and from the mixer stage an i.f. of 10.7 MHz is taken. This i.f. contains all the frequency deviation present in the received signal and after amplification is passed to a limiter stage. The limiter provides almost constant-amplitude output irrespective of differences in the amplitude of the received signal. Since the modulation component is contained in the frequency changes, any amplitude variations in the r.f. carrier can only be caused by noise and interference. The limiter reduces the effects of noise and interference.

An f.m. detector then recovers the audio signal from the modulated i.f. and this signal can be amplified and fed to the loudspeaker.

▶ *Automatic frequency control*
▶ *Automatic gain control*
▶ *Amplitude modulation*

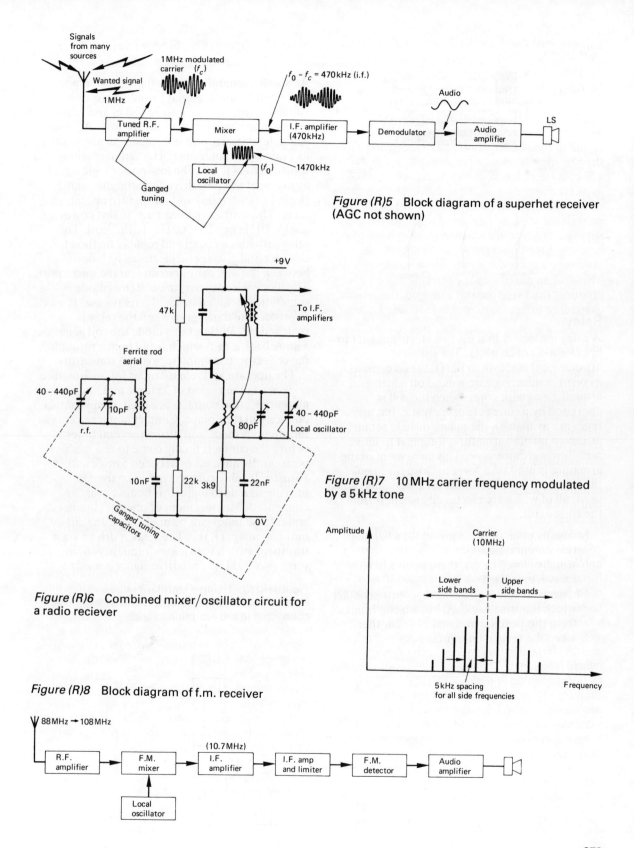

Figure (R)5  Block diagram of a superhet receiver (AGC not shown)

Figure (R)6  Combined mixer/oscillator circuit for a radio reciever

Figure (R)7  10 MHz carrier frequency modulated by a 5 kHz tone

Figure (R)8  Block diagram of f.m. receiver

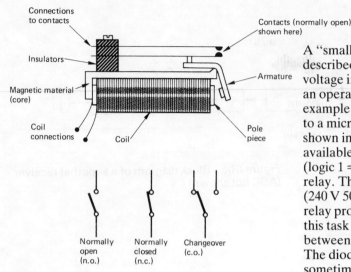

Connections to contacts

Insulators

Magnetic material (core)

Coil connections

Coil

Contacts (normally open) shown here)

Armature

Pole piece

Normally open (n.o.)

Normally closed (n.c.)

Changeover (c.o.)

## Relay

A relay is a device that uses an electromagnet to operate a set of contacts. The simplest arrangement is shown in fig. (R)9 and consists of a coil of conducting wire wound on a former around a magnetic core. When the coil is energised by a current (normally d.c. but a.c. types are available), the magnetic field set up attracts a pivoted armature, forcing it to move rapidly towards the core. This movement of the armature is used via a lever to close (or open) the contacts. Several contact arrangements can be used, all of which are electrically isolated from the coil circuit:

Normally open (n.o.): these close when the relay is energised.
Normally closed (n.c.): these open when the relay is energised.
Changeover (c.o.): these have a centre contact which is normally closed but which "breaks" from this position and makes to another when the relay is energised.

A "small" general-purpose relay of the type described (armature relay) would have a coil voltage in the range 5 V to 24 V d.c. and require an operating current of about 100 mA. A good example of its use is in interfacing an a.c. load to a microcomputer port. The arrangement is shown in fig. (R)10. The low-power logic signal available at the port drives a darlington switch (logic 1 = switch on) and this in turn operates the relay. The contacts connect a.c. mains power (240 V 50 Hz or 115 V 60 Hz) to the load. The relay provides a cheap and reliable method for this task and gives good electrical isolation between the a.c. mains circuit and the micro port. The diode wired across the coil (this diode is sometimes included inside the relay case) is used to protect the darlington when the relay is switched off. Without this diode the coil generates a large back e.m.f. which could punch through the collector/base junctions of the transistors.

The operate and release times for an armature relay are in the region of 15 msec and, unless mercury wetted contacts are used, some contact bounce will take place. Contact bounce occurs as the circuit is made and broken several times before the contacts finally come to rest in a new position. Bounce takes between 5 msec and 10 msec. If relay contacts are to be used to provide signals to logic, this bounce must be eliminated. Measurement of operate, bounce, and release times can be made using the circuit outlined in fig. (R)11. The relay is driven via a transistor switch with a low-frequency square wave (say 20 Hz). This drive signal is used to

*Figure (R)10*  Using a standard relay as an interface between a micro port and a load connected in the a.c. mains supply

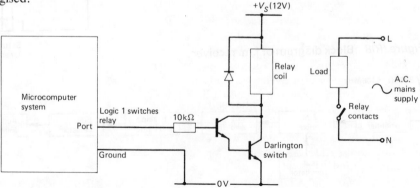

$+V_S$ (12V)

Relay coil

Load

L

A.C. mains supply

Microcomputer system

Port

Logic 1 switches relay

10kΩ

Darlington switch

Relay contacts

N

Ground

0V

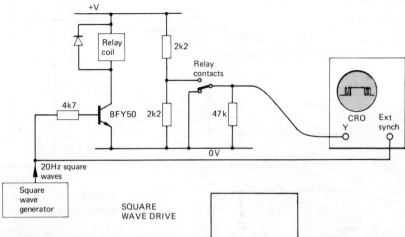

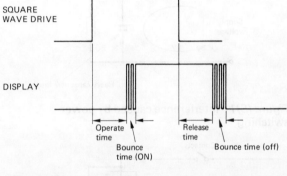

trigger the c.r.o. The contacts, changeover types, switch between 0 V and $\frac{1}{2}V_{CC}$ and the waveform is displayed on the c.r.o. as shown. Alternative switch connections can be used to display bounce only.

The life of a relay is typically $10^6$ to $10^7$ operations and depends to a great extent on the type of contact material used and the required switching action. Switching heavy inductive d.c. loads is the worst case since the back e.m.f. generated at switch-off will cause arcing at the contacts and consequent pitting and wear. Suppression circuits are then required.

A *latching relay* is one that, when forced to operate, remains on when the coil drive is removed. This bistable-type relay has two coils one for Set the other for Reset. A normal relay can be made to self-latch by wiring one of its own contacts in series with the supply; thus when the start switch (set) is made, the relay operates and its own contacts bypass the switch to the supply [fig. (R)12].

A *reed relay* [fig. (R)13] which is much faster in switching than an armature relay also uses the principle of magnetic attraction for its operation. The most basic reed relay is made from two ferromagnetic reeds encapsulated in a glass envelope. The energising coil is wound on a former and fits round this glass envelope. When the coil is energised, the magnetic field set up forces the reeds to flex towards each other and the contacts close. Operate and release times are typically 1 msec to 2 msec but again contact bounce at a fairly high frequency will occur. Both armature and reed relays can be obtained for p.c.b. mounting.

▶ *Interfacing* ▶ *Suppression*

*Figure (R)11* Circuit for measuring operate, bounce and release times

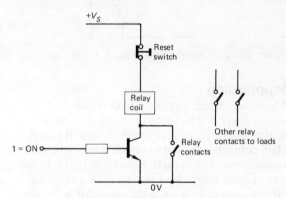

*Figure (R)12* Self-latching relay circuit

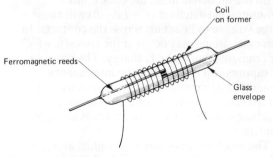

*Figure (R)13* Reed relay

281

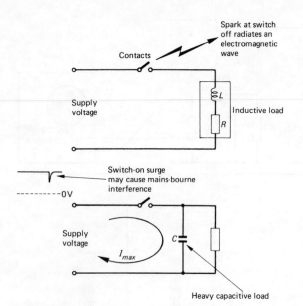

Figure (S)1 Interference caused by power switching

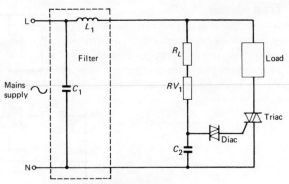

Figure (S)3 Using a filter to prevent switching spikes being fed into the mains supply

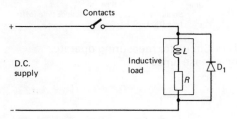

Figure (S)2a Suppression using a diode (d.c. only)

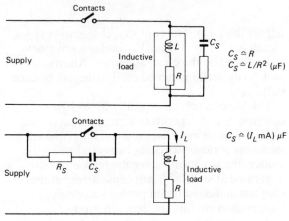

Figure (S)2b Suppression using an RC network

## Suppression

Whenever the current in a circuit is switched rapidly there will be some radiated or mains-bourne interference generated. A typical case is when contacts are used to switch power to either a capacitive or an inductive load [fig. (S)1]. In the case of a capacitive load there will be a surge of current at switch-on since the capacitor is uncharged; this may cause the contacts to weld. With the inductive load, the back e.m.f. generated at switch-off ($e = L di/dt$) will cause a large voltage to be set up across the contacts. In this case, arcing may occur at the contacts which will radiate a burst of r.f. energy. The purpose of suppression is to reduce the interference generated to a minimum. The main causes of interference are switch contacts, rotating machines, and thyristor/triac circuits using phase control.

The most common cause of contact arcing is when an inductive load is switched. There are

several methods of suppression that can be used. If the supply is d.c., a diode can be wired across the load [Fig. (S)2a]. When the switch is opened, the diode keeps current flowing and prevents a large back e.m.f. from being generated. Alternative circuits for suppression of the switch-off transient are shown in fig. (S)2b and consist of wiring a series $CR$ network across either the load or the switch. A series resistor must be included in order to limit the capacitive current.

To suppress transients in thyristor and triac circuits, some form of filter network is required. Take the case of a simple light dimmer which, at half power, switches the mains at peak value. The filter circuit, consisting of an inductor and capacitor connected as shown in fig. (S)3, acts as a low-pass filter in the line and reduces the amplitude of transients which are fed back into the power source.

## Transducer

A transducer is defined as any device that converts energy from one form into another, such as light energy into mechanical energy. In electronics, the transducers are the sensors used within a system to change input energy (heat, force, position) into electrical energy to give the small signals to drive the system or to give controlling feedback. Typical sensors can be grouped as follows;

| | |
|---|---|
| *Temperature sensors:* | Thermistor |
| | Thermocouple |
| | Platinum resistance |
| | Semiconductor junction |
| *Opto-sensors:* | Photoconductive cell |
| | Photodiode and |
| | Phototransistor |
| *Position and force sensors:* | Relective/slotted opto-switch |
| | Potentiometer |
| | Linear variable differential transformer |
| | Strain gauge |
| | Piezoelectric devices |

A *thermistor*, constructed of sintered oxides of nickel and manganese, is a sensor that exhibits a large change of resistance with temperature. The material is formed into rods or beads with the smallest volume required for sensing purposes

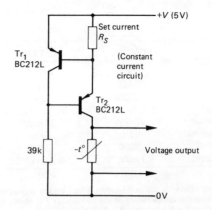

Figure (T)1b   Getting a useful signal from a thermistor

to achieve reasonably fast response. The bead is then sealed inside a glass envelope or into a stainless steel probe [fig. (T)1a]. Most types have a negative temperature characteristic following the law:

$$R_2 = R_1 e^{B(1/T_2 - 1/T_1)}$$

where $B$ = characteristic temperature constant (K)
$T$ = bead temperature (K)
$R_1$ = resistance at temperature $T_1$
$R_2$ = resistance at temperature $T_2$

$B$ is typically in the range 3000 to 3800 [see data given in fig. (T)2].

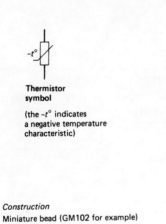

Thermistor symbol

(the $-t°$ indicates a negative temperature characteristic)

Construction
Miniature bead (GM102 for example)

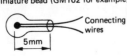

Connecting wires

5mm

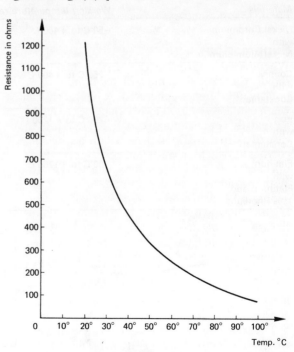

*Figure (T)1a*   Thermistor

**283**

|                        | GM 102            | GM 472            | GM 473            |
|------------------------|-------------------|-------------------|-------------------|
| Resistance 20°C        | 1 kΩ              | 4k7 Ω             | 47 kΩ             |
| $R_{min}$ (hot)        | 59 Ω              | 271 Ω             | 338 Ω             |
| Tolerance              | ±20%              | ±20%              | ±20%              |
| Temperature range      | −80°C to +125°C   | −80° to +125°C    | −60 to +200°C     |
| Maximum dissipation    | 70 mW             | 70 mW             | 120 mW            |
| Time constant (thermal)| 5 sec             | 5 sec             | 5 sec             |
| Temp. constant B       | 3000              | 3390              | 3930              |
| *Dissipation constant  | 0.7 mW/°C         | 0.7 mW/°C         | 0.7 mW/°C         |

*Dissipation constant refers to the self-heating effect and it represents the amount of power required to raise the temperature of the thermistor 1°C above ambient.

For example, suppose $I_t = 0.1$ mA and $R_t = 10$ kΩ.

Power dissipated by the thermistor is

$$P_t = I_t^2 R_t = (0.1 \times 10^{-3})^2 \times 10^4 = 0.1 \text{ mW}$$

In this case the dissipation would only cause a very slight difference between the temperature of the thermistor and that of the surroundings.

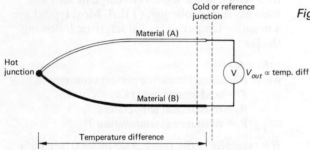

Cold or reference junction

Material (A)

Hot junction

$V$ $V_{out} \propto$ temp. diff

Material (B)

Temperature difference

Figure (T)3a Thermocouple temperature sensor

| Materials                                          | Working temperature range |
|----------------------------------------------------|---------------------------|
| Nickel Chromium with Nickel Aluminium              | −50°C to +400°C           |
| Copper with Constantan                             | −250°C to +400°C          |
| Iron with Constantan                               | −200°C to +1200°C         |
| Platinum with Platinum and 13% Rhodium             | −50°C to +1750°C          |

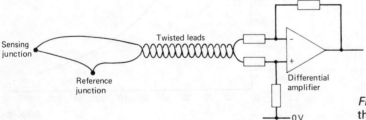

Sensing junction

Twisted leads

Reference junction

Differential amplifier

0 V

Figure (T)3b Typical circuit containing thermocouple

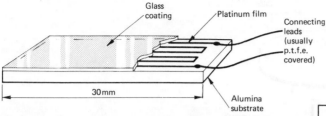

Figure (T)4  Platinum resistance temperature detector

In order to convert the changes in resistance into a voltage signal, the thermistor can be either fitted into a bridge circuit or supplied with a constant current [fig. (T)1*b*].

A *thermocouple* is a sensor made of two junctions of dissimilar metals with an e.m.f. generated proportional to the temperature difference between these two junctions. One junction is used for temperature measurement while the other is held at a reference temperature [fig. (T)3*a*]. An amplifier is required to boost the small (typically $40\,\mu\text{V}/°\text{C}$) output from a thermocouple. A circuit is shown in fig. (T)3*b*.

The *platinum resistance temperature detector* (PRTD) is constructed as a coil of platinum wire on an insulating former or as a film of platinum on an alumina substrate [fig. (T)4]. Usually the unit is made to have a resistance of $100\Omega$ at $0°\text{C}$ and operates over the range $-50°$ to $+500°\text{C}$. Platinum has a positive temperature coefficient of $+0.385\Omega/°\text{C}$. As with the thermistor, an output voltage can be obtained either by passing a

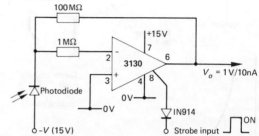

Figure (T)6  Light-detecting circuit using a photodiode to drive a MOSFET

constant current through the sensor or by fitting it in one arm of a bridge.

*Photodiodes* and *phototransistors* [fig. (T)5] are semiconductor devices fitted with a window to a junction. Any light falling on the junction creates hole-electron pairs which set up a current flow. The diode is operated with reverse bias so that its dark current is very low, typically a few tens of nano-amps. As the light level being measured increases, the reverse current also increases. Sensitivity is about $1\,\mu\text{A}/\text{mW}/\text{cm}^2$. Fig. (T)6 shows a circuit for detecting light level with a photodiode driving a MOSFET op-amp type 3130.

With a phototransistor, any light falling on the base/emitter junction is amplified to give a large collector current. The phototransistor is therefore a much more sensitive device then the diode. Both the photodiode and phototransistor are fast operating devices with rise times of better than $1\,\mu\text{sec}$.

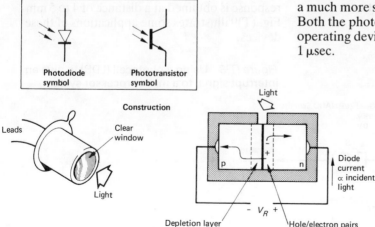

Figure (T)5  Photodiodes and phototransistors

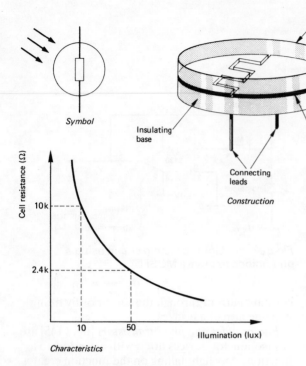

*Symbol*

*Insulating base*

*Connecting leads*

*Clear window*

*Semiconductor material (CdS)*

*Construction*

*Characteristics*

**Figure (T)7** The photoconductive cell

Another commony used light sensor is the *photoconductive cell* or *light-dependent resistor* (LDR). A film of light-sensitive material, such as cadmium sulphide, is deposited onto an insulating substrate and the unit is encapsulated leaving a clear end window. When light falls on the device, the cell resistance falls [fig. (T)7]. These types of cell are particularly suited to the task of detecting slowly varying light levels, as smoke detectors and lighting controllers. Fig. (T)8 illustrates the use of an LDR in generating an interrupt to a

microprocessor. CMOS logic gates are used because of their high input resistance. While light falls on the LDR, making its resistance low, the input voltage to gate A is held below its threshold value. As the light level falls, the input voltage rises and a point is reached when the output of gate A is forced to switch. $C_2$ is discharged via $R_3$ and, after a delay of approximately 5 seconds, the output of gate B goes high to give the interrupt signal. The delay circuit $R_3 C_2$ is used to prevent short-duration changes in light level, such as shadows, from causing the circuit to switch.

A wide variety of transducers can be used for sensing position, force and pressure. The *slotted opto-switch* and the *reflective opto-switch* are sensors that give an on/off indication and are therefore particularly suited for limit detection, position sensing, batch counting and level indication. Both use an infra-red diode as the emitter and a silicon phototransistor detector in one package. The optical link is either across a slot, where the beam can be interrupted to give a pulse signal from the phototransistor, or by a reflecting surface between the LED and the phototransistor. With the latter, optimum response is obtained at a distance of 4 to 5 mm. Fig. (T)9 illustrates some applications of these devices.

**Figure (T)8** Using a photocell (LDR) to give an interrupt signal to a microprocessor system

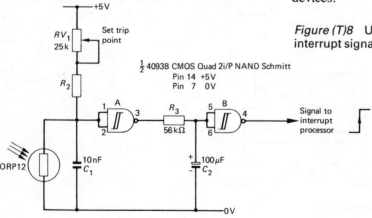

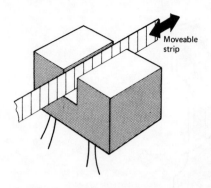

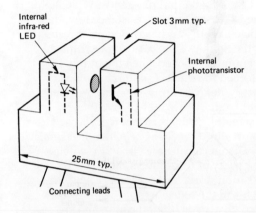

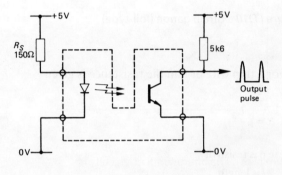

**SLOTTED OPTO-SWITCH**

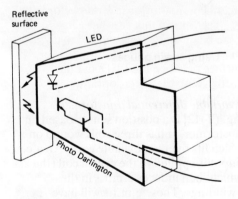

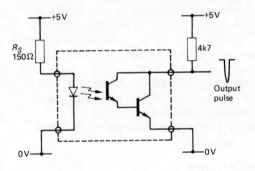

**REFLECTIVE OPTO-SWITCH**

*Figure (T)9* The slotted and reflective opto-switch

The *strain gauge* is one of the transducers commonly used for the detection and measurement of force, load, torque and strain. The basic unit consists of a resistive track on a flexible insulating base. The gauge is bonded to the mechanical part in which the strain is to be measured [fig. (T)10].

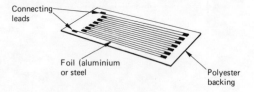

*Figure (T)10* Strain gauge (foil-type)

For a thin wire or film the resistance is given by

$$R = \rho \frac{l}{a}$$

where $\rho$ is resistivity

$l$ is length

$a$ is cross-sectional area.

A change in resistance due to strain can be expressed as

$$\frac{\Delta R}{R} = \frac{\Delta \rho}{\rho} + \frac{\Delta l}{l} - \frac{\Delta a}{a}$$

where $\frac{\Delta l}{l} = \varepsilon$, the strain in $\mu$m/m

Since the gauge is bonded to the member, any change of dimensions of the member will also cause the gauge resistance to change.

The main parameter for a strain gauge is called the gauge factor $K$.

$$K = \frac{\Delta R/R}{\Delta l/l} = \frac{\Delta R/R}{\varepsilon}$$

A typical value for $K$ is 2. Semiconductor gauges, which are much more sensitive, may have gauge factors of 100 or more.

Strain gauges are often used as a group of four in a Wheatstone bridge circuit. Two gauges are used for sensing, while the other two are mounted across the line of displacement and are used for temperature compensation. Fig. (T)11 illustrates the principle.

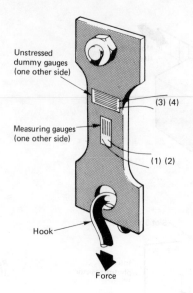

Unstressed dummy gauges (one other side)

(3) (4)

Measuring gauges (one other side)

(1) (2)

Hook

Force

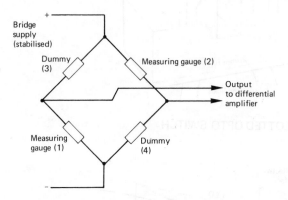

Bridge supply (stabilised)

+

Dummy (3)

Measuring gauge (2)

Output to differential amplifier

Measuring gauge (1)

Dummy (4)

−

*Figure (T)11* Using strain gauges

A *linear variable differential transformer* (LVDT) [fig. (T)12] is a position sensor based on the a.c. transformer. It has three coils wound on a former which fits over a moveable core. An a.c. reference input is applied to the centre coil (the primary) which then induces e.m.f.s in the secondary windings. These e.m.f.s will have amplitudes dependent upon the position of the core. When the core is dead centre, the secondary e.m.f.s will be equal and, since the windings are connnected in series, these e.m.f.s will cancel, giving zero output. As the core is moved away from the centre, the induced voltage will be greater in one secondary than in the other and an a.c. output voltage results. This a.c. output has an amplitude that is a linear function of the core displacement.

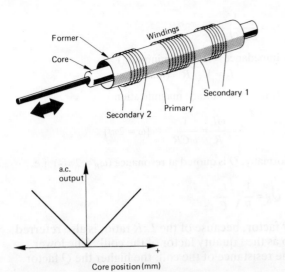

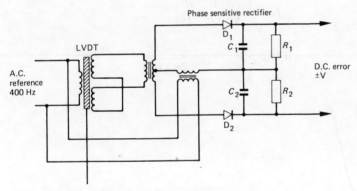

Typical characteristics

*Figure (T)12a* The linear variable differential transformer

Phase sensitive rectifier

*Figure (T)12b* Connecting an LVDT to a phase-sensitive rectifier

Non-linearity can be better than ±0.5% of maximum output and the measuring range of this type of sensor can be from 0.1 mm up to 75 mm. When the core moves through the centre position, the output goes through zero and then changes phase giving an a.c. output that is 180° out of phase with the reference a.c. A phase-sensitive rectifier is required to detect the position and direction of the core. Fig.(T)12 shows a typical circuit arrangement for use with this sensor.

▶ *Control systems*

# Tuned Circuits and Tuned Amplifiers

**Tuned Circuits**   These are circuits which are frequency-dependent and are used in filters, oscillators and radio frequency amplifiers. They are made up using an inductor and capacitor either connected in series or in parallel [fig. (T)13]. For both circuits there will be one frequency of the applied signal when the inductive reactance will be equal to the capacitive reactance. This condition is called *resonance*. At the resonant frequency $f_R$:

*a*)  The circuit acts as a pure resistance. For the series circuit, impedance is a minimum, while for the parallel circuit it is a maximum.

*b*)  The supply current is in phase with the supply voltage. For the series circuit, the supply current will be a maximum, while for the parallel circuit it is a minimum.

*Series circuit formulas*

$$Z = R + j\left(\omega L - \frac{1}{\omega C}\right)$$

At resonance, $\omega L = \dfrac{1}{\omega C}$

$$f_R = \frac{1}{2\pi\sqrt{(LC)}}$$

Impedance $= R\underline{/0^\circ}$   (minimum)
$\qquad\quad I_S = V_S / R$   (maximum)

$Q$ factor = Voltage magnification

$$= \frac{\omega L}{R} = \frac{1}{\omega CR} \qquad [\omega = 2\pi f]$$

Normally, $Q$ is quoted at resonance ($\omega_R = 2\pi f_R$), i.e.

$$Q_R = \frac{1}{R}\sqrt{\frac{L}{C}}$$

$Q$ factor, because of the $L{:}R$ ratio, is also referred to as the "quality factor of the coil". The lower the resistance of the coil, the higher the $Q$ factor.

$$\text{Bandwidth BW} = \frac{f_R}{Q_R} \qquad [\text{see fig. (T)14}]$$

*Parallel circuit formulas*

$$f_R = \frac{1}{2\pi}\sqrt{\left[\frac{1}{LC} - \frac{R^2}{L^2}\right]}$$

If $L^2 >> R^2$ $\qquad f_R \simeq \dfrac{1}{2\pi\sqrt{(LC)}}$

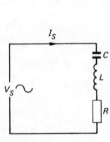

Series resonant circuit

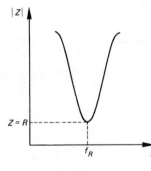

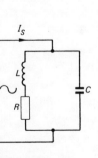

Parallel resonant circuit

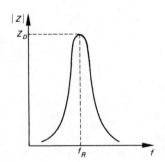

*Figure (T)13*   Series and parallel resonant circuits

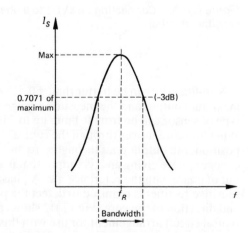

*Figure (T)14*   Bandwidth of a series tuned circuit

290

$Q$ factor = Ratio of circulating current to supply current (current magnification)

$$= \frac{\omega L}{R}$$

Dynamic impedance (impedance at $f_R$) is

$$Z_D = \frac{L}{CR} = Q\omega_R L = \frac{Q}{\omega_R C} \quad \text{(maximum)}$$

$$I_S = V_S/Z_D \quad \text{(minimum)}$$

$$BW = \frac{f_R}{Q}$$

The magnitude of the impedance just off resonance is given by:

$$|Z| = \frac{Z_D}{\sqrt{[1 + 4Q^2\delta^2]}}$$

where $\delta$ = fractional deviation = $(f - f_R)/f_R$

**Tuned Amplifiers**   This is an amplifier that has a very narrow bandwidth and is therefore highly selective. It uses a parallel tuned circuit as its load with the basic circuit, using a BJT, as shown in fig. (T)15a.

Since the dynamic impedance $Z_D$ of the tuned circuit can be very high, the voltage gain of the amplifier will be high at resonance while it will fall away sharply either side of resonance.

$$\text{At } f_R \qquad A_{VR} = \frac{h_{fe}Z_D}{h_{ie}(1 + h_{oe}Z_D)}$$

Note that $h_{oe}$ has the effect of reducing voltage gain and of reducing the $Q$ factor.

$$Q_L = \text{Loaded } Q \text{ factor} = \frac{Q}{1 + h_{oe}Z_D}$$

$$\text{and} \quad BW = \frac{f_R}{Q_L}$$

The effect of loading the amplifier is to "damp" the tuned circuit which reduces the gain and decreases selectivity [fig. (T)15b]. For a single stage, the effect of active device impedance and load resistance can be reduced by connecting the collector to a tapping on the coil [fig. (T)15c]. The coil effectively acts as a step-down transformer so that the reflected loading onto the tuned circuit is much reduced. For example with a turns ratio of 3:1, the reflected resistance ($R_L \| 1/h_{oe}$) becomes 9 times greater.

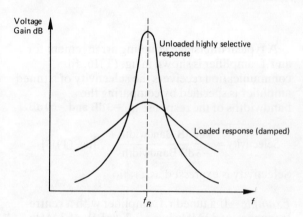

Figure (T)15b   Effect of loading a tuned amplifier

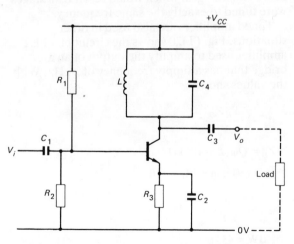

Figure (T)15a   Basic tuned transistor amplifier (CE)

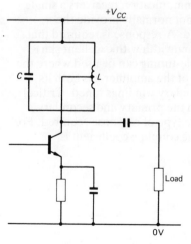

Figure (T)15c   One method of reducing the effect of loading

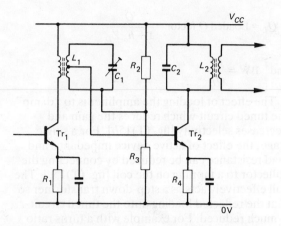

Figure (T)16   Interstage coupling in a tuned r.f. amplifier

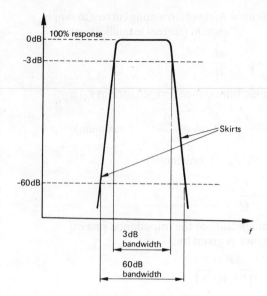

Figure (T)17   Selectivity specification

A typical interstage coupling arrangement for an r.f. amplifier is shown in fig. (T)16. In communication receivers the selectivity of a tuned amplifier is specified by comparing the bandwidths of the response at $-3\,\text{dB}$ and $-60\,\text{dB}$.

$$\text{Selectivity} = \frac{-60\,\text{dB Bandwidth}}{-3\,\text{dB Bandwidth}} \quad \text{[fig. (T)17]}$$

Selectivity is expressed as a ratio.

*Example*   If a tuned r.f. amplifier with a centre frequency of 32 MHz has a $-3$ db BW of 3 MHz and a $-60$ db BW of 9 MHz, then the selectivity ratio is 3:1. The smaller this ratio the higher the selectivity. In communication receivers a single tuned circuit cannot normally provide the required passband. A response is required that gives sufficient bandwidth with excellent skirt selectivity. Double-tuning can be used where the load transformer of the amplifier has both its primary and secondary windings tuned. Critical coupling between the primary and secondary windings gives the type of response required. For double-tuning, the coupling coefficient is

$$k = \frac{L_m}{\sqrt{(L_p L_s)}}$$

where $L_m$ = mutual inductance.

$$\text{Critical coupling } k_c = \frac{1}{\sqrt{(Q_p Q_s)}}$$

$k_c$ has a typical value of 0.025.

Alternatively stagger-tuning can be used where two or more amplifying stages have their resonant frequencies set at slightly different values. The overall response has increased bandwidth and good skirt selectivity [fig. (T)19]. Stagger tuning overcomes the problem of bandwidth shrinkage in cascaded stages, which would result if all stages were tuned to exactly the same frequency.

Tuned amplifiers are also used in other situations. Fig. (T)20 shows the circuit of a FET amplifier used to amplify the output of an a.c. bridge that has a supply frequency of 2 kHz. With the values shown,

$$A_v = \frac{Y_{fs} Z_D}{1 + Y_{os} Z_D}$$

$$Z_D = Q \omega_R L \simeq 125 \,\text{k}\Omega$$

$\therefore$   At resonance $A_v \simeq 110$

$$Q_L = \frac{Q}{1 + Y_{os} Z_D} \simeq 44$$

$\therefore$   BW = 45 Hz

▶ *Radio systems and circuits*

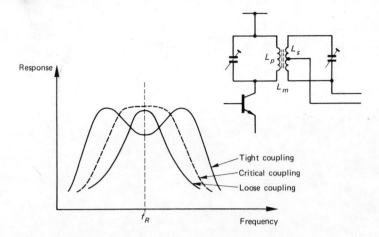

Response

Tight coupling
Critical coupling
Loose coupling

$f_R$    Frequency

*Figure (T)18*   Frequency response curve for double tuned circuits

Tuned stage A  →  Tuned stage B
$f_1$                    $f_2$

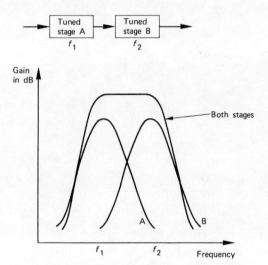

Gain in dB

Both stages

A          B

$f_1$      $f_2$      Frequency

*Figure (T)19*   Stagger tuning

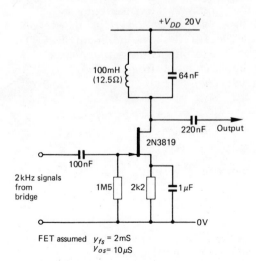

$+V_{DD}$ 20V

100mH (12.5Ω)    64nF

220nF  Output

2N3819

100nF

2kHz signals from bridge

1M5   2k2   1µF

0V

FET assumed   $y_{fs}$ = 2mS
                      $y_{os}$ = 10µS

*Figure (T)20*   JFET tuned amplifier